Spatial Networks

Marc Barthelemy

Spatial Networks

A Complete Introduction: From Graph
Theory and Statistical Physics to Real-World
Applications

 Springer

Marc Barthelemy (iD)
Institut de Physique Théorique
Commissariat à l'Energie Atomique
Gif-sur-Yvette, Essonne, France

Centre d'Analyse et de Mathématique
Sociales
Ecole des Hautes Etudes en Sciences
Sociales
Paris, France

ISBN 978-3-030-94108-6   ISBN 978-3-030-94106-2   (eBook)
https://doi.org/10.1007/978-3-030-94106-2

This Springer imprint is published by the registered company Springer Nature Switzerland AG
The registered company address is: Gewerbestrasse 11, 6330 Cham, Switzerland

We find beauty not in the thing itself but in the patterns of shadows, the light and the darkness, that one thing against another creates.

—Jun'ichirō Tanizaki

Preface

Watts and Strogatz, with the publication in 1998 of their seminal paper on small-world networks, opened the golden era of complex networks studies and showed, in particular, how statistical physics could contribute to the understanding of these objects. The first studies that followed considered the characterization of large graphs, their degree distribution, clustering coefficient or their average shortest path. New models of random graphs, beyond the well-known Erdos–Renyi archetype, were then proposed in order to understand some of the empirically observed features. However, many complex networks encountered in the real-world are embedded in space: nodes and links correspond to physical objects and there is usually a cost associated to the formation of edges. This aspect turns out to be crucial as it determines many features of the structure of these networks that we can call 'spatial'. It is difficult to consider that spatial networks actually form a subclass of complex networks, but rather constitute their own family specified by a set of properties that differ from the 'usual' complex networks. In particular, one of the most salient property in complex network is a broad degree distribution with the existence of hubs. This feature has a dramatic impact on dynamical processes occurring on these networks and is at the heart of studies on scale-free networks. The physical constraints in spatial networks prohibits in general the formation of hubs and their most interesting properties lie in their spatial organization and in the relation between space and topology.

Spatial networks—even if this was not the standard name at that time—were the subject of numerous studies in the 70s in regional science followed by quantitative geographers who were interested in characterizing the structure of transportation networks, from roads to subways and railways. Quantitative geographers realized the crucial importance of networks and produced a number of important results about these networks and their evolution. The recent revival of the interest in this subject, combined with an always larger amount of data allowed to make some progress in our understanding of these objects. The recent advances obtained in the understanding of spatial networks have generated an increased attention toward the potential implication of new theoretical models in agreement with data. Questions such as the structure and resilience of infrastructures, the impact of space on the formation of biological networks, or the impact of urban networks on epidemic spread are fundamental questions that we hope to solve in the near future.

In this book, we will discuss different aspects of spatial networks, focusing essentially on the characterization of their structure and on their modeling. Each chapter is as much as possible self-contained, and for the sake of clarity and readability, we tried to be as modular as possible in order to allow the reader interested in just one specific model or tool to focus essentially on the corresponding chapter.

Even if I started to work on complex networks and focused on purely topological quantities such as the degree distribution, I was always fascinated by this particular class of graphs that are embedded in space. The most common illustration is the example of maps and the intricate design of roads, railways or other features. I always felt that there is a richness in these graphs due to the interaction of connectivity constraints and space. After having worked on various aspects of this problem, I first published a review in Physics Reports followed by a book in the lecture notes series on Morphogenesis. However, a few years later, I realized that this book was incomplete and I took the opportunity to rewrite completely a new version with a new organization and new parts that I missed previously. All the material discussed in the previous book 'Morphogenesis of spatial networks' is included and revised in this new book, but many new chapters were also added. I also propose a new organization of the content and I now divide the book in two large parts: Part I is about characterizations of spatial networks and Part II is about their modeling. There are many other ways to organize this material, but I believe that in this way the potential reader could more easily find what she or he is looking for.

I tried to be exhaustive in both these parts and even if it is not possible to describe everything, I hope that I covered here pretty much of the material about the characterization of these particular networks and their modeling.

In the Chaps. 1–4 about characterization, I go over classical notions such as planar graphs or digraphs but I always try to present a particular point of view by insisting on aspects that are usually put aside by many authors. I also try to add an empirical grain of salt in all chapters. In this way, I can illustrate complex notions in a simple, concrete way and show that they are actually useful.

Shortest paths are crucial for understanding the structure of networks, and in particular, the betweenness centrality discussed in Chap. 5 is a simple tool that contains a lot of information and reveals quite a lot about the structure of a graph. The shape of shortest paths discussed in Chap. 6 is also an important topic in relation with first passage percolation, a standard problem in statistical physics. In Chap. 7 I discuss other types of paths: the simples paths which try to avoid turns as much as possible, and show how they bring us information about the mesoscopic structure of large networks. The entropy counts the complexity of paths in these networks and is also discussed in this chapter.

In Chap. 8, I will present various tools that allow to characterize the structure of spatial networks at a large scale. I will start with the notion of spatial dominance and marked point processes. For many systems, nodes have a mark (a real number such as the population of a city for example) and we have to characterize the interplay between the value of the mark and the spatial location of a node. We discuss for these objects a measure of spatial 'dominance' that was developed by Okabe and his collaborators and which is based on the Voronoi tessellation of points. The first passage times of random walkers can also be used to characterize spatial correlations and I will discuss such a recent method. I will also present new tools coming from topological data analysis such as the persistent homology. This tool was initially devised to characterize the 'shape' of sets of points but was adapted to the case of spatial networks such as streets or spider webs for example. I will end this chapter with a discussion on community detection whose results depend strongly on the existence of correlations between space and attribute, and on the choice of a null model.

In Chap. 9, I will present the different attemps for constructing a typology of planar graphs, including mathematical approaches, approximate methods, and recent machine learning results.

Finally, in the last Chapter 10 of this part, I address the important problem of time-evolving spatial networks and their characterization. I focus in this more empirically-oriented part on the evolution of the street network and the growth of subways. The large number of parameters and of possible measures is, maybe surprisingly, not very helpful and we will see how to identify the most relevant quantities for the characterization of the evolution of these systems.

The Part II concerns models of spatial networks. I completely reorganized the corresponding material and added many new results. I start with a short discussion about modelling (Part II) and I introduce in Chaps. 11 and 12 spatial generalizations of classical graphs such as the Erdos–Renyi and the Watts–Strogatz graph. In Chap. 13, I present several spatial generalizations of the classical preferential attachment model (also called Barabasi–Albert model).

In Chap. 14, I discuss tessellations and the associated Delaunay graph, and fragmentation models. There are many other possibilities to construct a graph over a set of points and I follow in the Chaps. 15 and 16, the classification coming mostly from mathematicians who distinguish proximity graphs (Chap. 15) and excluded volume graphs (Chap. 16). These chapters include, in particular, well-known graphs such as the random geometric graph, the Delaunay triangulation, β-skeletons, the Waxman model, etc., but also less known graphs such as the Gabriel graph, the bluetooth graph, or the radial graph for example.

I then discuss simple graphs constructed with branches and loops (Chap. 17) as their appear in a variety of contexts and often allow for analytical calculations. Chapters 18 and 19 concern then optimal networks. In Chap. 18, I discuss classical cases coming mostly from the mathematical and physics literature, while in Chap. 19 I discussed optimal networks from the more applied perspective of transportation networks. This allows me to connect the approaches discussed here from a statistical physics point of view with approaches such as location science coming from civil engineering.

In Chap. 20 I discuss growing network models where one adds at each time step a node that connects to the rest of the network according to the optimization of a specific cost-benefit function. This type of approach allows to investigate the relation between the properties of the network and those of the substrate where the network grows.

As can be seen in this short outline of the book, several disciplines are concerned. Scientists from statistical physics, random geometry, probability, computer sciences, and quantitative geography produced a wealth of interesting results and this book cannot cover all new studies about spatial networks. Owing to personal biases, space limitations and lack of knowledge, important topics might have been omitted, and I apologize in advance for omissions or errors and to those colleagues who feel that their work is not well represented here. I tried to minimize the possible gaps and I thank in advance colleagues who can point to me other related topics. Incomplete and imperfect as it is, I hope, however, that this book will be helpful to scientists interested in the formation and evolution of spatial networks, a fascinating subject at the crossroad of so many disciplines.

Paris, France   Marc Barthelemy
November 2021

Acknowledgements

My path in the network world started with my visit to Gene Stanley's group in Boston, where I worked in particular with Luis Amaral and Shlomo Havlin. I thank Gene for the freedom that he left me at that time and Luis and Shlomo for having introduced me to the analysis of empirical data. Back to Paris, I continued my exploration of networks with Alain Barrat and Alessandro Vespignani with whom we focused on the spread of epidemics and the impact of mobility on this process. I thank them both warmly for all the things I learned with them, from technical methods to the way of doing science. These studies on epidemic spread naturally led me to analyze transport networks at different spatial scales, and most importantly to understand the effect of space on the topology of these structures. These systems are indeed embedded in space and since the beginning of network studies, this aspect was mostly ignored. These different reasons, together with my fascination for maps, pushed me to look further about what we can now call spatial networks. In particular, I started to work on the most common example—road networks—and thanks to many discussions with Alessandro Flammini, we proposed a model for the formation and evolution of these systems. I then never stopped to explore this fascinating subject.

After some time, I joined the Institut de Physique Théorique in Saclay and I could continue in this interdisciplinary direction thanks to Henri Orland, who was the director at that time and thanks to his successors Michel Bauer and now Francois David, who provided such a great interdisciplinary environment for these funda-mental studies. In particular I could meet colleagues at the IPhT with a strong math-ematical background and from whom I could learn so much. In particular, I thank Jean-Marc Luck, Kirone Mallick, and Paul Krapivsky (a frequent visitor of our lab) for many discussions on many subjects in statistical physics, and Jeremie Bouttier, Emmanuel Guitter, and Philippe Di Francesco—their knowledge in combinatorics and planar maps helped me to understand small parts of this important topic in mathematical physics.

A constant interaction with other point of views, the need to explain yourself clearer are fundamental aspects in scientific research and I thank all my collab-orators, colleagues, together with my postdocs and Ph.D. students with whom I worked on different subjects related to networks. Another crucial aspect in this field is interdisciplinarity. This brought me to meet many scientists from whom I learned a lot about completely different aspects going from applied mathematics, probability, combinatorics, to economics, geography, and history. For all these discussions and interactions, I warmly thank: D. Aldous, E. Arcaute, A. Arenas, H. Barbosa, M. Batty, L. G. Benguigui, H. Berestycki, G. Bianconi, A. Blanchet, M. Boguna, P. Bordin, J.-P. Bouchaud, A. Bourges, A. Bretagnolle, M. Breuillé, G. Caldarelli, O. Cantu, G. Carra, A. Chessa, V. Colizza, J. Coon, Y. Crozet, M. De Nadai, J. Depersin, S. Derrible, C. P. Dettmann, S. Dobson, M. De Domenico, A. Flammini, S. Fortunato, M. Fosgerau, E. Frias, R. Gallotti, G. Ghoshal, J. Gleeson, M. C. Gonzalez, M. Gribaudi, J. Le Gallo, R. Le Goix, D. Helbing, R. Herranz, A. Kartun-Giles, E. Katifori, A. Kirkley, M. Kivela, P. Krapivsky, R. Lambiotte, V. Latora, F. Le Nechet, M. Lenormand, B. Lion, T. Louail, R. Louf, C. Mascolo, T. Di Matteo, Y. Moreno, R. Morris, I. Mulalic, J.-P. Nadal, V. Nicosia, A. Noulas, M. O'Kelly, J. Perret, S. Porta, M. A. Porter, D. Pumain, D. Quercia, C. Ratti, J. J. Ramasco, C. Roth, C. Rozenblat, M. San Miguel, F. Santambroggio, M. A. Serrano, S. Shai, E. Strano, E. Taillanter, M. Tomko, V. Verbavatz, A. Vignes, V. Volpati.

I also thank the Springer staff for its excellent support and reactivity. In particular, I thank Christoph Baumann for his help and constrant support about this project. I also thank Cecil Joselin Simon for coordinating this book.

For everything else, I thank Ralph and Shlomo and my loving family, Esther, Rebecca, and Catherine.

Contents

Part I   Characterization

1   From Complex to Spatial Networks 3
   1.1   Early Days 3
   1.2   Complex Networks 4
   1.3   Space Matters 5
   1.4   Definition and Representations 5
      1.4.1   Spatial Networks 5
      1.4.2   Representations of Networks 6
   References 8

2   Planar Graphs 9
   2.1   Graph Theoretical Tools 10
      2.1.1   Definition and Representation 10
      2.1.2 K_5 Is Not Planar 11
      2.1.3   Euler's Formula 12
      2.1.4   Distance from Planarity 13
   2.2   Planarity of Street Networks 18
   References 20

3   Directed and Mixed Graphs 23
   3.1   Theoretical Results 23
      3.1.1   Definitions 23
      3.1.2   Components of a Digraph 24
      3.1.3   Statistics of Loops 25
      3.1.4   Perturbation of the Shortest Path in a Directed
            Lattice 27
   3.2   Empirical Results: One-Way Streets 28
      3.2.1   Fraction of One-Ways 30
      3.2.2   Statistics of the Detour Index 31
      3.2.3   Asymmetry of Shortest Paths 33
   References 36

4   Simple Measures 39
   4.1   Irrelevant Measures for Spatial Networks 39
        4.1.1   Degree 40
        4.1.2   Length of Segments 42
        4.1.3   Cell Area 44
        4.1.4   Clustering and Assortativity 45
        4.1.5   Average Shortest Path and Diameter 47
        4.1.6   Empirical Illustrations 48
   4.2   Simple Characterizations 54
        4.2.1 α and γ Indices 54
        4.2.2   Organic Ratio and Ringness 55
        4.2.3   Edge Orientation Distribution 56
        4.2.4   Shape Factor 58
        4.2.5   Detour Index (or Stretch or Route Factor) 58
        4.2.6   Cost, Efficiency and Robustness 60
   References 63

5   Betweenness Centrality 65
   5.1   Definition 66
   5.2   General Properties 66
        5.2.1   Numerical Calculation: Brandes' Algorithm 67
        5.2.2   The Average BC 68
        5.2.3   Edge Versus Node BC 69
        5.2.4   Adding Edges 70
        5.2.5   Relation Between the BC and the Clustering
             Coefficient 72
        5.2.6   Scaling of the Maximum BC 72
   5.3   The BC for Simple Graphs 74
        5.3.1   Regular 1d and 2d Lattices 74
        5.3.2   Cayley Tree 77
        5.3.3   Branches and Ring 78
        5.3.4   Grid-Tree Network 83
   5.4   The BC in a Disk: The Continuous Limit 87
        5.4.1   The Infinite Density Limit 87
        5.4.2   Finite Density: A Perturbation Expansion 90
   5.5   Empirical Measures on Street Networks 97
        5.5.1   Spatial Patterns of Large BC Nodes 97
        5.5.2   The BC and Socio-Economic Indicators 101
        5.5.3   The BC Probability Distribution for Street
             Networks 101
        5.5.4   The Effect of One-Way Streets on the BC 105
   References 107

6   The Shape of Shortest Paths 109
6.1   (Euclidean) First Passage Percolation 109
6.1.1   Models and Definitions 109
6.1.2   Known Results About Exponents 112
6.1.3   Numerical Results 115
6.1.4   Travel Time and Transversal Fluctuations 117
References 118

7   Simplicity and Entropy 121
7.1   Simplicity 121
7.1.1   Simplest Paths 121
7.1.2   The Simplicity Index and the Simplicity Profile 124
7.1.3   A Null Model 125
7.1.4   Measures on Real-World Networks 125
7.2   Information Perspective 130
7.2.1   Entropy and Simplest Paths 131
7.2.2   Quantifying the Complexity 133
References 137

8   Large-Scale Tools 139
8.1   Spatial Dominance 140
8.1.1   Constructing the Dominance Tree 140
8.2   Class First-Passage Times 144
8.3   Community Detection in Spatial Networks 146
8.3.1   Modularity 147
8.3.2   A Null Model for Spatial Networks with Marks 148
8.3.3   Synthetic Spatial Network Benchmarks 152
8.3.4   Modifying the Modularity 153
8.4   Persistent Homology 157
8.4.1   Topological Data Analysis 157
8.4.2   Empirical Results 159
References 165

9   Typology of Planar Graphs 167
9.1   Area and Shape of Faces 167
9.1.1   Characterizing Blocks 168
9.1.2   A Typology of Planar Graphs 171
9.2   Approximate Mapping of a Planar Graph to a Tree 174
9.3   An Exact Bijection Between a Planar Graph and a Tree 178
9.4   Machine Learning Approaches 182
References 184

10   Measuring the Time Evolution of Spatial Networks 187
10.1   Road Networks 188
10.1.1   Organic Growth 188
10.1.2   Effect of Planning 196
10.1.3   Simplicity Measures 204

10.2   Subways 208
       10.2.1   Generalities 208
       10.2.2   Typical Numbers 210
       10.2.3   Network Evolution 212
       10.2.4   Standard Measures 213
       10.2.5   Efficiency 214
       10.2.6   Temporal Statistics: Bursts 217
       10.2.7   Core and Branches: Measures and Model 219
       10.2.8   Spatial Organization of the Core and Branches 227
References 231

Part II   Models

11   Spatial Generalizations of Random Graphs 235
       11.1   Spatial Version of Erdos-Renyi Graphs 235
              11.1.1   The Erdos-Renyi Graph 235
              11.1.2   Planar Erdos-Renyi Graphs 237
       11.2   The Hidden Variable Model for Spatial Networks 238
              11.2.1   Effect of Space 239
              11.2.2   Effect of Traffic 240
       References 241

12   Spatial Small-Worlds 243
       12.1   The Watts-Strogatz Model 243
       12.2   Spatial Generalizations in Dimension d 244
       12.3   Navigability in the Kleinberg Model 247
              12.3.1   Searchable Networks 247
              12.3.2   Sketch of Kleinberg's Proof 249
       References 252

13   Growing Spatial Networks 253
       13.1   Preferential Attachment and Space 253
              13.1.1   Preferential Attachment and Distance Selection 255
              13.1.2   Searching in Spatial Scale-Free Networks 262
       13.2   Attraction Potential Models 263
              13.2.1   The Connection Rule 264
              13.2.2   Uniform Distribution of Nodes 265
              13.2.3   Exponential Distribution of Centers 266
              13.2.4   Effect of Centrality and Density 267
              13.2.5   The Appearance of Core Districts 275
       References 276

14   Tessellations of the Plane 277
       14.1   The Voronoi Tessellation 277
              14.1.1   Average Properties of the Poisson-Voronoi
                       Tessellation 279
              14.1.2   Statistical Properties 281

14.1.3   Central Limit Theorem for the Total Length 285
14.2   Effect of the Density of Points 286
14.3   Crack and STIT Tessellations 288
14.4   Planar Fragmentation 289
References 292

15   Proximity Graphs 295
15.1   Random Geometric Graphs 296
15.1.1   The Hard Case 296
15.1.2   Soft Random Geometric Graphs 301
15.1.3   The Full Connectivity Probability 302
15.1.4   The Waxman Model 304
15.1.5   Random Geometric Graphs in Hyperbolic Space 308
15.2   Bluetooth Graph and Sparsification 309
15.3   The $k-$nearest Neighbour Model 310
15.3.1   Definition and Connectivity Properties 310
15.3.2   A Scale-Free Network on a Lattice 311
15.4   A Dynamical Proximity Model 313
15.4.1   The Model 313
15.4.2   Stationary State 314
15.4.3   Percolation Properties 315
15.4.4   Degree Distribution 315
15.5   Apollonian Networks 316
References 317

16   Excluded Volume Graphs 319
16.1   Delaunay Graph 319
16.2   Gabriel Graph 320
16.3   Relative Neighborhood Graph 323
16.4 β-Skeletons 323
References 326

17   Loops and Branches 327
17.1   Reducing the Complexity of a Spatial Network 327
17.2   A Loop and Branches Toy Model 330
17.2.1   Exact and Approximate Formulas for the BC 331
17.2.2   Threshold Value of w and Optimal ℓ 335
17.3   Analyzing the Impact of Congestion Cost 339
17.3.1   An Exactly Solvable Hub-and-Spoke Model 339
17.3.2   Congestion and Centralized Organization 343
References 346

18   Optimal Networks 347
18.1   Optimization, Complexity, and Efficiency 347
18.1.1   Complexity 347
18.1.2   Efficiency of Transport Network 348
18.2   Minimum Spanning Tree 350

	18.2.1	Minimum Spanning Tree on a Complete Graph	351
	18.2.2	Properties of the Euclidean Minimum Spanning Tree	353
18.3	Geometric t-Spanners		359
	18.3.1	Definition	359
	18.3.2	The Theta Graph	360
18.4	Optimal Trees: Generalization		361
18.5	Beyond Optimal Trees: Noise and Loops		366
References			369

19 Optimal Transportation Networks and Network Design 373
19.1	Empirical Motivation: The Structure of Subway Networks		374
19.2	Hub-and-Spoke Structure		376
19.3	One-Dimensional Problems		379
	19.3.1	A Single Open Line	379
	19.3.2	Transition for a Cyclic Service Line	383
19.4	Multiple Transit Lines		386
	19.4.1	Parallel Transit Lines	386
	19.4.2	Radial Lines	388
19.5	The Optimal Subway Problem		392
	19.5.1	A First Simplification: Optimal Placement	393
	19.5.2	The Minimum Average Distance to the Center	397
	19.5.3	Average Minimum Time Between All Pairs of Points	400
References			405

20 Greedy Models 407
20.1	A Model for Distribution Networks		408
20.2	Cost-Benefit Analysis		410
	20.2.1	Theoretical Formulation	411
	20.2.2	Crossover Between the Star Graph and the MST	412
	20.2.3	Spatial Hierarchy and Scaling	415
	20.2.4	Understanding the Scaling with a Toy Model	418
	20.2.5	Efficiency	419
20.3	Cost-Benefit Analysis: General Scaling Theory		422
	20.3.1	Theoretical Framework	424
	20.3.2	Subways	425
	20.3.3	Railways	429
References			433

Index 435

Acronyms

BA	Barabasi–Albert network
BC	Betweenness Centrality
CBA	Cost Benefit Analysis
dEMST	Dynamical euclidean minimum spanning tree
DT	Delaunay triangulation
EMST	Euclidean minimum spanning tree
ER	Erdos–Renyi graph
GG	Gabriel graph
GNN	Graph neural network
GSCC	Giant strongly connected component
GT	Greedy triangulation
GWCC	Giant weakly connected component
k-NNG	k Nearest Neighbor graph
ML	Machine learning
MST	Minimum spanning tree
NNG	Nearest Neighbor graph
OTT	Optimal traffic tree
PD	Persistent diagram
PH	Persistent homology
RNG	Relative neighborhood graph
SCC	Strongly connected component
SPT	Shortest path tree
STIT	Stability under iteration
TDA	Topological data analysis
WS	Watts–Strogatz graph

Part I
Characterization

Chapter 1
From Complex to Spatial Networks

The study of spatial networks—networks embedded in space—started essentially with quantitative geographers in the 60–70s who studied the structure and the evolution of transportation systems. The interest for networks was revived by Watts and Strogatz who opened the way to a statistical physics type of analysis and modelling of large networks. This renewed interest, together with an always growing availability of data, led to many studies of networks and their structures. Most of these studies focused on the topological properties of networks, leaving aside their spatial properties. It is only recently that researchers realized the importance of geometry—as opposed to topology—for spatial networks. In this chapter, we describe briefly the evolution of these fields and representations of spatial networks.

1.1 Early Days

The research activity on networks was intense these last two decades [1–4] but spatial networks were already the subject of many papers and books more than 40 years ago [5, 6]. In particular, in their great book [5], Haggett and Chorley explored the topology and the geometry of transportation networks (road networks, subways and railways). In the last chapter of their book, they addressed the problem of patterns of spatial evolution, a subject that is still at the heart of modern studies. In another study, Kansky [7] defined many indicators to characterize highways and roads, and in [8], Taafe, Morrill and Gould proposed a model for the evolution of road networks in cities, followed by many others (see [5]). Despite these various empirical and theoretical studies, the subject of the structure of these spatial networks was only revived later, first by geographers and then by physicists. The important difference between now and the 70s is certainly the availability of data, the existence of large

© The Author(s), under exclusive license to Springer Nature Switzerland AG 2022
M. Barthelemy, *Spatial Networks*,
https://doi.org/10.1007/978-3-030-94106-2_1

computer capacities and most importantly, a better knowledge of the structure of large networks.

In these early days, theoretical approaches were very simple and relied heavily on (basic) graph theory. The topological classification of networks amounted then essentially to distinguish between planar and non-planar, and for planar networks, betweeen trees and graphs with loops. For these networks, characterizations were basic and the indicators were mostly various combinations of the number of nodes N, the number of edges E and for planar graphs, the number of faces F. All these measures – the cyclomatic number, the Alpha, Beta and Gamma indices, the average degree, the average shortest path, etc. do not take into account the spatial nature of these networks, and therefore represent only one specific aspect of these objects. An important goal for geographers was then to understand the evolution of these systems and how these different network measures depend on socio-economical indicator. For example Kansky [7] discussed the relation of the Beta index (given by $\beta = E/N$) for railways with the gross energy consumption in different countries. Of course, they also investigated some spatial aspects such as the shape of the network, density of roads, flow properties, etc. We refer the interested reader to this work for more details.

1.2   Complex Networks

Independently from studies in quantitative geography or in graph theory, physicists started from 1998 to work intensively on networks. With the first paper on small-world networks by Watts and Strogatz [9], the statistical physics community realized that their tools for empirical analysis and modeling could be useful in other fields, even far from traditional objects of study in physics. This seminal work triggered a wealth of analysis of all possible networks available at that time. In particular, it was realized that many complex systems are very often organized under the form of networks and that these tools and models, and new ones, could have a strong impact across many disciplines. An important change of paradigm occurred when we realized [10, 11] that the usual Erdos-Renyi random network was not representative of most networks observed in real-world, and that we had to include large fluctuations of the degree. These strong fluctuations have a crucial impact on the dynamics that take place on these networks and many studies were devoted to this phenomenon [2].

This intense activity on complex systems thus led the researchers to think about the characterization of large networks. All the information is a priori encoded in the adjacency matrix but it is usually far too large and difficult to use under this form. In order to extract a smaller set of informations easily usable and that characterize the network, scientists introduced a number of measures that describe the statistical features of large networks. For example, for the class of networks with strong degree fluctuations, the degree distribution, the diameter, the clustering coefficient and the assortativity give a reasonable coarse-grained picture of the network and are in general enough to describe the dynamics on these networks. It appears that degree fluctuations are essential and govern many processes, but we note here that other quantities (such as correlations for example) can also play a critical role in the dynamics on networks (see for example the case of epidemics [12]).

1.3 Space Matters

These various studies on complex networks largely ignored space and considered that these networks were living in some abstract space with no metrics. In many cases indeed, the network is introduced as a simplified way to describe interactions between elements. Transportation and mobility networks, Internet, mobile phone networks, power grids, social and contact networks, neural networks, are however all embedded in space, and for these networks, space is relevant and topology alone does not contain all the information. In other words, in order to completely characterize these networks, we need not only the adjacency matrix but also the list of the positions of the nodes. An important consequence of space on networks is that there is naturally a cost associated to the length of edges which in turn has dramatic effects on the topological structure of these networks [13]. Characterizing and understanding the structure and the evolution of these 'spatial networks' became an important subject with many consequences in various fields ranging from epidemiology, neurophysiology, to ICT, urbanism and transportation studies. These networks are usually very large and we need statistical tools in order to describe the most accurately possible the salient aspects of their organization by taking into account both the topological and spatial aspects. In the first part of this book we will review the most important tools for their characterization.

1.4 Definition and Representations

The representation of a network is not unique and we introduce here the main definitions and representations used in the framework of spatial networks.

1.4.1 Spatial Networks

A graph $G = (V, L)$ is usually defined as a combination of a set V of N nodes and a set L of E links connecting these nodes. The $N \times N$ adjacency matrix A is then simply defined by $a_{ij} = 1$ if there is a link between nodes i and j and $a_{ij} = 0$ otherwise. This definition can be extended to weighted networks with $a_{ij} = w_{ij}$ where the weight w_{ij} denotes any quantity that flows on this link (i, j). Note that for directed networks, the matrix A is not symmetric and usually a_{ij} denotes the link that goes from j to i. Also, for practical purposes it is not necessary to store the whole

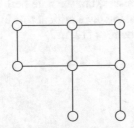
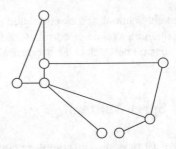

Fig. 1.1 These two networks have the same adjacency matrix and are topologically equivalent. However as shown here they can have a very different spatial representation and this information is encoded in the list of nodes' positions

matrix A as networks are often sparse and many elements are zeros. The convenient way to store the network is then to introduce the adjacency list which contains all the neighbors of a given node. While the full matrix necessitates to store N^2 elements, the adjacency list requires only to store a number at most equal to $N \times \Delta(G)$ where $\Delta(G)$ is the largest degree in the graph and is in general much smaller than N.

This standard representation of a graph is however not enough to describe a spatial network. The same graph can indeed be embedded in a plane in an infinite number of ways (see Fig. 1.1 for a simple example) and if we are interested in spatial features of the graph we need to specify this embedding. The minimum information needed for describing this aspect is the list of position of nodes: we denote $X = \{x_i\}$ this list. We will consider in most of the book that the quantity x_i for node i is a two-dimensional vector but for three-dimensional networks (such as the neural network for example), x_i is a 3d vector. Once we have $G = (V, L)$ and X, everything is known in principle about this spatial network and the purpose of simple characterizations is to extract useful, coarse-grained information from these large datasets.

1.4.2   Representations of Networks

Spatial networks can be represented directly by their embedding which is specified by the graph and the position of nodes, and which forms a 'map'. However in some cases (in particular for transportation systems, or the road network) it is useful to define other types of graphs. A specific example is the so-called dual network where we first identify 'lines' in the network (which are straight lines in the road network case or lines in the usual sense for subways). These lines (see Fig. 1.2 for a simple example) will be chosen as the nodes of the dual network and we connect two lines if they intersect. Note that the dual here is not the same as the dual graph used in the Voronoi-Delaunay construction for example (see chapter 14).

Concerning the important case of transportation networks, [15], Kurant and Thiran discuss very clearly the different representations of these systems (Fig. 1.3). The

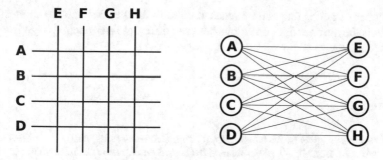

Fig. 1.2 (Left) Primal and (right) dual networks for a square lattice. In this example, the lattice in primal space has $N = 8$ routes. Each route has $k = N/2 = 4$ connections, so the total number of connections is $K_{tot} = k^2 = 16$. In the dual network, the four East-West routes (A, B, C, D) and the four North-South routes (E, F, G, H) form a bipartite graph $K_{4,4}$ with a diameter equal to 2. Figure taken from [14]

Fig. 1.3 a Direct representation of the routes (here for three different routes x, y, and z). **b** Space-of-changes (sometimes called P space [15]). A link connects two nodes if there is at least one vehicle that stops at both nodes. **c** Space-of-stops. Two nodes are connected if they are consecutive stops of at least one vehicle. **d** Space-of-stations. Here two stations are connected only if they are physically connected (without any station in between) and this network reflects the real physical network. Figure taken from [16]

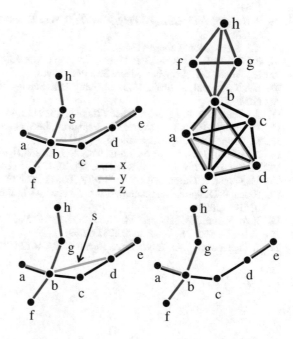

simplest representation is obtained when the nodes represent the stations and links the physical connections. One could however construct other networks such as the space-of-stops or the space-of-changes (see Fig. 1.3). One also finds in the literature on transportation systems, the notions of L and P-spaces [17–20] where the L-space connects nodes if they are consecutive stops in a given route. The degree in L-space is then the number of different nodes one can reach within one segment and the shortest path length represents the number of stops. In the P-space, two nodes are connected if there is at least one route between them so that the degree of a node is the number of nodes that can be reached directly. In this P-space the shortest path represents the number of connections between different lines needed to go from one node to another.

References

1. M. Newman, *Networks: An Introduction* (Oxford University Press, 2010)
2. A. Barrat, M. Barthélemy, A. Vespignani, *Dynamical Processes on Complex Networks*, reprint. (Cambridge University Press, Cambridge, 2012)
3. V. Latora, V. Nicosia, G. Russo, *Complex Networks: Principles, Methods and Applications* (Cambridge University Press, 2017)
4. F. Menczer, S. Fortunato, C.A. Davis, *A First Course in Network Science* (Cambridge University Press, 2020)
5. P. Haggett, R.J. Chorley, *Network Analysis in Geography*, vol. 67 (Edward Arnold London, 1969)
6. P. Haggett, A.D. Cliff, A. Frey, Tijdschrift Voor Economische En Sociale Geografie **68**(6) (1977)
7. K.J. Kansky, Ph.D. thesis (1963)
8. E.J. Taaffe, R.L. Morrill, P.R. Gould, Geograph. Rev. **53**(4), 503 (1963)
9. D.J. Watts, D.H. Strogatz, Nature **393**, 440 (1998)
10. L.A.N. Amaral, A. Scala, M. Barthelemy, H.E. Stanley, Proc. Nat. Acad. Sci. **97**(21), 11149 (2000)
11. A.L. Barabási, R. Albert, Science **286**(5439), 509 (1999)
12. M. Boguñá, R. Pastor-Satorras, A. Vespignani, Lect. Notes Phys. **625**, 127 (2003)
13. M. Barthelemy, Phys. Rep. **499**(1), 1 (2011)
14. R. Gallotti, M.A. Porter, M. Barthelemy, Sci. Adv. **2**(2), e1500445 (2016)
15. M. Kurant, P. Thiran, Phys. Rev. Lett. **96**(13), 138701 (2006)
16. M. Kurant, P. Thiran, Phys. Rev. E **74**(3), 036114 (2006)
17. P. Sen, S. Dasgupta, A. Chatterjee, P.A. Sreeram, G. Mukherjee, S.S. Manna, Phys. Rev. E **67**, 036106 (2003)
18. K.A. Seaton, L.M. Hackett, Physica A **339**, 635 (2004)
19. Y. Hu, D. Zhu, Physica A **388**, 2061 (2009)
20. J. Sienkiewicz, J.A. Hołyst, Phys. Rev. E **72.4**, 046127 (2005)

Chapter 2
Planar Graphs

In real-world applications, the spatial networks we encounter are in general planar. By definition, these networks can be represented in a two-dimensional plane without any edge crossings. Planar graphs pervade many different fields from abstract mathematics [1, 2], to quantum gravity [3], botanics [4, 5], geography and urban studies [6]. In particular, planar graphs are central in biology where they can be used to describe veination patterns of leaves or insect wings and display an interesting architecture with many loops at different scales [4, 5]. In the study of urban systems, planar networks are extensively used to represent, to a good approximation, various infrastructure networks [6] such as transportation networks [7] and streets patterns [8–26]. The planarity properties imply a certain number of relations—including the famous Euler relation—which explains their interest for theoretical studies.

However, in practical applications, some graphs are almost planar and it will be interesting to characterize the deviation from planarity both from an empirical and theoretical point of view. Mathematicians developped tools that quantifies how far from planarity a graph is and we will discuss some of them here. More precisely, if a graph G is not planar, the natural question that arises [27, 28] is how far away is G from planarity? We will also discuss planarization which is a related problem that is considered in graph algorithms and applications. Its formal definition is the following one [27]: Given a graph G, we consider operations on G that transform it into a planar graph.

From an empirical point of view, Boeing [29] analyzed the planarity of street networks and we will discuss his results at then end of this chapter.

M. Barthelemy, *Spatial Networks*,
https://doi.org/10.1007/978-3-030-94106-2_2

2.1 Graph Theoretical Tools

2.1.1 Definition and Representation

A graph is planar when there is at least one plane embedding such that no edges cross each other (see for example the graph textbook [30]). However if a certain embedding displays edge crossing, it does not necessarily mean that the graph is non-planar. Standard graph theory shows that a necessary and sufficient condition for planarity is the absence of subgraphs homeomorphic to the two graphs: K_5 or $K_{3,3}$ (see Fig. 2.1 where the complete graph K_n with $n = 5$ nodes and the complete bipartite graph $K_{n,m}$ with $n = 3$ and $m = 3$ are shown). This is the Kuratowsky theorem (see for example the textbook [30]) and there are efficient algorithms that can test this in $\mathcal{O}(N)$ time (see for example [31]).

Basically, a planar graph is thus a graph that can be drawn in the plane in such a way that its edges do not intersect. Not all drawings of planar graphs are without intersection and a drawing without intersection is sometimes called a plane graph or a planar embedding of the graph (the term planar map is also frequently used in combinatorial studies). In real-world cases, these considerations actually do not apply since the nodes and the edges represent in general physical objects.

A planar graph can also be embedded in the plane with edges mapped to straight line segments (Fig. 2.2) and this drawing is called a planar straight line graph. A theorem proves such an embedding exists for every planar graph [32]. This is an important object in computational geometry [33] and is useful for the representation of maps in geographical information systems. Such a graph without dead ends (i.e. vertices of degree 1) defines therefore a subdivision of the 2d space into polygonal regions. Many graphs belong to this class of planar straight line graphs, in particular, triangulations which are in addition maximal in the sense that any added edge will destroy planarity.

Fig. 2.1 Complete graphs K_5 and $K_{3,3}$. The Kuratowsky theorem states that all non-planar graphs have subgraphs homeomorphic to one (or both) of these graphs

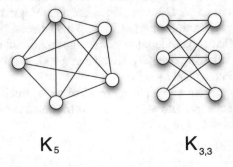

$$K_5 \qquad\qquad\qquad K_{3,3}$$

Fig. 2.2 Example of a planar straight line graph: the planar embedding is drawn with straight lines for edges

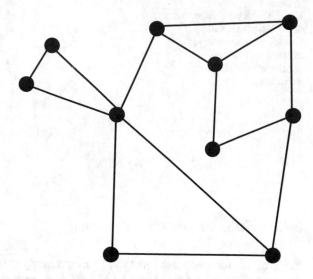

2.1.2 K_5 Is Not Planar

We note here that it is not trivial to demonstrate that a graph is non-planar and the demonstration is simplified by invoking the Jordan curve theorem (see for example [34]) which asserts that a continuous, non self-intersecting closed loop divides the plane into an 'interior' and an 'exterior' that can be connected by a continuous path that has to intersect the loop somewhere. In order to illustrate a non-planarity demonstration, we follow here [30] in the case of the complete graph K_5. We assume that K_5 is planar and will reach a contradiction. We denote its vertices by v_1, v_2, v_3, v_4, v_5 and since they are all connected to each other, the loop $C = v_1 v_2 v_3 v_1$ exists and is a Jordan curve separating an inside from an exterior domain. The node v_4 does not lie on C and we assume that it is in the inside domain (there is a similar argument in the other case where v_4 is outside). The interior of C is then divided into three different domains Ω_i for $i = 1, 2, 3$ delimited by the Jordan curves $C_1 = v_1 v_2 v_4 v_1$, $C_2 = v_4 v_2 v_3 v_4$ and $C_3 = v_4 v_3 v_1 v_4$ (see Fig. 2.3). The remaining node v_5 must then lie in either $\Omega_1, \Omega_2, \Omega_3$, or in the exterior of C. If $v_5 \in ext(C)$ then since $v_4 \in int(C)$ the Jordan theorem implies that the edge $v_4 v_5$ must cross C. If $v_5 \in \Omega_1$ (the two other cases are similar), we note that v_3 is exterior to this domain and according to the Jordan curve theorem the edge $v_3 v_5$ must cross the curve C_1. We thus find that K_5 cannot be planar.

Fig. 2.3 Demonstration that
K_5 is non planar. Case
considered in the text: v_4 is
inside C and divides the
interior of C in three
domains

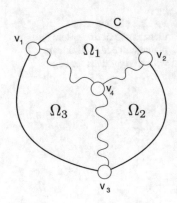

2.1.3   Euler's Formula

Basic results for planar networks can be found in any graph theory textbook (see for example [30]) and we recall here briefly the most important ones.

We start with very general facts that can be demonstrated for planar graphs, and among them Euler's formula is probably the best known. Euler showed that a finite connected planar graph satisfies the following formula

$$N - E + F = 2 \tag{2.1}$$

where N is the number of nodes, E the number of edges, and F is the number of faces. This formula can be easily proved by induction by noting that removing an edge decreases F and E by one, leaving $N - E + F$ invariant. We can repeat this operation until we get a tree for which $F = 1$ (the space around the tree counts as one face) and $N = E + 1$ leading to $N - E + F = E + 1 - E + 1 = 2$. This argument can be repeated in the case where the graph is made of C disconnected components and the Euler relation reads in this case

$$N - E + F = C + 1 \tag{2.2}$$

Moreover for any finite connected planar graph we have a bound for the average degree $\langle k \rangle$. Indeed, any face is bounded by at least three edges and every edge separates two faces at most which implies that $E \geq 3F/2$ (in other words $2E/3$ is an upper bound of the number of faces). From Euler's formula we then obtain

$$E \leq 3N - 6 \tag{2.3}$$

In other words, planar graphs are always sparse with a bounded average degree

$$\langle k \rangle = \frac{2E}{N} \leq 6 - \frac{12}{N} \tag{2.4}$$

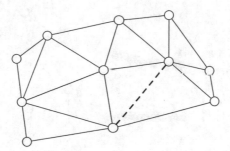

Fig. 2.4 Example of a triangulation constructed on a small set of point. If a face is not a triangle, we can always divide it in smaller triangles and preserve the planarity (we represented such an additional division by the dashed line in this figure)

which is therefore always smaller than 6. In other words, if a graph has an average degree $\langle k \rangle > 6$ it cannot be planar.

We end this part with a particular class of planar graphs that are constructed on a set of points distributed in the plane (see the Chap. 14 about tessellations). The maximal planar graph is obtained on a set of points if we cannot add another edge without violating the planarity. Such a planar graph is necessary a triangulation where all faces are triangles (indeed if a face is not a triangle we can always 'break' it into smaller triangles while preserving planarity—see the example of the dashed edge in Fig. 2.4). Such a planar network is useful in practical applications in order to assess for example the efficiency of a real-world planar network and provides an interesting null model.

For such a triangulation we have the equality $3F = 2E$ and using Euler's relation we obtain that the number of edges and faces are maximal and are equal to the bounds $E = 3N - 6$ and $F = 2N - 4$, respectively.

Obviously an important aspect of spatial, planar networks is the shape of faces that will contribute to the whole visual pattern. In the Chap. 4 we discuss the distribution of the area and the shape of faces and in the Chap. 9 we will discuss how these measures can be used for constructing a typology of planar graphs.

2.1.4 Distance from Planarity

2.1.4.1 The Crossing Number

As we have seen, we have to carefully distinguish the planarity of the graph—a topological notion—and the planarity of the embedding. When a graph is non-planar, it is impossible to find a two-dimensional representation without edge crossings. In contrast, even if a graph is planar, we can have embeddings that are not planar (see Fig. 2.5). The planarity is thus a topological concept and edge-crossing is a geometrical feature. It is therefore not obvious to relate the non-planarity of a graph

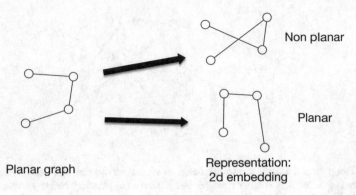

Fig. 2.5 A graph can be planar and have either non planar or planar 2d representations. In contrast, a non planar graph cannot have a 2d representation without crossings

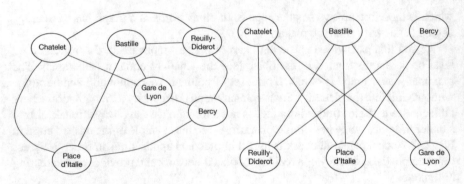

Fig. 2.6 (Left) 'Map' representation where the relative position of the nodes are respected. (Right) Same graph but which displays more clearly the $K_{3,3}$ structure. Despite this structure we observe that in the real graph there is only one edge crossing

and the local edge-crossing of the spatial network (which can be seen as an embedding of the graph). We can illustrate this on the case of the Paris subway. The planarity testing algorithm gives the subgraph shown in Fig. 2.6(left). We first represent the 'map' where we respect the relative spatial locations of the nodes. This is in contrast with the usual graph representation shown in Fig. 2.6(right) where we recognize the complete bipartite graph $K_{3,3}$. This example shows clearly the difference between planarity in the topological sense with the presence of subgraphs that are either $K_{3,3}$ or K_5, and the existence of edge-crossings in the real spatial network. Here we obwerve that we have only one crossing between the two lines Chatelet-Gare de Lyon and Bastille-Place d'Italie. In this respect, even if the graph is non-planar, the number of planarity violations in the spatial network is limited to one such event.

This notion of edge crossing has been formalized in graph theory with the crossing number $cr(G)$ of a graph G, defined as the lowest number of edge crossings of a plane drawing of G. This number is of practical importance: for example a circuit laid out on a chip corresponds to drawing a graph in 2d and wire crossings can cause potential problems and their number should be minimized.

For planar graphs, we thus have $cr(G) = 0$. We also note (a simple drawing can show that) that we have $cr(K_5) = cr(K_{3,3}) = 1$ suggesting that these graphs are the 'elementary bricks' for non-planarity and provides an intuitive reason why they appear in the Kuratowski theorem.

In general, the crossing number is very difficult to compute (and might be a NP-complete problem) and the interested readers can find some discussions about this problem in the excellent graph theory book [35] or in the more specialized survey [36]. A simple lower bound can however be constructed in the following way. We start from a graph G with N nodes, E edges, and $cr(G)$ crossings, and construct the graph H with the following rule: the nodes of H are those of G together with all crossing points that are now considered as nodes. The edges are the original ones together with the pieces of edges going from original nodes to crossing points. This new graph H has $N + cr(G)$ nodes, $E - 2cr(G) + 4cr(G) = E + 2cr(G)$ edges (each new vertex has a degree 4) and is planar. Using then Eq. (2.3) for the new graph H, we obtain

$$E + 2cr(G) \leq 3(n + cr(G)) - 6 \tag{2.5}$$

that is

$$cr(G) \geq E - 3N + 6 \tag{2.6}$$

This bound is good enough when E is linear in N (that is for a constant average degree) but for denser graphs where E is growing faster than N we need better bounds (see below).

Mathematicians worked in particular on the complete graph K_n where we have the scaling $cr(K_n) \sim n^4$. This scaling can be understood by computing the largest number of crossings [35]: if we place the vertices on a circle, in order to produce a crossing we have to choose 4 vertices, and if we assume that we always create a crossing, we obtain

$$cr(K_n) \leq \binom{n}{4} \sim n^4 \tag{2.7}$$

Actually better bounds can be found (see [35])

$$\frac{1}{80}n^4 + \mathcal{O}(n^3) \leq cr(K_n) \leq \frac{1}{64}n^4 + \mathcal{O}(n^3) \tag{2.8}$$

We also have results for the complete bipartite $K_{n,m}$ (for a survey of various results, see [36])

$$\frac{m(m-1)}{5} \left\lfloor \frac{n}{2} \right\rfloor \left\lfloor \frac{n-1}{2} \right\rfloor \leq cr(K_{m,n}) \leq \left\lfloor \frac{n}{2} \right\rfloor \left\lfloor \frac{n-1}{2} \right\rfloor \left\lfloor \frac{m}{2} \right\rfloor \left\lfloor \frac{m-1}{2} \right\rfloor \quad (2.9)$$

(where $\lfloor x \rfloor$ is the lowest nearest integer of x and $\lceil \cdot \rceil$ denote the nearest upper integer), demonstrating a scaling of the form

$$cr(K_{m,n}) \sim m^2 n^2 \quad (2.10)$$

Despite these results, crossing numbers are however not well known and only few general results are available. In particular, there is a theorem (Ajtai-Chvatal-Newborn-Szemeredi and Leighton, see [35]), that states that for a simple graph G with $E \geq 4N$ (which means an average degree $\langle k \rangle \geq 8$), the following bound holds

$$cr(G) \geq \frac{1}{64} \frac{E^3}{N^2} \quad (2.11)$$

We provide now a simple demonstration of this result which illustrates the fact that probability makes counting sometimes easy [37]. We consider a minimal drawing of a graph G and let p be a number in $[0, 1]$. We generate a subgraph of G denoted by G_p by selecting the nodes with probability p independently from each other. In this subgraph, we have N_p nodes, E_p edges, and X_p crossings. Since the bound Eq. (2.6) is true for all graphs, we certainly have the following inequality

$$\langle X_p - E_p + 3N + p \rangle \geq 0 \quad (2.12)$$

where $\langle \cdot \rangle$ denotes the average over all graphs constructed for a given value of p. It is easy to see that $\langle N_p \rangle = Np$ and $\langle E_p \rangle = p^2 E$ since an edge appears in G_p only if the two end nodes are present. A crossing is present only if all the four distinct vertices involved are present which implies that $\langle X_p \rangle = p^4 cr(G)$. We then obtain

$$cr(G) \geq \frac{pE - 3N}{p^3} \quad (2.13)$$

We now assume that $p = cN/E$ which implies that $E \geq cN$ and the previous equation becomes

$$cr(G) \geq \frac{c - 3}{c^3} \frac{E^3}{N^2} \quad (2.14)$$

For $c = 4$ we recover the result Eq. (2.11), but we note that the maximum constant prefactor is $4/243$ obtained for $c = 4.5$ (see [38] for a longer discussion on bounds).

2.1.4.2 Thickness, Skewness

If a graph G is not planar, the natural question that arises [27, 28] is how far away is G from planarity? The crossing number is a first indication, but there are other measures. For instance, the thickness of a graph, denoted by $\theta(G)$ is the minimum number of planar subgraphs into which the graph can be decomposed. The thickness of a planar graph is then equal to 1 and the thickness of a nonplanar graph is at least 2. More generally it has been proven that determining the thickness of a graph is a NP-complete problem [39]. We know however some results and bounds for the thickness (see [40] and references therein) and here is a partial list:

- For complete graphs, $\theta(K_n) = \lfloor \frac{n+7}{6} \rfloor$ except for $\theta(K_9) = \theta(K_{10}) = 3$. In particular the long debated problem Harari [40] whether $\theta(K_{16})$ is 3 or 4 is solved with $\theta(K_{16}) = 3$. It is interesting to note that $n = 16$ is the largest integer such that $\lfloor \frac{n+7}{6} \rfloor = 3$.
- For the complete bipartite graphs $K_{m,n}$, the problem is solved for almost all values of (m, n): $\theta(K_{m,n}) = \lceil mn/(m + n - 2) \rceil$ except when m and n are odd and there is an integer k such that $n = \lfloor 2k(n - 2)/(n - 2k) \rfloor$. If $m = n$, there is a shorter result: $\theta(K_{n,n}) = \lfloor \frac{n+5}{4} \rfloor$.

The skewness is another measure given by the number of edges that one has to remove from a graph G to obtain the maximum planar subgraph G' (i.e. the planar subgraph with the largest number of edges). If we denote by E the number of edges of G and E' of G', the skewness is then $E - E'$. The skewness of a graph is 0 if and only if the graph is planar. Unfortunately, finding the maximum planar subgraph is a NP-hard problem [27] which limits its scope of application.

2.1.4.3 Planarization: Splitting Number

There are various ways to transform a graph G into a planar one and a survey of results can be found in [27]. We will here discuss briefly vertex splitting operations as a mean to planarize a given graph.

First, we need to define what is meant by vertex splitting. We choose a vertex v and split it in two new vertices v_1 and v_2, and all the links originally connected to v are now connecting v_1 and v_2. The vertex splitting operation can be seen as the almost exact reverse of the following process. We start by deleting an edge $v_1 v_2$ that belongs to a graph G' and add a vertex v. The edges connected to v_1 and v_2 are then connected to v in this new graph G. We then say that the graph G' is obtained from G by splitting the vertex v. More generally, if a graph G' is obtained from a graph G by a sequence of vertex splitting operations, G' is a splitting of G. Note that going from G' to G is unique but the reverse is not true: given a graph G and a vertex v, there are many ways to split this vertex.

This splitting operation is central in many parts of graph theory such as the characterization of various classes of graphs (see for example [27] and references therein). It allows to define the splitting number $\sigma(G)$ of a graph G as the smallest number k

such that G can be obtained from planar graph G' by k vertex identification (which is the inverse of vertex splitting). Of course, $\sigma(G) = 0$ if G is planar. As for the other measures, determining the splitting number of a graph is a NP-hard problem and theoretically we have only bounds. For example, for a graph with $N \geq 3$, we have [27]

$$\sigma(G) \geq \lceil \frac{1}{3}E - N + 2 \rceil \tag{2.15}$$

2.2 Planarity of Street Networks

Street networks are embedded in space and represent the archetype of spatial networks. In the primal representation, the nodes are the intersections and links represent road segments between consecutive intersections. As discussed above, we have to distinguish the graph and its embedding. Along with its topology a spatial network has a geometry defined by geographical coordinates, lengths, shapes, and angles. A street network might then be non planar due to its embedding in space with overpasses or underpasses, while the graph representing it could be planar. In most studies of street networks, authors considered them as planar. Strictly speaking, this is obviously not true but this approximation is usually considered as an harmless simplification. Indeed, it can be argued that if there is a non-planarity in the street network (which means that two roads A and B crosses at a point which is not an intersection), there must be a way to go from one road to the other one. For example, on highways we have interchange roads and planarity holds at a coarser level: we can expect that in general the detour to go from road A to road B is small compared to the system size and that we could add a node at the crossing between A and B without modifying the large scale structure of the network.

Empirically, Goeffrey Boeing considered this problem seriously and proposed a quantitative assessment of this problem [29]. Boeing considered the data for consistently sized square-mile network at the centers of 50 major cities worldwide belonging to all continents. Boeing first tested the planarity of these networks, and defined a spatial planarity ratio ϕ which represents the ratio of the number of intersections $i_n = N(k \geq 2)$ observed in the spatial network to the number of all crossing i_p ($N(k)$ is the number of intersections with degree k)

$$\phi = \frac{i_n}{i_p} \tag{2.16}$$

The number $i_p - i_n$ thus represents the number of nonplanar edge crossings (i.e. over- or underpasses). The quantity $\phi \leq 1$ and the equality is reached for a spatial network without any over- or underpasses.

When we assume that the graph is planar we underestimate the average edge length by neglecting over/underpasses and we can then define two average length:

the average length ℓ_p of planar edges (considering all crossings as nodes) to the average length ℓ_n of nonplanar edges. We then construct the edge length ratio λ given by the ratio of these two lengths

$$\lambda = \frac{\ell_p}{\ell_n} \tag{2.17}$$

For a spatially planar street network, $\lambda = 1$ and in general $1 - \lambda$ counts the percentage of how much error we do by identifying all the crossings as nodes.

The results of this analysis are shown in the Table 2.1 for the drivable street networks and for the walkable network. Basically, among the drivable networks considered in this study, only a minority (20%) is formally planar. In addition and as expected, not all formally planar street network are spatially planar: for example Toronto is formally planar but $\phi = 92\%$ and $\lambda = 95\%$. The quantity ϕ and λ vary from 100% to values as low as $\phi = 54\%$ and $\lambda = 60\%$ for Moscow. On average, planar models that neglect nonplanarities, overcount intersections by 16% and underestimate street segment length by 11% (and similar numbers for the walkable street network, for more results see [29]).

More generally, it seems that older European cities are more planar than more recent american or chinese metropolises. These results display a large variability in the impact of considering a planar simplification and additional analysis should be conducted in each case in order to assess the potential errors by neglecting nonplanarities.

Table 2.1 Table of different planarity indicators for 50 major worldwide cities. From [29]

Country	City	Drive			Walk		
		Planar	ϕ	λ	Planar	ϕ	λ
Argentina	Buenos Aires	Yes	1.000	1.000	No	0.939	0.941
Australia	Sydney	No	0.729	0.735	No	0.902	0.884
Brazil	Sao Paulo	No	0.771	0.772	No	0.824	0.803
Canada	Toronto	Yes	0.922	0.946	No	0.838	0.824
	Vancouver	No	0.930	0.948	No	0.923	0.920
Chile	Santiago	No	0.873	0.885	No	0.967	0.965
China	Beijing	No	0.818	0.846	No	0.842	0.842
	Hong Kong	No	0.838	0.823	No	0.823	0.794
	Shanghai	No	0.682	0.708	No	0.660	0.641
Denmark	Copenhagen	Yes	0.992	0.988	No	0.994	0.985
Egypt	Cairo	No	0.897	0.913	No	0.899	0.886
France	Lyon	No	0.995	0.991	No	0.958	0.953
	Paris	No	0.982	0.987	No	0.928	0.907
Germany	Berlin	No	0.939	0.945	No	0.939	0.931
India	Delhi	Yes	1.000	1.000	Yes	0.997	0.989
Indonesia	Jakarta	Yes	0.990	0.994	No	0.969	0.965

(continued)

Table 2.1 (continued)

		Drive			Walk		
Iran	Tehran	No	0.962	0.973	No	0.953	0.951
Italy	Bologna	Yes	1.000	1.000	Yes	0.996	0.996
	Florence	Yes	0.993	0.994	No	0.980	0.974
	Milan	Yes	1.000	1.000	No	0.849	0.832
Japan	Osaka	No	0.868	0.868	No	0.953	0.950
	Tokyo	No	0.925	0.919	No	0.924	0.912
Kenya	Nairobi	No	0.974	0.971	No	0.949	0.938
Mexico	Mexico City	No	0.940	0.949	No	0.912	0.917
Nigeria	Lagos	No	0.960	0.972	No	0.990	0.988
Peru	Lima	No	0.941	0.951	No	0.923	0.921
Philippines	Manila	No	0.940	0.947	No	0.897	0.883
Russia	Moscow	No	0.540	0.596	No	0.842	0.833
Singapore	Singapore	No	0.864	0.869	No	0.891	0.880
Somalia	Mogadishu	Yes	1.000	1.000	Yes	1.000	1.000
South Africa	Johannesburg	No	0.847	0.877	No	0.997	0.997
Spain	Barcelona	Yes	1.000	1.000	No	0.925	0.917
Switzerland	Geneva	No	0.985	0.981	No	0.834	0.807
Thailand	Bangkok	No	0.993	0.979	No	0.990	0.984
Turkey	Istanbul	No	0.965	0.964	No	0.973	0.964
UAE	Dubai	No	0.679	0.668	No	0.852	0.837
UK	Edinburgh	No	0.974	0.965	No	0.987	0.983
	London	No	0.976	0.980	No	0.853	0.836
USA	Atlanta	No	0.720	0.765	No	0.805	0.788
	Chicago	No	0.748	0.786	No	0.792	0.787
	Cincinnati	No	0.723	0.746	No	0.929	0.922
	Dallas	No	0.584	0.639	Yes	0.961	0.956
	Los Angeles	No	0.581	0.627	No	0.784	0.785
	Miami	No	0.641	0.657	No	0.962	0.961
	New York	No	0.879	0.878	No	0.928	0.923
	Phoenix	No	0.949	0.958	No	0.977	0.972
	San Francisco	No	0.935	0.937	No	0.942	0.935
	Seattle	No	0.728	0.771	No	0.931	0.924
	Washington DC	No	0.948	0.956	No	0.964	0.961
Venezuela	Caracas	No	0.953	0.957	Yes	1.000	1.000

References

1. W. Tutte, Canadian J. Math. **15**, 249 (1963)
2. J. Bouttier, P. Di Francesco, E. Guitter, Electron. J. Combin. **11**(1), R69 (2004)

3. J. Ambjørn, T. Jónsson, *Quantum Geometry: A Statistical Field Theory Approach* (Cambridge University Press, 1997)
4. Y. Mileyko, H. Edelsbrunner, C.A. Price, J.S. Weitz, PloS one **7**(6), e36715 (2012)
5. E. Katifori, M.O. Magnasco, PloS one **7**(6), e37994 (2012)
6. M. Barthelemy, Phys. Rep. **499**(1), 1 (2011)
7. P. Haggett, R.J. Chorley, *Network Analysis in Geography*, vol. 67 (Edward Arnold London, 1969)
8. B. Hillier, J. Hanson, *The Social Logic of Space*, vol. 1 (Cambridge University Press Cambridge, 1984)
9. S. Marshall, *Streets and Patterns* (Routledge, 2004)
10. B. Jiang, C. Claramunt, Environ. Plann. B **31**(1), 151 (2004)
11. M. Rosvall, A. Trusina, P. Minnhagen, K. Sneppen, Phys. Rev. Lett. **94**(2), 028701 (2005)
12. S. Porta, P. Crucitti, V. Latora, Environ. Plann. B **33**, 705 (2006)
13. S. Porta, P. Crucitti, V. Latora, Physica A: Stat. Mech. Appl. **369**(2), 853 (2006)
14. S. Lämmer, B. Gehlsen, D. Helbing, Physica A: Stat. Mech. Appl. **363**(1), 89 (2006)
15. P. Crucitti, V. Latora, S. Porta, Phys. Rev. E **73**(3), 036125 (2006)
16. A. Cardillo, S. Scellato, V. Latora, S. Porta, Phys. Rev. E **73**(6), 066107 (2006)
17. F. Xie, D. Levinson, Geograph. Anal. **39**(3), 336 (2007)
18. B. Jiang, Physica A: Stat. Mech. Appl. **384**(2), 647 (2007)
19. A. Masucci, D. Smith, A. Crooks, M. Batty, Eur. Phys. J. B-Condens. Matter Complex Syst. **71**(2), 259 (2009)
20. S.H. Chan, R.V. Donner, S. Lämmer, Eur. Phys. J. B-Condens. Matter Complex Syst. **84**(4), 563 (2011)
21. T. Courtat, C. Gloaguen, S. Douady, Phys. Rev. E **83**(3), 036106 (2011)
22. E. Strano, V. Nicosia, V. Latora, S. Porta, M. Barthelemy, Sci. Rep. **2** (2012)
23. M. Barthelemy, P. Bordin, H. Berestycki, M. Gribaudi, Sci. Rep. **3** (2013)
24. M.P. Viana, E. Strano, P. Bordin, M. Barthelemy, Sci. Rep. **3** (2013)
25. R. Louf, M. Barthelemy, Sci. Rep. **4** (2014)
26. S. Porta, O. Romice, J.A. Maxwell, P. Russell, D. Baird, Urban Stud. 0042098013519833 (2014)
27. A. Liebers, in *Graph Algorithms and Applications 2* (World Scientific, 2004), pp. 257–330
28. M. Chimani, C. Gutwenger, Disc. Math. **309**(7), 1838 (2009)
29. G. Boeing, Environment and planning B: urban analytics and city science **47**(5), 855 (2020)
30. J. Clark, D. Holton, *A First Look at Graph Theory* (World Scientific, 1991)
31. D. Jungnickel, *Algorithm and Computation in Mathematics*, vol. 5 (Springer, Heidelberg, 1999)
32. F. István, Acta scientiarum mathematicarum **11**(229–233), 2 (1948)
33. F.P. Preparata, M.I. Shamos, *Computational Geometry: an Introduction* (Springer Science & Business Media, 2012)
34. T.C. Hales, Am. Math. Mon. **114**(10), 882 (2007)
35. D.B. West, *Introduction to Graph Theory*, vol. 2 (Prentice Hall, Upper Saddle River, 2001)
36. M. Schaefer, Electron. J. Comb. **1000**, DS21 (2013)
37. M. Aigner, G.M. Ziegler, *Proofs from The Book* (Berlin, Germany, 1999)
38. L.A. Székely, Disc. Math. **276**(1–3), 331 (2004)
39. A. Mansfield, in *Mathematical Proceedings of the Cambridge Philosophical Society*, vol. 93 (Cambridge University Press, 1983), pp. 9–23
40. E. Mäkinen, T. Poranen, *An Annotated Bibliography on the Thickness, Outerthickness, and Arboricity of a Graph* (2009)

Chapter 3
Directed and Mixed Graphs

Most studies focusing on directed networks are not considering spatial networks. However, directed spatial networks are not that seldom. Street networks with one-ways provide a natural example, but we can also think of power grids, river networks, and naturally transportation networks etc. [1] where there is a flow indicating a preferred direction. In this case, the edges can represent a direction from node i to j that is allowed while the opposite direction from j to i is not necessarily present. In this chapter, we will discuss some general properties of digraphs and then consider the important empirical case of street networks with one-ways. This will lead us in particular to the notion of mixed graphs where links can be either directed or not.

3.1 Theoretical Results

3.1.1 Definitions

Formally, a directed graph, or 'digraph', $D = (V, A)$ consists of a set of vertices V and a set A of ordered pairs of vertices called arcs in graph theory (or directed edges in less formal settings). We can represent a digraph by drawing the vertices as points and arcs as arrows from a point to another. All the usual quantities defined for graphs such as paths, cycles, etc. apply naturally to digraphs. The important point to remember is that a path from a node i to another node j can only use arcs along their direction. One could for example think of a street network with one-ways: a car driver can go from a point of the network to another if there is a path that respects one-ways. A path from a node v_1 to a node $v_{\ell+1}$ in a digraph is then a sequence $v_1, a_1, v_2, a_2, \ldots, v_\ell, a_\ell, v_{\ell+1}$ where a_i is the arc from v_i to v_{i+1}. The length of this path is here ℓ. A *cycle*–also sometimes called a *loop*–is a closed path $(i_0 = i_n)$ in

© The Author(s), under exclusive license to Springer Nature Switzerland AG 2022
M. Barthelemy, *Spatial Networks*,
https://doi.org/10.1007/978-3-030-94106-2_3

which all vertices and all edges are different. It is clear that for a given number of nodes the number of loops increases with the number of edges.

Starting from a digraph D with arcs between nodes, when an arc exists between nodes i and j, and also from j to i (this would correspond to a two-ways road for example), we can replace the two arcs by an undirected edge. The resulting object constructed over a set of vertices joined by one-way arcs or undirected edge is called a mixed graph.

For digraphs, we define the incoming degree (also called the in-degree) $k_{in}(i)$ of a node i as the number of incoming arcs. Similarly, the outgoing degree (or out-degree) $k_{out}(i)$ is the number of outgoing links. Since each directed link contributed to both total incoming and outgoing degree, we have the trivial relation between the average quantities

$$\langle k_{in} \rangle = \langle k_{out} \rangle \tag{3.1}$$

which is valid for any digraph.

3.1.2 Components of a Digraph

The notion of path in digraphs allows us to discuss the reachability of vertices, i.e. the possibility of going from one vertex to another following the connections given by the arcs in the network.

A graph is called *connected* if there exists a path connecting any two vertices in the graph. A *component* $\mathscr{C}$ of a graph is defined as a connected subgraph. Two components $\mathscr{C}_1 = (\mathscr{V}_1, \mathscr{E}_1)$ and $\mathscr{C}_2 = (\mathscr{V}_2, \mathscr{E}_2)$ are disconnected if it is impossible to construct a path $\mathscr{P}_{i,j}$ with $i \in \mathscr{V}_1$ and $j \in \mathscr{V}_2$. In a connected network every vertex is reachable from any other vertex. The connected components of a graph thus define many properties of its physical structure.

An interesting property of graphs is then the distribution of components, and in particular the existence of a *giant component* $\mathscr{G}$, defined as the component whose size scales with the number of vertices of the graph, and therefore diverges in the limit $N \to \infty$. The presence of a giant component implies that a macroscopic fraction of the graph is connected, in the sense that it is possible to find a way across a certain number of edges, joining any two vertices.

In a digraph, the structure of the components is somewhat more complex as the presence of a path from the node i to the node j does not necessarily guarantee the presence of a corresponding path from j to i. Therefore, the definition of a giant component needs to be adapted to this case. In general, the component structure of a directed network can be decomposed into a giant weakly connected component (GWCC), corresponding to the giant component of the same graph in which the edges are considered as undirected, plus a set of smaller disconnected components (DC) (see Fig. 3.1). The GWCC is itself composed of several parts due to the directed nature of its edges: The giant strongly connected component (GSCC), in which there

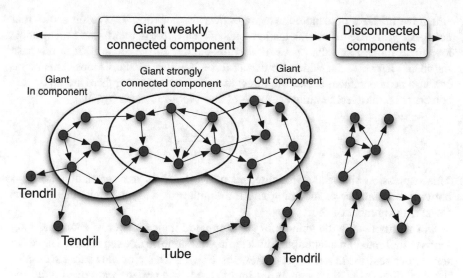

Fig. 3.1 General component structure of a directed graph

is a directed path joining any pair of nodes. The giant IN-component (GIN), formed by the nodes from which it is possible to reach the GSCC by means of a directed path. The giant OUT-component (GOUT), formed by the nodes that can be reached from the GSCC by means of a directed path. Last but not least there are the tendrils that contains nodes that cannot reach or be reached by the GSCC (among them, the tubes that connect the GIN and GOUT) that form the rest of the GWCC.

3.1.3 Statistics of Loops

The structure of digraphs can be characterized by the statistics of cycles, or loops . In undirected networks, there are usually many short loops leading to a large clustering which together with a small diameter leads to the small-world structure. The statistics of loops in directed networks has been studied in [2] and we discuss here some of these results. In particular, the main result is that for a large class of directed networks including foodwebs, power-grids, metabolic, neural, transcription and WWW networks, the number of loops is much smaller than what is expected for a random graph. For example, the directed neural network of the C. Elegans has less than 50% of the short loops expected for a random graph constructed over the same vertices and with the same degree sequence.

For an undirected random graph, the average number $N_u(L)$ of short loops of length L is given by [3, 4]

$$N_u(L) = \frac{1}{2L} \left(\frac{\langle k^2 \rangle - \langle k \rangle}{\langle k \rangle} \right)^L \tag{3.2}$$

where the brackets $\langle \cdot \rangle$ denote the average over the network. This result shows that networks with large degree fluctuations such as scale-free networks have many short loops and much more than networks without hubs. In particular, this is the case for spatial networks for which we expect a low number of short loops. This result can also be generalized to random directed networks and when there are no degree correlations, the average number $N_d(L)$ of short loops is given by [2]

$$N_d(L) = \frac{1}{L} \left(\frac{\langle k_{in} k_{out} \rangle}{\langle k_{in} \rangle} \right)^L \tag{3.3}$$

This expression shows that for directed networks, it is the correlation $\langle k_{in} k_{out} \rangle$ between the number of incoming and the number of outgoing links that governs the average number of short loops.

For real networks the authors of [2] compared their number of short and long loops to their random counterpart (by keeping the same degree sequence). The result for various real-world networks is shown in Fig. 3.2 This plot shows that six out of the nine directed networks analyzed in this study are under-shortlooped, with the only exceptions of the metabolic and transcription networks, which are marginally

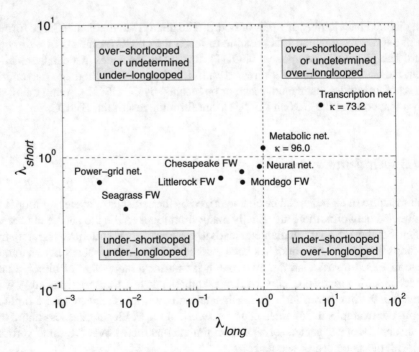

Fig. 3.2 Overrepresentation of short and long loops in real-world networks. When $\lambda_{short} \gg 1$ short loops are over-represented (compared to a random network with the same degree distribution), and when $\lambda_{short} \ll 1$ short loops are underrepresentated (the same discussion applies for long loops characterized by λ_{long}). Figure taken from [2]

over-shortlooped. The only spatial network analyzed here is the texas power-grid network and clearly displays a lack of long loops which is expected for a distribution network. However, more measures on various spatial networks are needed at this point in order to confirm this feature.

3.1.4 *Perturbation of the Shortest Path in a Directed Lattice*

In directed lattices, changing the direction of an edge can have an impact at a much larger scale. In particular, shortest paths can be largely affected by such a direction flip. We represent in Fig. 3.3 the impact of flipping one edge on a directed shortest path. After this flip, there is a new shortest path and there are essentially two main quantities of interest that were studied in [5]. First, the variation of the length of the shortest path and the area between the old and the new shortest paths (shaded area in the figure).

This problem is obviously interesting in general and has been studied in the particular case of a two-dimensional square lattice at critical probability p_c for bond percolation in [5] and we will discuss their results.

The bond percolation model can be generalized on directed lattices (see for example the Chap. 3 and also [6–11]) and the authors of [5] focused on directed networks where all the directions are equiprobable, leading to a critical system on the square lattice [6]. They study here the shortest path ℓ (with the smallest amount of steps) connecting two opposite sides of the lattice of size L. They found numerically that

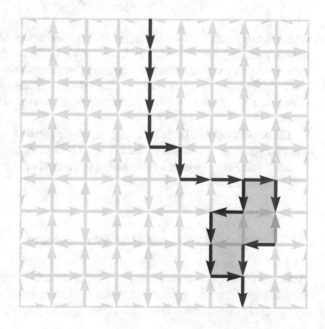

Fig. 3.3 The original shortest path is in red. After inverting one of the links on the path (green arrow), the shortest path changes. The new shortest path is shown in blue. It is chosen out of all possible shortest paths such that it minimizes the area of the shaded region. Figure from [5]

the distribution $p(\ell, L)$ for the shortest path has the same scaling found for bond percolation [12] given by

$$p(\ell, L) = g(\ell/L^{d_{min}})/\ell, \tag{3.4}$$

where $d_{min} \approx 1.13$ is the fractal dimension of the shortest path [13, 14] which is consistent with the fact that this model is in the same universality class as standard percolation [11, 15].

One then chooses randomly one of the shortest paths (if there is more than one) and a randomly selected edge along this path is flipped. The new shortest path ℓ_{new} that encloses the smallest area together the original one is selected. Note that the length change of the shortest path does not necessarily to be non-negative and a new path can be shorter than the original one (but the new shortest path cannot pass through the flipped edge [5]). If the shortest paths have the same endpoints they are called 'convergent' and 'divergent' otherwise (terminology used in [5]). We will present here the results for convergent paths which represent the majority of cases (for large L, convergent paths with non-zero change in length represent 73% of all cases).

Numerical results indicate that flipping an edge lying on the shortest path has a nonlocal effect in the form of power-law distributions for both the differences in shortest path lengths and for the minimal enclosed areas (Fig. 3.4). The fits (using maximum likelihood estimation and extrapolation) lead to the exponent values $\alpha = 1.36 \pm 0.01$ for the path length differences. The data collapse indicate tha the area difference follows the scaling law

$$p(A) = L^{-2\beta} f(A/L^2) \tag{3.5}$$

where the exponent is $\beta = 1.186 \pm 0.001$ for the enclosed areas. It is interesting to note that the exponent β agrees within error bars with the exponent $\beta = 1.16 \pm 0.03$ for the size distribution of the areas enclosed by watersheds from landscapes that only differ slightly at one location reported in [16] in the case of uncorrelated landscapes with Hurst exponent $H = -1$. There are however no theoretical explanations for this so far.

These behaviors indicate that very small perturbations made to the shortest path can cause nonlocal changes in the system. Of course, these results are valid for the critical lattice and it would be interesting to understand what happens in general (for non critical lattices or spatial networks).

3.2 Empirical Results: One-Way Streets

Almost all studies on street networks [17–26] describe them as undirected graphs but formally they are a mixture of both undirected links and one-way streets (represented by directed edges) which is by definition a mixed graph [27, 28].

Fig. 3.4 Data collapse for
the distribution of non-zero
differences in shortest path
lengths $p(\Delta l)$ for convergent
paths using data for different
lattice sizes L. The dashed
line represents a power law
curve with exponent
$\alpha = 1.36$. (b) Data collapse
for the distribution of
enclosed areas $p(A)$ using
data for different lattice sizes
L. The dashed line represents
a power law fit to the data
with exponent $\beta = 1.186$.
Figures from [5]

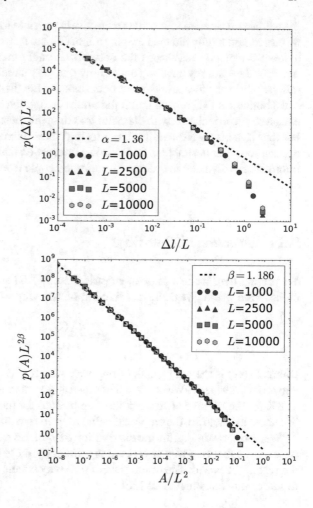

Despite their relevance for practical applications [1], there are very few results available for directed street networks, except for the following one: Robbins' theorem [29] states that it is possible to choose a direction for each edge—called hereafter a strong orientation—of an undirected graph G turning it into a directed graph that has a path from every vertex to every other vertex, if and only if G is connected and has no bridge (i.e. an edge whose deletion increases the graph's number of connected components). Indeed, if there is a bridge in the graph it is by definition the only link that connects a part A to another part B. If this link has a direction, say from A to B, there are no paths from any node in B to A.

Robbins' seminal result can be extended to mixed graphs [30], stating that if G is a strongly connected mixed graph, then any undirected edge of G that is not a bridge may be made directed without changing the connectivity of G. Hence, it is possible to turn streets into one-ways as long as their removal does not disconnect the whole street network. It is thus recursively possible for any bridgeless network to be turned into a fully directed graph. In most cities, it should then be possible to find a street-orientation that keeps the network strongly connected. Robbins' theorem however does not say anything about how one-way streets modify shortest paths. In this respect, very few results were obtained: for the diameter for example, Chvatal and Thomassen [31] proved that if the undirected graph has a diameter d, then there exist a strong orientation with diameter less than the (best possible) bound $2d + 2d^2$, but that it is also a NP-hard problem to find. It is interesting to note that for some applications, it is desirable to find a strong orientation that is not efficient, i.e. doesn't minimize the diameter in order to discourage people from driving in certain sections [1].

3.2.1 Fraction of One-Ways

We present here the results of an empirical study [32] about one-ways in about 140 cities in the world. We define the fraction of one-way streets as

$$p = \frac{L_1(G)}{L(G)} \tag{3.6}$$

where $L_1(G)$ is the total length of one-way streets and $L(G)$ the total length of the network G of size N. We observe that this fraction ranges from very low values such as 8% for the average of African cities up to 31% for the average of European ones. We show in Table 3.1 the empirical value of p in five different cities.

We also show in Fig. 3.5 the distribution of p in different continents. In particular, we observe that one-way streets are significantly more common in Europe than in the rest of the world. The occurrence of one-way streets seems thus to be connected to more complex street plans [26].

Table 3.1 Empirical fraction Eq. (3.6) of one-way streets in five different cities

City	Country	One-way share (%)
Beijing	China	37
Casablanca	Morocco	19
Paris	France	66
New York City	USA	55
Buenos Aires	Argentina	71

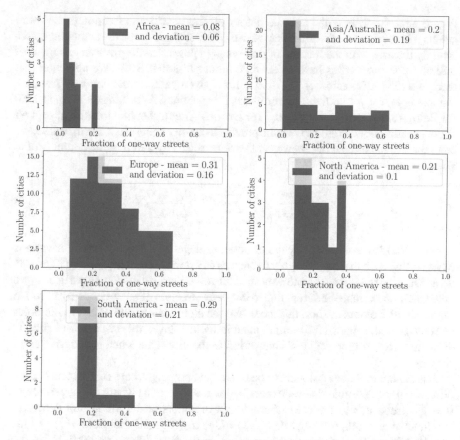

Fig. 3.5 Distribution of the fraction p of one-way streets for the five continents (the fraction is defined as the total length of one-way streets over the total length of the network, Eq. (3.6). Figure from [32]

3.2.2 Statistics of the Detour Index

The presence of one-ways leads to detours and we can characterize this effect quantitatively. We denote by $d_G(i, j)$ the shortest path distance from node i to node j on the undirected graph G and $d_{\vec{G}}(i, j)$ the corresponding quantity for the mixed graph denoted by $\vec{G}$ (when one-ways are taken into account). The average detour due to one-ways is then defined as (see Chap. 4 for a discussion about the detour index in general)

$$\bar{\eta} = \frac{1}{N(N-1)} \sum_{(i,j) \in G} \frac{d_{\vec{G}}(i, j)}{d_G(i, j)} - 1 \qquad (3.7)$$

Figure 3.6a shows how the average detour increases with the fraction of one-way streets p in the dataset of world cities we use. We first observe that the detour increases roughly linearly with the fraction of one-ways (a power law fit gives an exponent of 0.8) and that most cities have an average detour less than 10%. We also note that there is a large dispersion of this detour for a given value of the one-way fraction. For example, for $p \approx 0.6$ the detour varies from about 6% for Singapore up to 15% for Beirut (and even 5% for $p = 0.7$ for Buenos Aires), showing that the impact on shortest paths depends strongly on the precise location of one-ways. Furthermore, we can separate the impact of one-ways at various scales by defining the detour profile given by

$$\eta(d) = \frac{1}{N(d)} \sum_{(i,j) \ s.t. \ d_G(i,j)=d} \frac{d_{\vec{G}}(i,j)}{d_G(i,j)} - 1 \qquad (3.8)$$

(where $N(d)$ is the number of pairs of nodes at distance d). We observe for various cities on Fig. 3.6b that $\eta(d)$ roughly decreases as a power law of the form $\eta(d) \sim d^{-\theta}$ demonstrating the impact of one-way streets even for large distance. In particular, we note that if on average the detour due to one-way streets is of the order of 10%, which seems small, detours at short distances may be significantly higher (up to the order of 100%). Also, even if 10% is small at an individual level, this has a non-neglibible effect in terms of time cost and congestion at the city scale when summed over all car users.

The exponent θ does not seem to be universal and ranges between 0.2 and 0.8 for different cities. We note that we expect in general $\theta \in [0, 1]$ where the upper-bound $\theta = 1$ corresponds to the case where one-way streets create a constant detour in the directed network, implying $d_{\vec{G}}(i, j) = C + d_G(i, j)$ and therefore $\eta(d) \sim 1/d$. The case $\theta = 0$ corresponds to the situation where the detour is proportional to the distance traveled: $d_{\vec{G}}(i, j) \propto d_G(i, j)$ implying $\eta(d) \sim const$. In any case, this slow decrease of $\eta(d)$ with d signals the long-range effect of one-ways on shortest paths (Fig. 3.6).

In order to get a theoretical understanding of these results, we consider the average detour on the honeycomb lattice and observe that it increases with p (see Fig. 3.7 for Paris). We also see in Fig. 3.7 that the real detour is below the result obtained for a random distribution of one-way streets (similar results are obtained for other cities). This confirms the importance of the precise location of one-ways that can affect in very different ways the shortest paths statistics.

For the honeycomb lattice (Fig. 3.8), the average detour $\eta(d)$ due to directed links for a trip of distance d scales as a power-law of d with $\eta(d) \sim d^{-\theta}$ (the quantity d is here normalized by its maximum value). We find $\theta = 0.5 \pm 0.1$ as shown in the data collapse of Fig. 3.8(a). More precisely, we also show that the relation is of the form $\eta(d) = A(p)d^{-1/2}$ that remains valid for all p and with $A(p) \sim p^{\nu}$ with $\nu \approx 2.3$ (see Fig. 3.8b). This result in $1/\sqrt{d}$ suggests the possibility of an argument relying on the sum of random quantities leading to $d_{\vec{G}}(i, j) - d_G(i, j) \sim \sqrt{d}$.

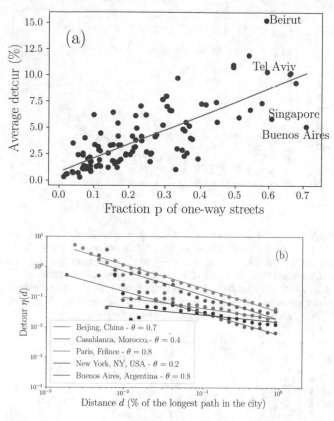

Fig. 3.6   a Distribution of the average detour (%) as a function of the fraction p of one-way streets for 146 world cities ($R^2 = 0.59$). **b** For five selected cities in the world, we plot the average detour $\eta(d)$ due to one-way streets for a trip of distance d as a function of d (normalized by the maximum distance obtained for each city). The detour can be fitted by a power law $\eta(d) \sim d^{-\theta}$. We find that θ differs from one city to another and ranges roughly from 0.2 to 0.8. In particular, small exponent values (such as in the case of NYC) might be correlated with the presence of very long one-way streets leading to a large detour even at very large spatial scales. We have $R^2 = 0.87$ for Beijing, $R^2 = 0.25$ for Casablanca, $R^2 = 0.99$ for Paris, $R^2 = 0.12$ for NYC and $R^2 = 0.90$ for Buenos Aires. Figure from [32]

3.2.3   Asymmetry of Shortest Paths

In [33] the authors investigate the asymmetry of shortest paths in real city networks. More precisely, for different pairs of origin-destination, they analyze the log-ratio

$$r = \ln \frac{\ell_D}{\ell_O} \tag{3.9}$$

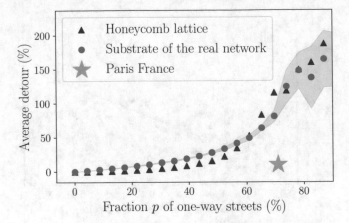

Fig. 3.7 Average detour $\overline{\eta}$ as a function of the fraction p of randomly chosen one-way streets in the city of Paris (France). In this statistical process, the detour increases with the fraction p. We note, however, that the empirical detour in the real world (indicated by a star symbol) remains below the result expected from a random uniform distribution of one-way streets. This indicates that the actual choice of one-way streets in Paris is far from what would be obtained by a random choice of one-way streets and favors small detours. We compare these results to those obtained for a honeycomb lattice, whose degree distribution is close to the Paris one

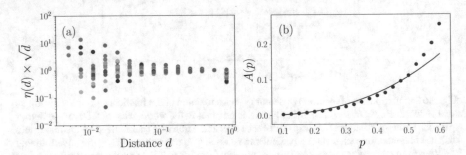

Fig. 3.8 Average detour for the honeycomb lattice. **a** We show the quantity $\eta(d)\sqrt{d}$ versus d normalized by its value at $d = 0.1$. The curves collapse onto a single one independent from d and the observed discrepancies for d close to 1 and d small come from finite-size effects. **b** The collapse suggest a form $\eta(d) = A(p)d^{-1/2}$, and a power law fit gives $A(p) \sim p^{\gamma}$ with $\gamma \approx 2.3$ ($R^2 = 0.94$). We, however, observe discrepancies at large p

where ℓ_O is the length of the shortest route from O to D (and from D to O for ℓ_D). When the distances are close to each other $\ell_O \approx \ell_D$ this log-ratio is approximately equal to $r \approx \ell_D - \ell_O/\ell_0$ and there measures the relative differences between the distances ℓ_D and ℓ_O.

The authors of [33] analyzed ten large cities worldwide (with more than one million inhabitants) and found that the average of r is zero. This means that on average $\langle \ell_0 \rangle = \langle \ell_D \rangle$ as observed numerically for the city of Paris for distances larger than 1km (Fig. 3.9a).

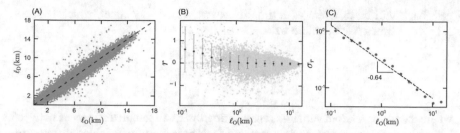

Fig. 3.9 Fluctuations of r for the city of Paris. **a** We have on average $\ell_O \approx \ell_D$ but there are non-neglibible fluctuations around this behavior. **b** The average of r (red circles) converges to zero for distances larger than $\sim$ 1km. The fluctuations are here indicated by the vertical bars. **c** However, the fluctuations σ_r, indicated by the bars, tend to decrease smoothly with ℓ_O and well fitted by a power with exponent 0.64 for Paris. Figure taken from [33]

The average of r is zero, but its standard deviation defined by

$$\sigma_r = \sqrt{\langle r^2 \rangle - \langle r \rangle^2}$$ (3.10)

decays smoothly with ℓ_O. In Fig. 3.9c, it is shown that this decay is consistent with a power-law behavior of the form

$$\sigma_r \sim r^{-\beta}$$ (3.11)

with β than ranges from 0.4 to 0.64 for the set of cities studied in [33]. There is apparently no relation between the value of this exponent and the fraction of one-way streets.

An interesting quantity is the fraction of one-way streets used in the shortest paths that is not necessarily equal to the fraction of one-ways in the entire network. This fraction $\langle f \rangle$ average over all shortest path of a given length ℓ behaves as

$$\langle f \rangle(\ell) \sim \frac{1}{\ell^\alpha}$$ (3.12)

For Paris for example, the exponent is $\alpha \approx 0.18 \pm 0.01$. This behavior indicates that the longer the route and the lower the fraction of the path that used one-way streets and the larger the fraction of main two-ways streets.

The authors propose the following argument to relate these two exponents α and β. When $\ell_O \approx \ell_D$, we have by definition $\ell_D \approx \ell_O + r\ell_O$ and which leads to

$$\sigma_r = \frac{\sigma_{\ell_D}}{\ell_O}$$ (3.13)

where σ_{ℓ_D} is the dispersion of ℓ_D (in all this argument ℓ_O is fixed and the goal is to get σ_r as a function of this variable). In this limit, we can estimate the scaling of σ_r by dividing a shortest path in $N \propto \ell_O$ equal parts and by writing that the shortest path back will be different each time it crosses a one-way street, which happens $N\langle f\rangle(\ell_O)$ times. We therefore write

$$\ell_D = \ell_O + \sum_i^{N\langle f\rangle(\ell_O)} \eta_i \tag{3.14}$$

where the η_i are random variables representing the small length differences between a shortest path and its way back. Assuming that these random variables are uncorrelated, one then finds that

$$\sigma_{\ell_D} = \ell_0\langle f\rangle(\ell_O) \tag{3.15}$$

Combining these different results then leads to

$$\sigma_r \sim 1/\ell_O^{(1+\alpha)/2} \tag{3.16}$$

which implies that $\beta = (1+\alpha)/2$. This relation has been tested on the dataset used in [33] revealing a good agreement with however some large deviations for cities with a small fraction of one-ways (Berlin, Fortaleza, and San Francisco). For the case where one-ways are distributed at random, we expect $\langle f\rangle \approx const.$ (for sufficiently large ℓ), hence $\alpha = 0$. We should then observe in this case $\beta = 0.5$ which is indeed recovered for the randomly reshuffled case of Paris where the value $\beta_{reshuf} = 0.57$ is measured. The deviations from $\beta = 0.5$ are therefore resulting from underlying spatial correlations between the locations of one way streets.

References

1. F.S. Roberts, *Graph Theory and its Applications to Problems of Society* (SIAM, 1978)
2. G. Bianconi, N. Gulbahce, A.E. Motter, Phys. Rev. Lett. **100**(11), 118701 (2008)
3. Z. Burda, J. Jurkiewicz, A. Krzywicki, Phys. Rev. E **70**(2), 026106 (2004)
4. G. Bianconi, M. Marsili, J. Stat. Mech. **2005**, P06005 (2005)
5. F. Hillebrand, M. Luković, H.J. Herrmann, Phys. Rev. E **98**(5), 052143 (2018)
6. S. Redner, J. Phys. A: Math. Gen. **14**(9), L349 (1981)
7. S. Redner, Phys. Rev. B **25**(5), 3242 (1982)
8. S. Redner, Phys. Rev. B **25**(9), 5646 (1982)
9. S.R. Broadbent, J.M. Hammersley, in *Mathematical Proceedings of the Cambridge Philosophical Society*, vol. 53 (Cambridge University Press, 1957), pp. 629–641
10. A.W. De Noronha, A.A. Moreira, A.P. Vieira, H.J. Herrmann, J.S. Andrade Jr., H.A. Carmona, Phys. Rev. E **98**(6), 062116 (2018)
11. H.K. Janssen, O. Stenull, Phys. Rev. E **62**(3), 3173 (2000)
12. S. Havlin, B. Trus, G. Weiss, D. Ben-Avraham, J. Phys. A: Math. Gen. **18**(5), L247 (1985)
13. Z. Zhou, J. Yang, Y. Deng, R.M. Ziff, Phys. Rev. E **86**(6), 061101 (2012)
14. A. Aharony, A.B. Harris, Physica A: Stat. Mech. Appl. **191**(1–4), 365 (1992)
15. Z. Zhou, J. Yang, R.M. Ziff, Y. Deng, Phys. Rev. E **86**(2), 021102 (2012)
16. E. Fehr, D. Kadau, J.S. Andrade Jr., H.J. Herrmann, Phys. Rev. Lett. **106**(4), 048501 (2011)

17. B. Jiang, C. Claramunt, Environ. Plann. B **31**(1), 151 (2004)
18. J. Buhl, J. Gautrais, N. Reeves, R. Solé, S. Valverde, P. Kuntz, G. Theraulaz, Eur. Phys. J. B **49**, 513 (2006)
19. E. Strano, M. Viana, A. Cardillo, L.D.F. Costa, S. Porta, V. Latora, (2012). arXiv:1211.0259
20. F. Xie, D. Levinson, Geograph. Anal. **39**(3), 336 (2007)
21. S. Lämmer, B. Gehlsen, D. Helbing, Physica A: Stat. Mech. Appl. **363**(1), 89 (2006)
22. E. Strano, V. Nicosia, V. Latora, S. Porta, M. Barthelemy, Sci. Rep. **2** (2012)
23. P. Crucitti, V. Latora, S. Porta, Phys. Rev. E **73**(3), 036125 (2006)
24. R. Louf, M. Barthelemy, Sci. Rep. **4** (2014)
25. M. Barthelemy, *Morphogenesis of Spatial Networks* (Springer, 2018)
26. G. Boeing, Appl. Netw. Sci. **4**(1), 1 (2019)
27. E.M. Arkin, R. Hassin, Disc. Appl. Math. **116**(3), 271 (2002)
28. M. Beck, D. Blado, J. Crawford, T. Jean-Louis, M. Young, Graph. Comb. **31**(1), 91 (2015)
29. H.E. Robbins, Am. Math. Mon. **46**(5), 281 (1939)
30. F. Boesch, R. Tindell, Am. Math. Mon. **87**(9), 716 (1980)
31. V. Chvátal, C. Thomassen, J. Comb. Theory Ser. B **24**(1), 61 (1978)
32. V. Verbavatz, M. Barthelemy, Phys. Rev. E **103**(4), 042313 (2021)
33. H.P. Melo, D.P. Mota, J.S. Andrade Jr., N.A. Araújo (2021). arXiv:2111.07434

Chapter 4
Simple Measures

Many studies on complex networks focus on how to characterize them and what are the most relevant measures for understanding their structure. In particular, the degree distribution and the existence of the second moment for an infinite network was shown to be critical when studying dynamical processes on networks. These behaviors are strongly connected to degree fluctuations and to the existence of hubs. In the case of spatial networks, the physical constraints are usually large and prevent the appearance of such hubs. These constraints also impact other quantities that are non-trivial for complex networks but that become irrelevant for spatial networks. We review here these measures that are essentially useless for spatial networks and we then discuss useful, simple measures that were mostly introduced in the context of quantitative geography.

4.1 Irrelevant Measures for Spatial Networks

Quantities that depend very much on the spatial constraints turn out to be irrelevant for spatial networks in general. The prime example is the degree distribution which in many complex networks was found to be a broad law and in some cases well fitted by a power law of the form

$$P(k) \sim k^{-\gamma} \tag{4.1}$$

with $1 < \gamma < 3$ [1]. In this case of 'scale-free network', degree fluctuations are very large which has a direct impact on many dynamical processes that take place on the network, such as epidemics for example [2]. For spatial networks however the degree has to satisfies steric constraints. If we consider a road network, nodes represent the intersections and the degree of a node is the number of streets starting

© The Author(s), under exclusive license to Springer Nature Switzerland AG 2022   39
M. Barthelemy, *Spatial Networks*,
https://doi.org/10.1007/978-3-030-94106-2_4

from it and is therefore clearly limited as a result of space. As a consequence, the degree distribution for most spatial networks is not broad but displays a fast decaying tail (such as an exponential for example).

Other irrelevant parameters include the clustering coefficient or the assortativity. Indeed as shown below the clustering coefficient is always large: if a node is connected to two other nodes there are usually located in its neighborhood which in turn increases the probability that they are connected to each other, leading to a large clustering coefficient. In the following, we will discuss in more detail these different measures.

4.1.1 Degree

We recall here that a graph G with N nodes and E edges can be described by its $N \times N$ adjacency matrix A which is defined as

$$A_{ij} = \begin{cases} 1 & \text{if } i \text{ and } j \text{ are connected} \\ 0 & \text{otherwise} \end{cases} \tag{4.2}$$

If the graph is undirected then the matrix A is symmetric. The degree of a node is by definition the number of its neighbors and is given by

$$k_i = \sum_j A_{ij} \tag{4.3}$$

The first simple indicator of a graph is the average degree

$$\langle k \rangle = \frac{1}{N} \sum_i k_i = \frac{2E}{N} \tag{4.4}$$

where here and in the following the brackets $\langle \cdot \rangle$ denote the average over the nodes of the network. In particular the scaling of $\langle k \rangle$ with N indicates if the network is sparse (which is the case when $\langle k \rangle \to const.$ for $N \to \infty$).

In [3, 4] measurements for street networks in different cities in the world are reported. Based on the data from these sources, the authors of [5] plotted (Fig. 4.1a) the number of roads E (edges) versus the number of intersections N. The plot is consistent with a linear fit with slope ≈ 1.44 (which is consistent with the value $\langle k \rangle \approx 2.5$ measured in [4]). The quantity $e = E/N = \langle k \rangle / 2$ displays values in the range $1.05 < e < 1.69$, in between the values $e = 1$ and $e = 2$ that characterize tree-like structures and $2d$ regular lattices, respectively. Few exact values and bounds are available for the average degree of classical models of planar graphs. In general it is known that $e \leq 3$ (see Chap. 2), while it has been shown that $e > 13/7$ for planar Erdös-Rényi graphs ([6].

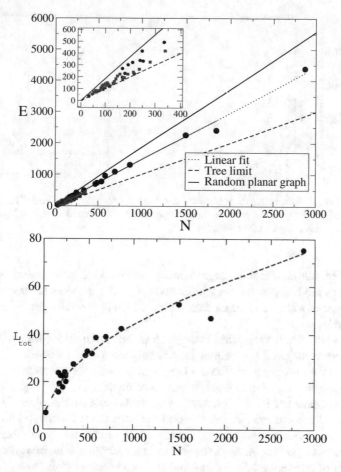

Fig. 4.1 **a** Numbers of roads versus the number of nodes (i.e. intersections) for data from [3] (circles) and from [4] (squares). In the inset, we show a zoom for a small number of nodes. **b** Total length L_{tot} versus the number of nodes. The line is a fit which predicts a growth as $\sqrt{N}$ (data from [3] and figures from [5])

The distribution of degree $P(k)$ is usually a quantity of interest in complex networks and can display large heterogeneities such as it is observed in scale-free networks (see for example [7]). We observe that for spatial networks such as airline networks or the Internet the degrees can be very heterogeneous (see [2]). However, when physical constraints are strong or when the cost associated with the creation of new links is large, a cut-off appears in the degree distribution [8] and in some case the distribution can be very peaked. This is the case for the road network for example and more generally in the case of planar networks for which the degree distribution $P(k)$ is then of little interest. For example, in a study of 20 German cities, Lämmer et al. [9] showed that most nodes have four neighbors (the full degree distribution is

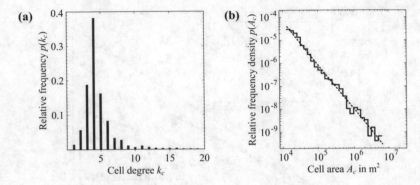

Fig. 4.2 a Degree distribution of degrees for the road network of Dresden. **b** The probability distribution of the cell's surface areas A_c obeys a power law with exponent $\alpha \approx 1.9$ (for the road network of Dresden). Figure taken from [9]

shown in Fig. 4.2a) and that the degree rarely exceeds 5 for various world cities [3]. These values are however not very indicative: planarity imposes severe constraints on the degree of a node and on its distribution which is generally peaked around its average value.

We note here that in real-world cases such as the road network for example, it is natural to study the usual (or 'primal') representation where the nodes are the intersections and the links represent the road segment between the intersection. However another representation, the dual graph can be of interest (see [10] and for the road network it is constructed in the following way: the nodes are the roads and two nodes are connected if there exist an intersection between the two corresponding roads. One can then measure the degree of a node which represents the number of roads which intersect a given road. Also, the shortest path length in this network represents the number of different roads one has to take to go from one point to another. Even if the road network has a peaked degree distribution its dual representation can display broad distributions [11]. Indeed, in [11] measurements were made on the dual network for the road network in the US, England, and Denmark and showed large fluctuations with a power law distribution with exponent $2.0 < \gamma < 2.5$.

4.1.2   Length of Segments

In Fig. 4.1b, we see that the total length L_{tot} of the network versus N for the cities considered in [3]. Data are well fitted by a power function of the form

$$L_{tot} = \mu N^{\beta} \tag{4.5}$$

with $\mu \approx 1.51$ and $\beta \approx 0.49$. In order to understand this result, one has to focus on the street segment length distribution $P(\ell_1)$. This quantity has been measured

Fig. 4.3 Length distribution $P(\ell_1)$ for the street network of London (and for the model GRPG proposed in this study). Figure taken from [12])

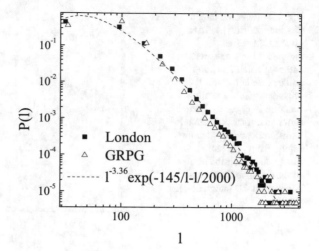

for London in [12] and is shown in Fig. 4.3. This figure shows that the distribution decreases rapidly and the fit proposed by the authors of [12] suggests that

$$P(\ell_1) \sim \ell_1^{-\gamma} \tag{4.6}$$

with $\gamma \simeq 3.4$ which implies that both the average and the dispersion are well-defined and finite. If we assume that this result extends to other cities, it means that we have a typical distance ℓ_1 between nodes which is meaningful. This typical distance between connected nodes then naturally scales as

$$\ell_1 \sim \frac{1}{\sqrt{\rho}} \tag{4.7}$$

where $\rho = N/L^2$ is the density of vertices and L the linear dimension of the ambient space. This implies that the total length scales as

$$L_{tot} \sim E\ell_1 \sim \frac{\langle k \rangle}{2} L\sqrt{N} \tag{4.8}$$

This simple argument reproduces well the $\sqrt{N}$ behavior observed in Fig. 4.1b and also the value (given the error bars) of the prefactor $\mu \approx \langle k \rangle L/2$.

At a global level, Strano et al. [13] used a database comprising all major roads on Earth. They found that the road length distribution with urban areas and croplands are not distinguishable from each other, once rescaled by the average length in each area. This length distribution can be written as

$$P_i(\ell_1) = \frac{1}{\ell_1} F\left(\frac{\ell_1}{\langle \ell_i \rangle}\right) \tag{4.9}$$

Fig. 4.4 Rescaled length distribution $P(\ell_1|A)$ for a given surface area A computed for urban areas and croplands. The good collapse demonstrates the universality of the scaling function F Eq. (4.9). The different curves correspond to different surface area bins. Figure from [13] and distributed under the terms of the Creative Commons Attribution 4.0 International License

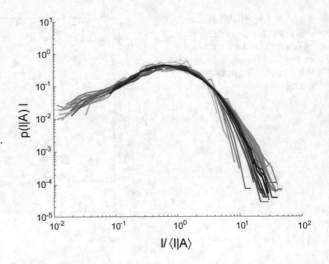

where i indicates if the distribution refers to urban areas ($i = U$) or croplands ($i = C$) and $\langle \ell_i \rangle$ is the corresponding average length. The scaling function F appears however to be the same for both types of areas. In contrast, the average quantities $\langle \ell_i \rangle$ depend on the zone considered and scale differently with the surface area in urban areas or in croplands (see [13])).

In addition to the universality of the scaling form, these results also show that the length is distributed according to a relatively peaked distribution (at least for large values). The authors of [13] also considered the total length and found that it varies linearly for large areas (for both area types), except in denser urban areas where it scales as $L_{tot} \sim A^\beta$ with $\beta \approx 0.6$. Unfortunately, they didn't consider the rescaled quantity $L_{tot}/\sqrt{A}$ versus the number of nodes in order to test the simple argument leading to Eq. (4.8).

4.1.3 Cell Area

Planar graphs naturally produces a set of non-overlapping cells (or faces, or blocks) and covering the embedding plane. In the case of the road network, the distribution of the area A of these cells has been measured for the city of Dresden in Germany (Fig. 4.2b) and has the form

$$P(A) \sim A^{-\alpha} \tag{4.10}$$

with $\alpha \simeq 1.9$, which was confirmed by measures on other cities [14]. This broad law is in sharp contrast with the simple picture of an almost regular lattice which would predict a distribution $P(A)$ peaked around ℓ_1^2.

It is interesting to note that if we assume that $A \sim 1/\ell_1^2 \sim 1/\rho$ and that the density ρ is distributed according to a law $f(\rho)$ (with a finite $f(0)$), a simple calculation gives

$$P(A) \sim \frac{1}{A^2} f(1/A) \qquad (4.11)$$

which behaves as $P(A) \sim 1/A^2$ for large A. This simple argument thus suggests that the observed value ≈ 2.0 of the exponent is universal and essentially reflects the random variation of the density. This possible universality thus shows that the area distribution is unable to distinguish different spatial networks and cannot be used as an interesting characterization.

4.1.4   Clustering and Assortativity

Complex networks are essentially characterized by a small set of parameters which are not all relevant for spatial networks. For example, the degree distribution which has been the main subject of interest in complex network studies is usually peaked for planar networks, due to the spatial constraints. We will see here that it is the same for the clustering and the assortativity which don't bring useful informations about the structure of spatial networks.

4.1.4.1   Clustering Coefficient

The clustering coefficient of a node i of degree k_i is defined as

$$C(i) = \frac{E_i}{k_i(k_i - 1)/2} \qquad (4.12)$$

where E_i is the number of edges among the neighbors of i. This quantity gives some information about local clustering and was the object of many studies in complex networks. The clustering coefficient depends on the number of triangles or cycles of length 3 and can also be computed by using powers of the adjacency matrix. For instance the quantity $\frac{1}{6}\mathrm{Tr}(A^3)$ is the number C_3 of cycles of length tree and is related to the clustering coefficient. Analogously we can define and count cycles of various lengths (see for example [15, 16] and references therein) and compare this number to the ones obtained on null models (lattices, triangulations, etc.).

For the Erdos-Renyi (ER) random graphs with finite average degree $\langle k \rangle$, the average clustering coefficient is simply given by

$$\langle C \rangle = p \sim \frac{\langle k \rangle}{N} \qquad (4.13)$$

where the brackets $\langle \cdot \rangle$ denote the average over the nodes of the network. In contrast, for spatial networks, closer nodes have a larger probability to be connected, leading to a large clustering coefficient (see for example the calculation in the random geometric graph case, see Chap. 15). Here also, this means that due to spatial constraints, the clustering coefficient doesn't contain interesting information about the graph structure.

Many studies define the clustering coefficient per degree classes which is given by

$$C(k) = \frac{1}{N(k)} \sum_{i/k_i=k} C(i) \tag{4.14}$$

The behavior of $C(k)$ versus k thus gives an indication how the clustering is organized when we explore different classes of degrees. However, in order to be useful, this quantity needs to be applied to networks with a large range of degree variations which is usually not the case in spatial networks.

4.1.4.2   Assortativity

In general the degrees at the two end nodes of a link are correlated and to describe these correlations one needs the two-point correlation function $P(k'|k)$. This quantity represents the probability that any edge starting at a vertex of degree k ends at a vertex of degree k'. Higher order correlation functions can be defined and we refer the interested reader to [17] for example. The function $P(k'|k)$ is however not easy to handle and one can define the assortativity [18, 19]

$$k_{nn}(k) = \sum_{k'} P(k'|k)k' \tag{4.15}$$

A similar quantity can be defined for each node as the average degree of the neighbors

$$\langle k_{nn}(i) \rangle = \frac{1}{k_i} \sum_{j \in \Gamma(i)} k_j \tag{4.16}$$

where $\Gamma(i)$ denotes the set of neighbors of node i. There are essentially two classes of behaviors for the assortativity. If $k_{nn}(k)$ is an increasing function of k, vertices with large degrees have a larger probability to connect to similar nodes with a large degree. In this case, we speak of an *assortative* network and in the opposite case of a *disassortative* network. It is expected in general that social networks are mostly assortative, while technological networks are disassortative. However for spatial networks spatial constraints usually implies a flat function $k_{nn}(k)$.

4.1.5 *Average Shortest Path and Diameter*

Usually, there are many paths between two nodes in a connected networks and if we keep the shortest one it defines a distance on the network

$$\ell(i, j) = \min_{paths(i \to j)} |path| \tag{4.17}$$

where the length $|path|$ of the path is defined as its number of edges. The average shortest path is then obtained by averaging $\ell(i, j)$ over all pairs of nodes in order to characterize the 'size' of the network. Indeed for a d-dimensional regular lattice with N nodes, this average shortest path $\langle \ell \rangle$ scales as

$$\langle \ell \rangle \sim N^{1/d} \tag{4.18}$$

In a small-world network (see [20] and Chap. 10) constructed over a $d-$dimensional lattice $\langle \ell \rangle$ has a very different behavior

$$\langle \ell \rangle \sim \log N \tag{4.19}$$

The crossover from a large-world behavior $N^{1/d}$ to a small-world one with $\log N$ can be achieved for a density p of long links (or 'shortcuts') [21] such that

$$pN \sim 1 \tag{4.20}$$

The effect of space could thus in principle be detected in the behavior of $\langle \ell \rangle (N)$. It should however be noted that if the number of nodes is too small this can be a tricky task. In the case of brain networks for example, a behavior typical of a three-dimensional network in $N^{1/3}$ could easily be confused with a logarithmic behavior if N is not large enough.

Using this distance $\ell(i, j)$ on the graph, we define the eccentricity $\varepsilon(i)$ of a vertex i by

$$\varepsilon(i) = \max_{j \in V} \ell(i, j) \tag{4.21}$$

(where V is the set of vertices of the graph) which characterizes how 'central' the node i is: if the eccentricity is 'small' all vertices are not too far from i. With this eccentricity, we then define the radius $r(G)$ of the graph G as the minimum eccentricity over all vertices

$$r(G) = \min_{i \in V} \varepsilon(i) \tag{4.22}$$

and the vertex i_c that realizes this value (i.e. such that $r(G) = \varepsilon(i_c)$) is called the center of the graph.

In contrast, the diameter $d(G)$ of the graph is the maximum eccentricity and equivalently the largest possible distance between pairs of nodes in the graph

$$d(G) = \max_i \varepsilon(i) = \max_{i,j} \ell(i, j) \tag{4.23}$$

A peripheral vertex i_p is then such that its eccentricity is maximum $\varepsilon(i_p) = d(G)$.

In most applications, the diameter is confused with the average shortest paths as they have usually the same scaling with the number of nodes N. It is difficult to compute mathematically the diameter for any graph, but from a physical point of view we know that shortcut links responsible for the small-world behavior ensures that the diameter scales as $\log N$ (which can be proven rigorously for Barabasi-Albert scale-free networks [22]). For planar graphs, some bounds can be obtained [23] using the inverse degree of a graph given by $1/K(G) = \sum_{i \in V} \frac{1}{k_i}$ where k_i i the degree of node i. The theorem proved in this paper is the following bound for the diameter $d(G)$ of any planar graph

$$d(G) < \frac{5}{2} \frac{1}{K(G)} \tag{4.24}$$

4.1.6 Empirical Illustrations

We discuss here some simple results obtained on transportation networks, that illustrate the fact that indeed some measures useful for understanding complex networks, are actually irrelevant in the case of spatial networks and do not convey interesting informations.

4.1.6.1 Power Grids and Water Distribution Networks

Power grids are one of the most important infrastructure in our societies. In modern countries, they have evolved for a rather long time (sometimes a century) and are now complex systems with a large variety of elements and actors playing in their functioning. This complexity leads to the relatively unexpected result that their robustness is actually not very well understood and large blackouts such as the huge August 2003 blackout in North America demonstrate the fragility of these systems.

The topological structure of these networks was studied in different papers such as [8, 24, 25]. In particular, in [8, 24], the authors consider the Southern Californian and the North American power grids. In these networks the nodes represent the power plants, distribution and transmission substations, and the edges correspond to transmission lines. These networks are typically almost planar (see for example the Italian case, Fig. 4.5) and we expect a peaked degree distribution, decreasing typically as an exponential of the form $P(k) \sim \exp(-k/\langle k \rangle)$ with $\langle k \rangle$ of order 3 in

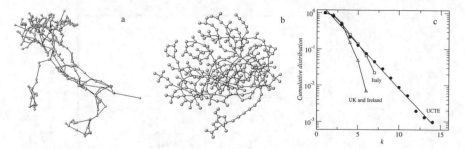

Fig. 4.5 a Map of the Italian power grid. **b** Topology of the Italian power grid. **c** Degree distribution for the European network (UCTE), Italy and the UK and Ireland. In all cases the degree distribution is peaked and can be fitted by exponential. Figure taken from [25]

Fig. 4.6 Cumulative degree distribution of substations for the North American power grid. The straight line represents an exponential fit (a similar result is obtained for Southern California [8]). Figure taken from [24]

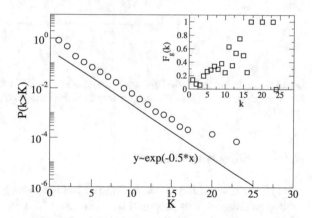

Europe and 2 in the US. The other studies on US power grids confirm that the degree distribution is exponential (see Fig. 4.6). In [24], Albert, Albert, and Nakarado also studied the load (a quantity similar to the betweenness centrality) and found a broad distribution. The degree being peaked we can then expect very large fluctuations of load for the same value of the degree, as expected in general for spatial networks. These authors also found a large redundancy in this network with however 15% of bridges (also called cut-edges).

Also, as expected for these networks, the clustering coefficient is rather large and even independent of k as shown in the case of the power grid of Western US (see Fig. 4.7).

Beside the distribution of electricity, our modern societies also rely on various other distribution networks. The resilience of these networks to perturbations is thus an important point in the design and operating of these systems. In [27], Yazdani and Jeffrey study the topological properties of the Colorado Springs Utilities and the Richmond (UK) water distribution networks. Both these networks (of size $N = 1786$ and $N = 872$, respectively) are sparse planar graphs with very peaked degree distributions (the maximum degree is 12).

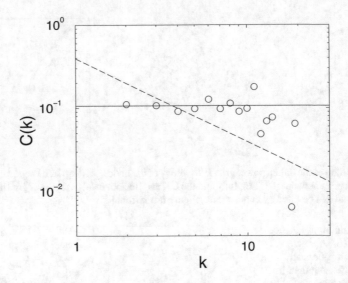

Fig. 4.7 Scaling of the clustering $C(k)$ for the Power Grid of the Western United States. The dashed line has a slope -1 and the solid line corresponds to the average clustering coefficient. Figure taken from [26]

4.1.6.2   Subways and Buses

One of the first studies (after the Watts-Strogatz paper) on the topology of a transportation network was proposed by Latora and Marchiori [28] who considered the Boston subway network. It is a relatively small network with $N = 124$ stations. The average shortest path is $\langle \ell \rangle \sim 16$ a value which is closer to the two-dimensional result $\sqrt{124} \approx 11$ than to the 'small-world' logarithmic behavior $\ln 124 \approx 5$.

In [29] Sienkiewicz and Holyst study a larger set made of public transportation networks of buses and tramways for 22 Polish cities and in [30] von Ferber et al. study the public transportation networks for 15 world cities. The number of nodes of these networks varies from $N = 152$ to 2811 in [29] and in the range [1494, 44629] in [30]. Interestingly enough the authors of [29] observe a strong correlation between the number of stations and the population which is not the case for the world cities studied in [30] where the number of stations seems to be independent from the population (see the Chap. 20 for a detailed discussion about the connection between socio-economical indicators and the properties of networks). For polish cities the degree has an average in the range [2.48, 3.08] and in a similar range [2.18, 3.73] for [30]. In both cases, the degree distribution is relatively peaked (the range of variation is usually of the order of one decade) consistently with the existence of physical constraints [8].

Due to the relatively small range of variation of N in these various studies [28–30], the behavior of the average shortest path is not clear and could be fitted by a logarithm or a power law as well. We however note that the average shortest path is usually large (of order 10 in [29] and in the range [6.4, 52.0] in [30]) compared to $\ln N$, suggesting that the behavior of $\langle \ell \rangle$ might not be logarithmic with N but more likely scales as $N^{1/2}$, a behavior typical of a two-dimensional lattice.

The average clustering coefficient $\langle C \rangle$ in [29] varies in the range [0.055, 0.161] and is larger than a value of the order $C_{ER} \sim 1/N \sim 10^{-3} - 10^{-2}$ corresponding to a random ER graph. The ratio $\langle C \rangle / C_{ER}$ is explicitly considered in [30] and is usually much larger than one (in the range [41, 625]). The degree-dependent clustering coefficient $C(k)$ seems to present a power law dependence, but the fit is obtained over one decade only.

In another study [31], the authors study two urban train networks (Boston and Vienna which are both small $N = 124$ and $N = 76$, respectively) and their results are consistent with these conclusions.

4.1.6.3 Railways

One of the first studies of the structure of railway networks [32] concerns a subset of the most important stations and lines of the Indian railway network and has $N = 587$ stations. In the P-space representation (see Chap. 1), there is a link between two stations if there is a train connecting them and in this representation, the average shortest path is of order $\langle \ell \rangle \approx 5$ which indicates that one needs 4 connections on average to go from one node to another one. In order to obtain variations with the number of nodes, the authors considered different subgraphs with different sizes N. The clustering coefficient varies slowly with N and is always larger than ≈ 0.7 which is much larger than a random graph value of order $1/N$. Finally in this study [32], it is shown that the degree distribution is behaving as an exponential and that the assortativity $k_{nn}(k)$ is flat showing an absence of correlations between the degree of a node and those of its neighbors.

In [33], Kurant and Thiran studied the railway system of Switzerland and major trains and stations in Europe (and also the public transportation system of Warsaw, Poland). The Swiss railway network contains $N = 1613$ nodes and $E = 1680$ edges (Fig. 4.8). All conclusions drawn are consistent with the various cases presented in this chapter. In particular, the average degree is $\langle k \rangle \approx 2.1$, the average shortest path is ≈ 47 (consistent with the $\sqrt{N}$ result for a two-dimensional lattice), the clustering coefficient is much larger than its random counterpart, and the degree distribution is peaked (exponentially decreasing).

In summary, these empirical results for different spatial networks show the following facts:

- The degree distribution $P(k)$ is peaked.
- The clustering is large compared to the one for the random case and the clustering profile $C(k)$ is essentially flat.
- The average shortest path is in general behaving as $\sqrt{N}$, as expected for a two-dimensional planar graph (although in some cases it is difficult to distinguish from the small-world behavior characterized by $\log N$).
- The assortativity is usually flat: $k_{nn}(k) \approx$ const.

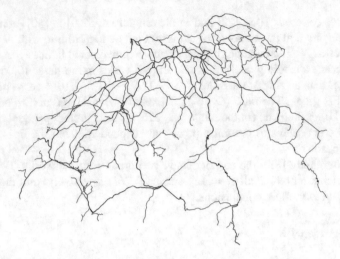

Fig. 4.8 Physical map of the Swiss railway networks. Figure taken from [33]

4.1.6.4   Neural Networks

The human brain with about 10^{10} neurons and about 10^{14} connections is one of the most complex network that we know of. The structure and functions of the brain are the subjects of numerous studies and different recent techniques such as electroencephalography, magnetoencephalography, functional RMI, etc., can be used in order to reconstruct networks for the human brain (see Fig. 4.9 and for a clear and nice introduction see for example [34] and [35]).

Brain regions that are spatially close have a larger probability of being connected than remote regions as longer axons are more costly in terms of material and energy [34]. Wiring costs depending on distance is thus certainly an important aspect of brain networks and we can expect spatial networks to be relevant in this rapidly evolving topic. So far, many measures seem to confirm a large value of the clustering coefficient, and a small-world behavior with a small average shortest-path length [36, 37]. It also seems that neural networks do not optimize the total wiring length but rather the processing paths thanks to shortcuts [38]. This small-world structure of neural networks could reflect a balance between local processing and global integration with rapid synchronization, information transfer, and resilience to damage [39].

In contrast, the nature of the degree distribution is still under debate and a recent study on the macaque brain [40] showed that the distribution is better fitted by an exponential rather than a broad distribution. Besides the degree distribution, most of the observed features were confirmed in later studies such as [41] where Zalesky et al. propose to construct the network with MRI techniques where the nodes are distinct grey-matter regions and links represent the white-matter fiber bundles. The spatial resolution is of course crucial here and the largest network obtained here is

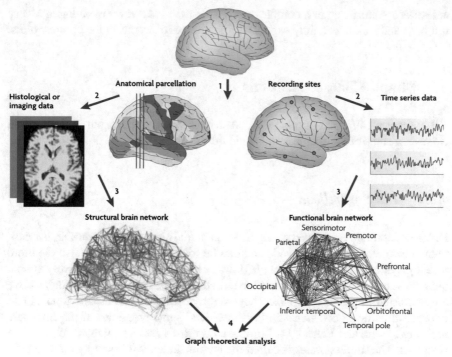

Fig. 4.9 Structural and functional brains can be studied with graph theory by following different methods shown step-by-step in this figure. Figure taken from [34]

of size $N \approx 4,000$. These authors find a large clustering coefficients with a ratio to the corresponding random graph value of order 10^2. Results for the average shortest path length $\langle \ell \rangle$ are however not so clear due to relatively low values of N. Indeed, for N varying from $1,000$ to $4,000$, $\langle \ell \rangle$ varies by a factor of order $1.7 - 1.8$ [41]. A small-world logarithmic behavior would predict a ratio

$$r = \frac{\langle \ell \rangle (N = 4000)}{\langle \ell \rangle (N = 1000)} \sim \frac{\log(4000)}{\log(1000)} \approx 1.20 \qquad (4.25)$$

while a 3-dimensional spatial behavior would give a ratio of order $r \approx 4^{1/3} \approx 1.6$ which is closer to the observed value. Larger sets would however be needed in order to be sure about the behavior of this network concerning the average shortest path and to distinguish a $\log N$ from a $N^{1/3}$ behavior expected for a three-dimensional lattice.

Things are however more complex than it seems and even if functional connectivity correlates well with anatomical connectivity at an aggregate level, a study [42] shows that strong functional connections exist between regions with no direct structural connections, demonstrating that structural and functional properties of neural networks are entangled in a complex way. This field is however evolving at a very fast pace and we can certainly expect a large number of results in the coming years.

4.2 Simple Characterizations

Throughout the first part of this book, we will see many metrics but in this section we review some of the simplest and standard measures.

4.2.1 α and γ Indices

Different indices were defined a long time ago mainly by scientists working in quantitative geography since the 1960s and can be found in [43–45] (see also the more recent paper by Xie and Levinson [46]). Most of these indices are relatively simple but give valuable information about the structure of the network, in particular if we are interested in planar networks. They were used to characterize the topology of transportation networks: Garrison [47] measured some properties of the Interstate highway system and Kansky [48] proposed up to 14 indices to characterize these networks. The simplest index is called the gamma index and is simply defined by

$$\gamma = \frac{E}{E_{max}} \tag{4.26}$$

where E is the number of edges and E_{max} is the maximal number of edges (for a given number of nodes N). For non-planar networks, E_{max} is given by $N(N-1)/2$ for non-directed graphs and for planar graphs we saw in the Chap. 2 that $E_{max} = 3N - 6$ leading to a modified γ index

$$\gamma_P = \frac{E}{3N - 6} \tag{4.27}$$

The gamma index is a simple measure of the density of the network but one can define a similar quantity by counting the number of elementary cycles instead of edges. The number of elementary cycles for a network is known as the cyclomatic number (see for example [49]) and is equal to

$$\Gamma = E - N + 1 \tag{4.28}$$

Using $E \leq 3N - 6$ for a planar graph, we find that this quantity is always less or equal to $\Gamma \leq 2N - 5$ which leads naturally to the definition of the alpha index (also coined 'meshedness' in [4])

$$\alpha = \frac{E - N + 1}{2N - 5} \tag{4.29}$$

This index lies in the interval $[0, 1]$ and is equal to 0 for a tree and equal to 1 for a maximal planar graph.

In the large N limit these indices are however not all independent. In particular, using the definition of the average degree $\langle k \rangle = 2E/N$ the quantities α and γ_P read for $N \gg 1$

$$\alpha \simeq \frac{\langle k \rangle - 2}{4} \tag{4.30}$$

$$\gamma_P \simeq \frac{\langle k \rangle}{6} \tag{4.31}$$

which shows that in fact for a large network these indice do not contain much more information than the average degree.

4.2.2 Organic Ratio and Ringness

Other interesting indices were proposed in order to characterize specifically road networks [46, 50]. For example, in some cities the degree distribution is very peaked around $3 - 4$ and the ratio

$$r_N = \frac{N(1) + N(3)}{\sum_{k \neq 2} N(k)} \tag{4.32}$$

can be defined [50] where $N(k)$ is the number of nodes of degree k. If this ratio is small the number of dead ends and of 'unfinished' crossing ($k = 3$) is small compared to the number of regular crossings with $k = 4$. In the opposite case of r_N close to 1, there is a dominance of $k = 4$ nodes which signals a more organized street network.

The authors of [50] also define the 'compactness' of a city which measures how much a city is 'filled' with roads. If we denote by A the area of a city and by ℓ_T the total length of roads, the compactness $\Psi \in [0, 1]$ can be defined in terms of the hull and city areas

$$\Psi = 1 - \frac{4A}{(L_{tot} - 2\sqrt{A})^2} \tag{4.33}$$

In the extreme case of one square city of linear size $L = \sqrt{A}$ with only one road encircling it, the total length is $L_{tot} = 4\sqrt{A}$ and the compactness is then $\Psi = 0$. At the other extreme, if the city roads constitute a square grid of spacing a, the total length is $\ell_T = 2L^2/a$ and in the limit of $a/L \to 0$ one has a very large compactness $\Psi \approx 1 - a^2/L^2$.

We end this section by mentioning the ringness. Arterial roads (including freeways, major highways) provide a high level of mobility and serve as the backbone of the road system [46]. Different measures (along with many references) are discussed and defined in [46]. In particular the ringness is defined as

$$\phi_{ring} = \frac{\ell_{ring}}{L_{tot}} \tag{4.34}$$

where ℓ_{ring} is the total length of arterials on rings and where the denominator ℓ_{tot} is the total length of all arterials. This quantity is in [0, 1] and is thus an indication of the importance of a ring and to what extent arterials are organized as trees.

4.2.3 Edge Orientation Distribution

Another simple interesting measure concerns the orientation of edges. For the sake of concretness we will consider the street network [51]. In this study, Boeing plots the polar histogram of each city's street orientation. In other words, in a polar representation the radius $r(\theta)$ is equal to the probability of finding a street with the angle θ (for practical applications we need to bin the polar histrogram and in his study, Boeing used 36 bins). Two examples are shown in the Fig. 4.10.

On the left, we see the street network of Manhattan and the corresponding polar histogram. The grid structure can clearly be seen in this histogram which shows

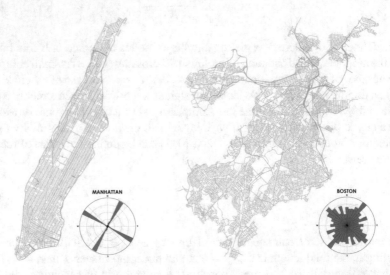

Fig. 4.10 Street networks and corresponding polar histograms for Manhattan and Boston. Figure from [51] and distributed under the terms of the Creative Commons Attribution 4.0 International License

that streets can have two directions perpendicular to each other only. On the right, the street network of Boston is shown. In this case, we have various patches with different structures. Some neighborhoods feature a grid (such as Back Bay or South Boston) but are aligned with each other, and together with (mostly older) streets in various directions results in a broad polar histrogram. As we will see also in Chap. 10, this type of polar histogram can also be used in order to characterize the temporal evolution of (street) networks.

This polar histogram can be used for classification purposes and Boeing plotted them according to a measure of their heterogeneity (the entropy here) and the result is shown in Fig. 4.11.

City Street Network Orientation

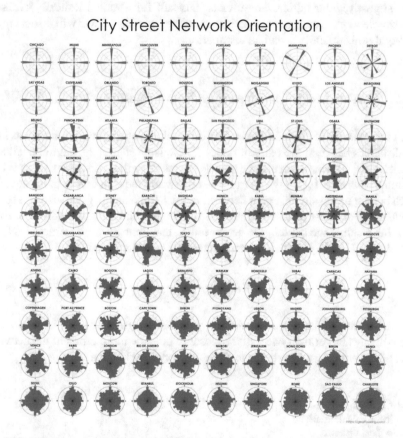

Fig. 4.11 Polar histograms of 100 world cities' street orientations, sorted according to their entropy. About half of these cities (49%) have an at least approximate north-south-east-west orientation trend. Figure from [51] and distributed under the terms of the Creative Commons Attribution 4.0 International License

4.2.4 Shape Factor

The authors of [9] measured the distribution of the form or shape factor defined as
the ratio of the area of the cell to the area of the circumscribed circle

$$\Phi = \frac{4A}{\pi D^2} \qquad (4.35)$$

(for practical applications, D can be also taken as the longest distance in the cell). For
german cities, the authors [9] found that most cells have a form factor between 0.3 and
0.6, suggesting a large variety of cell shapes, in contradiction with the assumption
of an almost regular lattice. These facts thus call for a model radically different
from simple models of regular or perturbed lattices. In Chap. 9, we will discuss more
thoroughly this quantity ϕ and its distribution.

4.2.5 Detour Index (or Stretch or Route Factor)

When the network is embedded in a two-dimensional space, we can define at least
two distances between the pairs of nodes. There is of course the natural euclidean
distance $d_E(i, j)$ which can also be seen as the 'as crow flies' distance. There is also
the 'route' distance $d_R(i, j)$ from i to j by computing the sum of length of segments
which belong to the shortest path between i and j (this quantity can also be seen as
the weighted shortest distance where the weight of a link is its length). The route
factor (also called the stretch, the detour index, the circuitry, or directness [52]) for
this pair of nodes (i, j) is then given by (see Fig. 4.12 for an example)

$$\eta(i, j) = \frac{d_R(i, j)}{d_E(i, j)} \qquad (4.36)$$

(in some works, an equivalent definition with $d_R(i, j)/d_E(i, j) - 1$ is used). This
ratio is always larger than one and the closer to one, the more efficient the network.
From this quantity, we can derive another one for a single node i defined by

Fig. 4.12 Example of a
detour index calculation. The
'as crow flies' distance
between the nodes A and B
is $d_E(A, B) = \sqrt{10}$ while
the route distance over the
network is $d_R(A, B) = 4$
leading to a detour index
equal to
$\eta(A, B) = 4/\sqrt{10} \simeq 1.265$

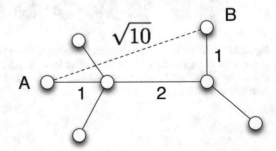

$$\eta(i) = \frac{1}{N-1} \sum_j \eta(i, j) \tag{4.37}$$

which measures the 'accessibility' for this specific node i. Indeed the smaller it is and the easier it is to reach the node i (Accessibility is a subject in itself—see for example [53]—and there are many other measures for this concept and we refer the interested reader to the articles [54–56]). This quantity $\eta(i)$ is related to the quantity so-called 'straightness centrality' [57] defined as

$$C^S(i) = \frac{1}{N-1} \sum_{j \neq i} \frac{d_E(i, j)}{d_R(i, j)} \tag{4.38}$$

If one is interested in assessing the global efficiency of the network, one can compute the average over all pairs of nodes (used in [58])

$$\langle \eta \rangle = \frac{1}{N(N-1)} \sum_{i \neq j} \eta(i, j) \tag{4.39}$$

The average $\langle \eta \rangle$ or the maximum η_{max}, and more generally the statistics of $\eta(i, j)$ are important and contain a lot of information about the spatial network under consideration. Aldous and Shun discuss the problems and avantages of these various quantities in [59] and we reproduce here some important elements of this paper. First, it is important to note that both $\langle \eta \rangle$ and η_{max} are unsatisfactory. For instance, it is unreasonable to characterize a network as inefficient because of the absence of a direct route between a given pair of nodes (leading to a very large η_{max}). In constrast, the disadvantage of $\langle \eta \rangle$ comes from the fact that for large networks, its value is governed by pairs of nodes that are far apart and therefore says nothing about the route length between nearby nodes. It is then interesting to define the following quantity [59]

$$\eta(d) = \frac{1}{N_d} \sum_{ij/d_E(i,j)=d} \eta(i, j) \tag{4.40}$$

(where N_d is the number of nodes such that $d_E(i, j) = d$) whose shape can help for characterizing combined spatial and topological properties (see also Chap. 7 for more empirical examples). This quantity $\eta(d)$ can be considered as an intermediate quantity that avoids the respective problems of $\langle \eta \rangle$ or η_{max}. Aldous and Shun further propose to consider the quantity

$$\eta^* = \max_d \eta(d) \tag{4.41}$$

which provides an intuitive way to compare the efficiency of different networks for providing short routes at any scale.

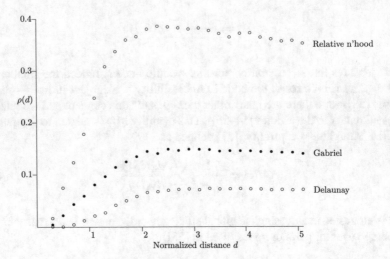

Fig. 4.13 The function $\eta(d)$ for three theoretical networks on random graphs (the distance d is normalized by). From top to bottom: RNG, GG, DT (see Chap. 16 for details about these graphs). The distance is normalized by the typical distance between nodes $\sqrt{A/N}$. Figure from [59]. Reprinted with the permission of the Institute of Mathematical Statistics

The general shape of $\eta(d)$ obtained for various graphs is shown in the Fig. 4.13. For various networks (the relative neighborhood graph, the Gabriel graph and the Delaunay triangulation), this profile has a characteristic shape with a maximum followed by a slow decrease. This typical shape can be interpreted as follows. For small distances, an efficient network will connect directly close nodes implying a small $\eta(d)$. For large distances, the presence if many alternate routes prevents the detour to grow to very large values. It is expected that this characteristic shape will hold for any 'reasonable' network model [59], although this is a conjecture at this stage.

4.2.6 Cost, Efficiency and Robustness

The minimum number of links to connect N nodes is $E = N - 1$ and the corresponding network is a tree. We can also look for the tree which minimizes the total length given by the sum of the length of all links

$$L_{tot} = \sum_{e \in E} d_E(e) \tag{4.42}$$

where $d_E(e)$ denotes the length of the link e. This procedure leads to the minimum spanning tree (MST) which has a total length L_{tot}^{MST} (see also Chap. 18). Obviously the tree with a very large detour index is not a very efficient network from the point of view of transportation for example, and usually more edges are added to the network, leading to an increase of accessibility but also of L_{tot}. A natural measure of the 'cost' of the network is then given by

$$C = \frac{L_{tot}}{L_{tot}^{MST}} \tag{4.43}$$

Adding links thus increases the cost but improves accessibility or the *transport performance* P of the network which can be measured as the average distance between all pairs of nodes, normalized to the same quantity but computed for the minimum spanning tree

$$P = \frac{\langle \ell \rangle}{\langle \ell_{MST} \rangle} \tag{4.44}$$

Another measure of efficiency was also proposed in [60, 61] and is defined as

$$\mathscr{E} = \frac{1}{N(N-1)} \sum_{i \neq j} \frac{1}{\ell(i, j)} \tag{4.45}$$

where $\ell(i, j)$ is the shortest path distance from i to j. This quantity $\mathscr{E}$ is zero when there are no paths between the nodes and is equal to one for the complete graph (for which $\ell(i, j) = 1$). The combination of these different indicators and comparisons with the MST or the maximal planar network can be constructed in order to characterize various aspects of the networks under consideration (see for example [4]).

Finally, adding links improves the resilience of the network to attacks or dysfunctions. A way to quantify this is by using the *fault tolerance (FT)* (see for example [62]) measured as the probability of disconnecting part of the network with the failure of a single link. The benefit/cost ratio is then estimated by the quantity FT/L_{tot}^{MST} which is a quantitative characterization of the trade-off between cost and efficiency [62].

Buhl et al. [4] measured different indices for 300 maps corresponding mostly to settlements located in Europe, Africa, Central America, India. They found that many networks depart from the grid structure with an α index usually low. For various world cities, Cardillo et al. [3] found that the α index varies from 0.084 (Walnut Creek) to 0.348 (New York City) which reflects in fact the variation of the average degree. Indeed for both these extreme cases, using Eq. (4.30) leads to $\alpha_{NYC} \simeq (3.38 - 2)/4 \simeq 0.345$ and for Walnut Creek $\alpha_{WC} \simeq (2.33 - 2)/4 \simeq 0.083$.

Measures of efficiency are relatively well correlated with the α index but displays broader variations demonstrating that small variations of the α index can lead to large variations in the shortest path structure. Cardillo et al. [3] plotted the relative efficiency.

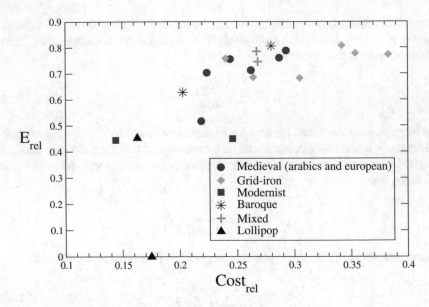

Fig. 4.14 Relative efficiency versus relative cost for the street network of 20 different cities in the world. In this plot the point $(0, 0)$ corresponds to the MST and the point $(1, 1)$ to the greedy triangulation. Figure taken from [3]

$$E_{rel} = \frac{\mathscr{E} - \mathscr{E}^{MST}}{\mathscr{E}^{GT} - \mathscr{E}^{MST}} \tag{4.46}$$

versus the relative cost

$$C_{rel} = \frac{C - C^{MST}}{C^{GT} - C^{MST}} \tag{4.47}$$

where GT refers to the greedy triangulation (the maximal planar graph). The cost is here estimated as the total length of segments $C \equiv L_{tot}$ and the obtained result is shown in Fig. 4.14 which demonstrates two things. First, it shows—as expected— that efficiency is increasing with the cost with an efficiency saturating at ~ 0.8. In addition, this increase is slow: typically, doubling the value of C shifts the efficiency from ~ 0.6 to ~ 0.8. Second, it shows that most of the cities are located in the high cost-high efficiency region. New York City, Savannah and San Francisco have the largest value of the efficiency (~ 0.8) with a relative cost value around ~ 0.35. It seems however at this stage difficult to clearly identify different classes of cities and further studies with a larger number of cities is probably needed in order to confirm the typology proposed in [3].

References

1. M. Newman, *Networks: An Introduction* (Oxford University Press, 2010)
2. A. Barrat, M. Barthélemy, A. Vespignani, *Dynamical Processes on Complex Networks*, reprint. (Cambridge University Press, Cambridge, 2012)
3. A. Cardillo, S. Scellato, V. Latora, S. Porta, Phys. Rev. E **73**(6), 066107 (2006)
4. J. Buhl, J. Gautrais, N. Reeves, R. Solé, S. Valverde, P. Kuntz, G. Theraulaz, Eur. Phys. J. B **49**, 513 (2006)
5. M. Barthelemy, A. Flammini, Phys. Rev. Lett. **100**(13), 138702 (2008)
6. S. Gerke, C. McDiarmid, Combinatorics. Probab. Comput. **13**, 165 (2004)
7. R. Albert, A.L. Barabasi, Rev. Mod. Phys. **74**, 47 (2002)
8. L.A.N. Amaral, A. Scala, M. Barthelemy, H.E. Stanley, Proc. Nat. Acad. Sci. **97**(21), 11149 (2000)
9. S. Lämmer, B. Gehlsen, D. Helbing, Physica: Stat. Mech. Appl. **363**(1), 89 (2006)
10. S. Porta, P. Crucitti, V. Latora, Environ. Plann. B **33**, 705 (2006)
11. V. Kalapala, V. Sanwalani, A. Clauset, C. Moore, Phys. Rev. E **73**, 026130 (2006)
12. A. Masucci, D. Smith, A. Crooks, M. Batty, Eur. Phys. J. B-Condens. Matter Complex Syst. **71**(2), 259 (2009)
13. E. Strano, A. Giometto, S. Shai, E. Bertuzzo, P.J. Mucha, A. Rinaldo, R. Soc. Open Sci. **4**(10), 170590 (2017)
14. M. Barthelemy, Phys. Rep. **499**(1), 1 (2011)
15. G. Bianconi, M. Marsili, J. Stat. Mech. **2005**, P06005 (2005)
16. H. Rozenfeld, J. Kirk, E. Bollt, D. ben Avraham, J. Phys. A **38**, 4589 (2005)
17. A. Barrat, M. Barthelemy, A. Vespignani, *Dynamical Processes in Complex Networks* (Cambridge University Press, Cambridge, UK, 2008)
18. R. Pastor-Satorras, A. Vázquez, A. Vespignani, Phys. Rev. Lett. **87**, 258701 (2001)
19. A. Vázquez, R. Pastor-Satorras, A. Vespignani, Phys. Rev. E **65**, 066130 (2002)
20. D.J. Watts, D.H. Strogatz, Nature **393**, 440 (1998)
21. M. Barthelemy, L.A.N. Amaral, Phys. Rev. Lett. **82**, 3180 (1999)
22. B. Bollobás, O. Riordan, Combinatorica **24**(1), 5 (2004)
23. R. Fulek, F. Morić, D. Pritchard, Disc. Math. **311**(5), 327 (2011)
24. R. Albert, I. Albert, G.L. Nakarado, Phys. Rev. E **69**, 025103 (2004)
25. R. Solé, M. Rosas-Casals, B. Corominas-Murtra, S. Valverde, Phys. Rev. E **77**, 026102 (2008)
26. E. Ravasz, A.L. Barabási, Phys. Rev. E **67**, 026112 (2003)
27. A. Yazdani, P. Jeffrey, in *submitted to Water Distribution System Analysis Conference, WDSA2010* (AZ, USA, 2010)
28. V. Latora, M. Marchiori, Physica A **314**, 109 (2001)
29. J. Sienkiewicz, J.A. Hołyst, Phys. Rev. E **72**(4), 046127 (2005)
30. C. Von Ferber, T. Holovatch, Y. Holovatch, V. Palchykov, in *Traffic and Granular Flow'07* (Springer, 2009), pp. 709–719
31. K.A. Seaton, L.M. Hackett, Physica A: Stat. Mech. Appl. **339**(3), 635 (2004)
32. P. Sen, S. Dasgupta, A. Chatterjee, P. Sreeram, G. Mukherjee, S. Manna, Phys. Rev. E **67**(3), 036106 (2003)
33. M. Kurant, P. Thiran, Phys. Rev. Lett. **96**(13), 138701 (2006)
34. E. Bullmore, O. Sporns, Nat. Rev. Neurosci. **10**, 186 (2009)
35. M. Chavez, M. Valencia, V. Latora, J. Martinerie, to appear in Int. J. Bif. Chaos (2010)
36. V.M. Eguiluz, D. Chialvo, G. Cecchi, M. Baliki, A. Apkarian, Phys. Rev. Lett. **94**, 018102 (2005)
37. M. Kaiser, C. Hilgetag, Neurocomputing **58–60**, 297 (2004)
38. M. Kaiser, C. Hilgetag, PLoS Comput. Biol. **2**, 0805 (2005)
39. L.F. Lago-Fernandez, R. Huerta, F. Corbacho, J.A. Siguenza, Phys. Rev. Lett. **84**, 2758 (2000)
40. D. Modha, R. Singh, Proc. Nat. Acad. Sci. (USA) **107**, 13485 (2009)
41. A. Zalesky, A. Fornito, I. Harding, L. Cocchi, M. Yucel, C. Pantelis, E. Bullmore, NeuroImage **50**, 970 (2010)

42. C. Honey, O. Sporns, L. Cammoun, X. Gigandet, J.P. Thiran, R. Meuli, P. Hagmann, Proc. Nat. Acad. Sci. (USA) **106**, 2035 (2010)
43. P. Haggett, R.J. Chorley, *Network Analysis in Geography*, vol. 67 (Edward Arnold London, 1969)
44. E. Taaffe, H.G. Jr., *Geography of Transportation* (Prenctice Hall, Englewood Cliffs, NJ, 1973)
45. J.P. Rodrigue, C. Comtois, B. Slack, *The Geography of Transport Systems* (Routledge, New York, NY, 2006)
46. F. Xie, D. Levinson, Geograp. Anal. **39**(3), 336 (2007)
47. W. Garrison, Reg. Sci. **6**, 121 (1960)
48. K. Kansky, *Structure of Transportation Networks: Relationships Between Network Geometry and Regional Characteristics* (University of Chicago Press, Chicago, 1969)
49. J. Clark, D. Holton, *A First Look at Graph Theory* (World Scientific, 1991)
50. T. Courtat, C. Gloaguen, S. Douady, Phys. Rev. E (2010)
51. G. Boeing, Appl. Netw. Sci. **4**(1), 1 (2019)
52. D. Levinson, A. El-Geneidy, Reg. Sci. Urban Econ. **39**, 732 (2009)
53. M. Batty, Env. and Plan. B: Plann. Des. **36**, 191 (2009)
54. S. Handy, D. Niemeier, Environ. Plann. A **29**, 1175 (1997)
55. D. Levinson, J. Econ. Geograp. **8**(1), 55–77 (2008)
56. A. El-Geneidy, D. Levinson, Netw. Spat. Econ. (2010)
57. P. Crucitti, V. Latora, S. Porta, Chaos **16**, 015113 (2006)
58. I. Vragovic, E. Louis, A. Diaz-Guilera, Phys. Rev. E **71**, 036122 (2005)
59. D.J. Aldous, J. Shun, Stat. Sci. **25**(3), 275 (2010)
60. V. Latora, M. Marchiori, Phys. Rev. Lett. **87**, 198701 (2001)
61. V. Latora, M. Marchiori, Eur. Phys. J. B **32**, 249 (2003)
62. A. Tero, S. Takaji, T. Saigusa, K. Ito, D. Bebber, M. Fricker, K. Yumiki, R. Kobayashi, T. Nakagaki, Science **327**, 439 (2010)

Chapter 5
Betweenness Centrality

There are many centralities that characterize the importance of a node (or an edge) in a large network. Among all these centralities, the betweenness centrality (BC) captures important aspects of the network and its structure. The BC is able to point to nodes with great importance in these networks. This is particularly true for resilience for example where the removal of high BC nodes can lead to a macroscopic breakdown of the system. Also, if one assume that

- a quantity is traveling on shortest paths in the network, and
- if the origin-destination matrix is flat (this corresponds to the uniform demand case where each pair of nodes constitutes an origin-destination couple)

the BC of a node (or an edge) then corresponds to the local traffic that can be found at this node. Highly congested points are then signalled by very large values of the BC. This is relevant for any type of transportation network where goods and individuals travel but also for networks such as the Internet where information packets can experience congestion problems at routers. In the particular case of spatial networks, the localization of these congested points can reveal some interesting features about the organization of the network and the flow pattern on it.

For complex networks, the BC generally scales with the degree, showing that in general central nodes are the hubs. In spatial networks however we do not have hubs and the degree is not equivalent to the BC. Generally speaking, for a lattice we expect the BC to decrease with the distance to the barycenter of all nodes, while the disorder introduces fluctuations that can dramatically alter this effect of distance and create non-trivial patterns. In particular, we observe the appearance of loops made of links that can have a BC larger than the barycenter of all nodes. In other words, disorder can completely modify the typical behavior observed for regular lattices. The pattern of the large BC nodes for spatial networks is then an interesting interplay between space and topology and contains a lot of information about the structure of these networks. In this chapter, we will first discuss general properties of the BC and then address more advanced questions.

M. Barthelemy, *Spatial Networks*,
https://doi.org/10.1007/978-3-030-94106-2_5

5.1   Definition

The importance of a node is characterized by its so-called centrality. There are many different centrality indicators such as the degree, the closeness, etc., but we will focus here on the betweenness centrality $g(i)$ which is defined for each node i as [1–5].

$$g(i) = \frac{1}{\mathcal{N}} \sum_{s \neq t} \frac{\sigma_{st}(i)}{\sigma_{st}} \tag{5.1}$$

where σ_{st} is the number of shortest paths going from s to t and $\sigma_{st}(i)$ is the number of shortest paths going from s to t through the node i. In general the summation is on $s \neq t$ and $s \neq v, t \neq v$ and this is the convention that we will adopt in this book. This quantity $g(i)$ thus characterizes the importance of the node i in the organization of flows in the network. Note that with this definition, the betweenness centrality of terminal nodes (i.e. with degree $k = 1$) is zero. The betweenness centrality can similarly be defined for edges

$$g(e) = \frac{1}{\mathcal{N}} \sum_{s \neq t} \frac{\sigma_{st}(e)}{\sigma_{st}} \tag{5.2}$$

where $\sigma_{st}(e)$ is the number of shortest paths going from s to t through the edge e.

The constant $\mathcal{N}$ is the normalisation and we will use here $\mathcal{N} = (N - 1)(N - 2)$ which counts the number of pairs and ensures that $g(v) \in [0, 1]$. Another way to understand this normalization is to note that it corresponds to the BC of the center of a star graph of N nodes (see below).

It is important to stress here that other types of centralities could be defined by choosing other type of paths. The most common choice is indeed the shortest path which leads to the usual BC, but we will also use the more general case of weighted shortest path, which corresponds to the quickest path if the weight of a link represents time.

5.2   General Properties

The betweenness centrality satisfies a certain number of properties (see for example [6]) that are often ignored by practioners of networks. These results can be obtained by standard methods of graph theory [7] or by using eigenvalues of the graph Laplacian [8] to estimate bounds. These general properties can however be useful to check analytical or numerical results and we reproduce here the most important ones. We refer the interested reader to the more mathematical presentation in [7].

5.2.1 Numerical Calculation: Brandes' Algorithm

The naive numerical computation of the BC can lead to very slow algorithms (typically of complexity $\mathcal{O}(N^3)$) and Brandes [9] proposed an algorithm of complexity $\mathcal{O}(N^2)$ (and $\mathcal{O}(N^2 \log N)$ for weighted networks) that became the standard way to compute this quantity. Since this algorithm is so important for network studies, we give a sketch here of its structure. Brandes used the dependencies defined as partial sums

$$\delta_s(v) = \sum_t \delta_{st}(v) \qquad (5.3)$$

where

$$\delta_{st}(v) = \frac{\sigma_{st}(v)}{\sigma_{st}} \qquad (5.4)$$

Brandes demonstrated that the quantities $\delta_s(v)$ obey the following recursion relation [9]

$$\delta_s(v) = \sum_{w:v \in P_s(w)} \frac{\sigma_{sv}}{\sigma_{sw}}(1 + \delta_s(w)) \qquad (5.5)$$

where the sum is over all w's that follow v, or in Brandes' terms, v belongs to the set of predecessors of w defined as

$$P_s(w) = \{u \in V | \{u, v\} \in E, \; d(s, v) = d(s, u) + \omega(u, v)\} \qquad (5.6)$$

where E is the set of edges and $\omega(u, v)$ is the weight on the link $u - v$ (and is equal to $\omega(u, v) = 1$ in the unweighted case). This relation can be understood with the following hand-waving argument. On a tree we have the relation $\delta_{st}(w) = \delta_{st}(v) - 1$ for each 'successor' of v leading to $\delta_{st}(v) = \delta_{st}(w) + 1$. For a general graph, only a fraction σ_{sv}/σ_{sw} goes to w via the node v and we have

$$\delta_{st}(v) = \sigma_{sv}/\sigma_{sw}(1 + \delta_{st}(w) \qquad (5.7)$$

leading to the general relation Eq. (5.5) above when summed over all w's (and t's).

Starting from a given node, we can then use the recursion Eq. (5.5) and then perform a loop on all nodes, leading to a complexity of order N^2.

5.2.2   The Average BC

The first simple bound of the BC is obtained by noting that its largest possible value for a network with N nodes is realized for the center of a star graph (also denoted by the bipartite complete graph $K_{1,N-1}$). In this case the BC is given

$$g_{K_{1,N-1}}(0) = (N-1)(N-2) \tag{5.8}$$

(up to a factor 2 depending on the summation used to define the BC). This quantity is thus the natural normalization in order to obtain a BC belonging to [0, 1].

We now compute the sum over all edges of the graph (more precisely, over the giant connected component of the graph)

$$\sum_{e \in G} g(e) \tag{5.9}$$

From the definition of the BC, this quantity reads

$$\sum_{e} g(e) = \frac{1}{\mathcal{N}} \sum_{st} \frac{1}{\sigma_{st}} \sum_{e} \sigma_{st}(e) \tag{5.10}$$

For a fixed pair of nodes s and t, if e belongs to a shortest path connecting these nodes, we have $\sigma_{st}(e) = 1$ and if not $\sigma_{st}(e) = 0$. This implies that

$$\sum_{e} \sigma_{st}(e) = \# \text{ of links belonging to } \mathcal{P}(s,t) \tag{5.11}$$

where $\mathcal{P}(s,t)$ is the set of shortest paths connecting s and t. All shortest paths from s to t have the same number of links (by definition) and is given by $d(s,t)$. We thus obtain

$$\sum_{e} \sigma_{st}(e) = d(s,t)\sigma_{st} \tag{5.12}$$

In summary, we have

$$\sum_{e} g(e) = \frac{1}{\mathcal{N}} \sum_{st} \frac{1}{\sigma_{st}} d(s,t)\sigma_{st}$$

$$= \frac{1}{\mathcal{N}} \sum_{st} d(s,t)$$

$$= \langle \ell \rangle \tag{5.13}$$

where $\langle \ell \rangle$ is the average shortest path in the graph.

It is tempting to try to extend this type of calculation to higher moments but there are problems that are difficult to solve. For example, for $\sum_e g(e)^2$ we would have to compute terms of the form

$$\sum_e \sigma_{st}(e)\sigma_{s't'}(e) \tag{5.14}$$

which leads to discuss the intersection of shortest paths $\mathscr{P}(s,t) \cap \mathscr{P}(s',t')$. It seems difficult to count that exactly, an important term is of course given by $s' = s, t' = t$ but other terms are certainly non-negligible. It is therefore unclear at this stage if we can show that higher moments of the BC (and hence the distribution) depends on global indicators only and not on the details of the graph.

5.2.3 Edge Versus Node BC

The BC for a link e is defined as follows

$$g(e) = \frac{1}{\mathscr{N}'} \sum_{s,t} \frac{\sigma_{st}(e)}{\sigma_{st}} \tag{5.15}$$

while we saw that for a node v it is defined as

$$g(v) = \frac{1}{\mathscr{N}} \sum_{s,t} \frac{\sigma_{st}(v)}{\sigma_{st}} \tag{5.16}$$

In these equations, $\mathscr{N}$ and $\mathscr{N}'$ are normalization constants, the sum is computed over nodes s and t, σ_{st} is the number of shortest paths from s to t, and $\sigma_{st}(e)$ is the number of these shortest paths going through e (and a similar definition for $\sigma_{st}(v)$). We will use the same normalization constant $\mathscr{N} = \mathscr{N}' = (N-1)(N-2) \sim N^2$ where N is the number of nodes.

We will now show that these centralities $g(e)$ and $g(v)$ are not independent. Indeed a path that goes through v has to pass necessarily through two of the neighbors of v

$$\sigma_{st}(v) = \frac{1}{2} \sum_{v' \in \Gamma(v)} \sigma_{st}(v, v') \tag{5.17}$$

where $\sigma_{st}(v, v')$ is the number of shortest paths that go through the link (v, v'). Here $\Gamma(v)$ denotes the set of neighboring nodes of v. We can then write for the node BC

$$g(v) = \frac{1}{2\mathscr{N}} \sum_{st} \sum_{v' \in \Gamma(v)} \frac{\sigma_{st}(v, v')}{\sigma_{st}}$$

$$= \frac{1}{2} \sum_{e \in \Gamma(v)} \frac{1}{\mathcal{N}} \sum_{st} \frac{\sigma_{st}(e)}{\sigma_{st}}$$

$$= \frac{1}{2} \sum_{e \in \Gamma(v)} g(e) \tag{5.18}$$

where $e \in \Gamma(v)$ denotes here also the set of links incident to v. This result shows that the fundamental quantity is in fact the edge BC, from which we can compute the node BC which is in some sense an aggregated quantity.

The average node centrality given by $1/N \sum_v g(v)$ is then equal to

$$\frac{1}{N} \sum_v g(v) = \frac{1}{2N} \sum_v \sum_{e \in \Gamma(v)} g(e) \tag{5.19}$$

$$= \frac{1}{N} \sum_e g(e) \tag{5.20}$$

$$= \langle \ell \rangle \tag{5.21}$$

and is also related to the average shortest path.

5.2.4 Adding Edges

Adding edges has a crucial impact on the BC, and even the addition of a single link can modify the whole pattern of the centrality. For example, in the simple case of a one-dimensional lattice $i = 1, \ldots, N$, the BC is given by $g_0(i) = (i-1)(N-i)$ with a maximum at $i = N/2$ (see Fig. 5.1). When one link is added between nodes a and b, the betweenness centrality of the nodes in the interval $]a, b[$ decreases. For example, if $a < N/2 < b$ it is not difficult to see that the variation of the centrality of the node $N/2$ is bounded by

$$\delta g(N/2) < -a(N-b) \tag{5.22}$$

which basically means that the shortest paths from a node $i \in [1, a]$ to a node $j \in [b, N]$ follow the shortcut and avoid the node $N/2$. In contrast, the betweenness centrality of the contact points increases and the betweenness centrality of node a is now

$$g(a) = g_0(a) + \delta g(a) \tag{5.23}$$

where $\delta g(a)$ is positive and essentially counts the new pairs of nodes which are connected by a shortest path going through the new link (a, b).

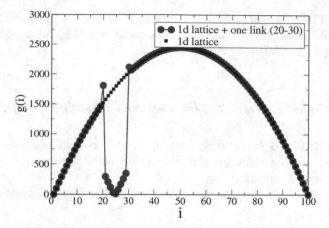

Fig. 5.1 Example of how the addition of a link perturbs the centrality. In black, the betweenness centrality for the 1d lattice (of size $(N = 100)$ has a maximum at the barycenter $N/2 = 50$. The addition of a link between $a = 20$ and $b = 30$ decreases the betweenness centrality between a and b and increases the betweenness centrality of nodes a and b

More generally, bounds on the average BC can be constructed when adding a link [7]. We denote the average BC of a graph G by

$$\overline{g}(G) = \frac{1}{N} \sum_i g(i) \tag{5.24}$$

and we recall (see the previous section) that this quantity is proportional to the average shortest path (which helps in constructing bounds and understanding the proofs).

We construct the graph G' by adding a link between two vertices u and v at distance $d = d(u, v) > 1$. It can then be shown that [7]

$$\overline{g}(G') \leq \overline{g}(G) - \frac{2(d - 1)}{N} \tag{5.25}$$

The main reason for such a relation is that distances in G' are lowered thanks to the shortcut and that in general we have the following inequality

$$\sum_{i,j} d_{G'}(i, j) \leq \sum_{i,j} d_G(i, j) - 2(d - 1) \tag{5.26}$$

where $d_{G'}(i, j)$ denotes the distance computed in G'. The immediate consequence of this bound is that if now $G' = (V, E')$ is a subgraph of $G = (V, E)$ and r denotes the number of links of G that are not in G', we obtain [7]

$$\overline{g}(G) \leq \overline{g}(G') - \frac{2r}{N} \tag{5.27}$$

We note that these results concern the average centrality but they don't imply that locally the BC of all nodes should decrease (indeed in the Fig. 5.1, we observe that the two endpoints a and b see their BC increase).

5.2.5   Relation Between the BC and the Clustering Coefficient

It has been noted that for networks with diameter 2 there is relation between the BC and the clustering coefficient and this relation has been empirically verified on various diameter 2 networks [10].

If the graph is a tree then the (not normalized) BC of a node is essentially given by

$$g_T(i) \sim N^2 \tag{5.28}$$

where the prefactor actually depends on the precise structure of the tree and the sizes of the different branches emanating from i. If we assume that we add loops to this tree, the BC at node i will decrease and it is natural to expand it in following way

$$g(i) = g_T(i) - \sum_{\ell \geq 1} N_\ell \tag{5.29}$$

where N_ℓ is the number of pairs of nodes connected by a path of length ℓ. For example if the neighbors of i are connected by paths of length one only, we obtain

$$g(i) = g_T(i) - g_T(i)C(i) \tag{5.30}$$

where $C(i)$ is the clustering coefficient of node i. This leads to a relation of the form

$$g(i) \sim N^2(1 - C(i)) \tag{5.31}$$

which was tested on different networks [10].

5.2.6   Scaling of the Maximum BC

In communication networks, it is intuitive to think that the traffic between nodes tends to go through a small core of nodes. In this case, the shortest paths are somehow curved inwards and it has been suggested that this is related to the global curvature of the network [11, 12]. A natural way to measure the impact of the structure on the load in the network is then to understand how the maximum traffic varies with various graph properties. The BC is a reasonable proxy for traffic and the theoretical question can be reduced to the scaling of the maximum BC, denoted by g_{max} with the size of the network measured by the number of nodes. In general, the relation is of the form

$$g_{max} \sim N^{\tau} \tag{5.32}$$

where the exponent τ depends on the network. Narayam and Saniee [11] studied empirically various networks and found essentially two families characterized by different values of $\tau = 1.5$ and $\tau = 2$ with typically, euclidean lattice networks displaying a value $\tau = 1.5$. These authors proposed the idea that this behavior is controlled by the curvature of the network and this was justified mathematically by Jonckheere et al. [12]. We discuss below a simplified version of their argument.[1] The idea is to write in the continuous limit the BC of a small region X as

$$g(X) \sim \int_{D \times D} \mathbb{1}(X \cap \mathscr{P}(s,t)) ds dt \tag{5.33}$$

here $\mathbb{1}$ is the indicator function and $X \subset D$ where D is the whole bounded domain (typically D is the ball $B_R(0)$ of radius R centered at the origin). In the euclidean case, the result is that

$$g(X) \sim R \frac{vol(X)}{vol(D)} \tag{5.34}$$

which means that the typical shortest path from s to t which goes through X is of length R times the probability to cross X which is $vol(X)/vol(D)$. We can illustrate this on the Fig. 5.2 where we assume that the euclidean lattice is of size $N = R^d$ and where we assume that the region of size ε around 0 contain the nodes with large BC. The probability that a shortest path (which are here straight lines) goes through this $\varepsilon-$region is

$$p = \frac{\varepsilon}{R^{d-1}} \tag{5.35}$$

The BC at 0 is then $R^d \times R^d \times p$ leading to

$$g_{max} \sim R^{d+1} \tag{5.36}$$

For euclidean-like networks we therefore have the scaling

$$g(X) \sim N^{1+\frac{1}{d}} vol(X) \tag{5.37}$$

and for $d = 2$, we recover the $N^{1.5}$ behavior (and for $d = 1$ a behavior in N^2).

For non-euclidean lattices, we can define various measures of curvature in graphs and we refer to [12] for more details. In particular, a graph is said to be Gromov

---

[1] I thank J.-M. Luck for discussions on this point.

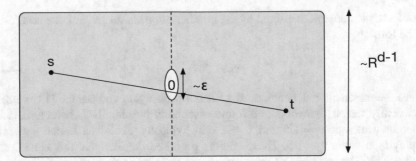

Fig. 5.2 Illustration of the simple argument for the scaling of the maximum BC for euclidean like networks

hyperbolic if there exists $\delta < \infty$ such that for any geodesic triangle abc (where ab, bc and ac are geodesic paths), there is an inscribed triangle xyz with $x \in bc$, $y \in ac$, $z \in ab$ of perimeter $d(x, y) + d(y, z) + d(z, x) \leq \delta$. In this hyperbolic case, the maximum BC essentially scales as the volume of X and we have

$$g(X) \sim (vol(D))^2 \sim N^2 \tag{5.38}$$

This result shows that for hyperbolic networks, the geodesic path between any two nodes always go through the center.

Planar graphs are in general flat and we expect a scaling of the maximum with $\tau = 1.5$. For spatial networks that are not planar, it would be interesting to measure this exponent. More generally, the behavior of the maximum BC might be a good indicator for classifying different types of networks [11].[2]

5.3 The BC for Simple Graphs

The BC cannot be computed analytically in general but some simple graphs can be studied and solved. These simple cases provide some sort of elementary 'bricks' that can help for understanding more complex networks.

5.3.1 Regular 1d and 2d Lattices

For one-dimensional lattices with N nodes, it is easy to see that the BC of node i ($i \in [\![1, N]\!]$) is given by

---

[2] I thank Saray Shai for discussions on this point.

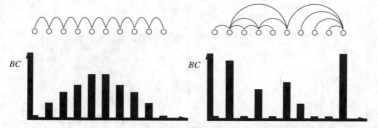

Fig. 5.3 Betweenness centrality for a one-dimensional lattice. (Left) When there is no disorder, the barycenter is the most central nodes. (Right) In the case of a disordered network, the degree becomes relevant and the most central nodes have large degrees

$$g(i) = \frac{i}{N}\left(1 - \frac{i}{N}\right) \tag{5.39}$$

The barycenter of all nodes $i_b = N/2$ is then also the most central node (see Fig. 5.3). When we introduce disorder—by removing or rewiring links—the BC becomes important at nodes that can be far away from the barycenter (see Fig. 5.3). In the extreme case where space doesn't play a role anymore such as in scale-free networks, the average BC per degree classes $g(k)$ scales as [13]

$$g(k) \sim k^{\mu} \tag{5.40}$$

where μ is an exponent that depends on the structure of the graph. Even if there are fluctuations around this scaling it shows that essentially the degree controls the BC in these graphs.

In the two-dimensional case we expect a similar behavior where the maximum is at the center of the network. More precisely, it is easy to express the BC of a node as a sum of combinatorial factors that count the number of paths. The number of paths between points (a, b) and (i, j) with $a < i$ and $b < j$ being $\binom{i+j-a-b}{i-a}$, the centrality of the node (i, j) on the grid $[\![-L, L]\!] \times [\![-L, L]\!]$ is:

$$g(i, j) = \frac{1}{4L^2} \sum_{\sigma \in \{-1,1\}} \sum_{a=-L}^{i} \sum_{b=-L}^{j}$$

$$\sum_{c=i}^{L} \sum_{d=j}^{L} \frac{\binom{j-b+\sigma(i-a)}{\sigma(i-a)}\binom{d-j+\sigma(c-i)}{\sigma(c-i)}}{\binom{c+d-a-b}{\sigma(c-a)}} \tag{5.41}$$

where $\sigma = \pm 1$ corresponds to nodes with $j < 0$ and $j > 0$, respectively. This expression is difficult to analyze, but we can resort to the simple approximation described in Fig. 5.4a. We assume that the number of paths going through the node $(i, 0)$ is proportional to the product of areas described in gray in Fig. 5.4, normalized by the total number of paths (we have to multiply the result by a factor 2 by symmetry). We thus obtain

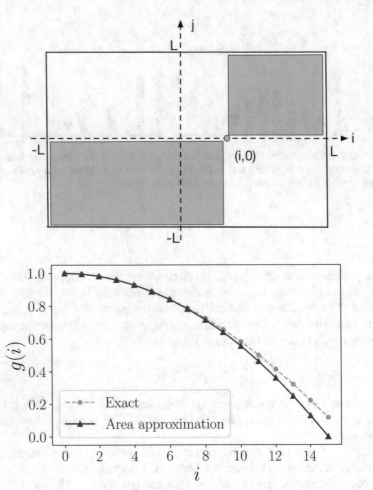

Fig. 5.4 **a** Approximation for computing the BC for the 2d grid: the BC at node $(i, 0)$ is proportional to the product of shaded areas (up to a factor two accounting for the white squares). **b** Comparison of the approximation of Eq. (5.42) to the analytical formula of Eq. (5.41) computed numerically for $L = 15$

$$g(i, 0) \propto L(L - i) \times (i + L)L/L^4$$

$$= 1 - \left(\frac{i}{L}\right)^2 \tag{5.42}$$

We compare this approximation to the exact numerical result showing a very good agreement (Fig. 5.4b). The discrepancy appears essentially for $i \approx L$ where the approximation predicts $g(L) = 0$ which is exact for $L \gg 1$. Using this same argument, we find that for any node located at (i, j), we find

$$g(i, j) \simeq (1 - i^2)(1 - j^2) \tag{5.43}$$

We compare this expression to the exact result and we observe that the most important errors appear at the boundary of the square but, despite its simplicity, the approximation is very good in the bulk of the square [18].

5.3.2 Cayley Tree

We follow here the derivation of [41] and consider a Cayley tree of size N with fixed branching ratio k and all leaves are at the same depth. We adopt the convention where $l = L$ for the leaf level and $l = 0$ for the root. A node on the l-th level has $k - 1$ branches directly below it at the $(l + 1)$-th level, and each of these children as a number M_{l+1} of children itself (its descendant). The total number of descendant of a node a level $l + 1$ (including the parent node) is then

$$M_{l+1} = \sum_{l'=l+1}^{L} k^{l'-l-1}$$
$$= \frac{k^{L-l} - 1}{k - 1} \tag{5.44}$$

The betweenness $g(l)$ of a node v at level l is then given by (by symmetry of the problem all nodes at the same level have the same BC)

$$g(l) = \binom{k-1}{2} M_{l+1}^2 + \binom{k-1}{1} M_{l+1}(N - M_l) \tag{5.45}$$

The first term corresponds to all paths connecting nodes being descendant of different children of node v, while the second term corresponds to path going from a descendant of v to another node that is not a descendant of v. This last term is the dominant contribution and we have for large N (and therefore large L)

$$g(l) \approx (k - 1)N M_{l+1} \approx A k^{-l} \tag{5.46}$$

where the prefactor reads $A = N k^L$. Using the relation

$$P(g) = \sum_l P(g|l)P(l). \tag{5.47}$$

where the probability to have a node at level l is given by $P(l) = k^l/N$ and where $P(g|l) = \delta(g - Ak^{-l})$ since g is completely determined by the level l in which it lies in the tree. We then obtain [41]

$$P(g) = \frac{A}{g} \tag{5.48}$$

which implies a broad distribution of the BC (there is in general a cut-off at large BC due to the finite size).

5.3.3 Branches and Ring

As discussed above, we observe non-trivial objects such as high BC loops in random graphs. It is important to understand the formation of these structures and the conditions for their existence. In particular, it seems that randomness can induce very large perturbation in the spatial distribution of the BC and we can be in a situation where the barycenter is not the most central node, or equivalently, that the BC is not a simple decreasing function of the distance to the barycenter anymore. In order to understand this phenomenon and the condition for the appearance of central loops, we discuss here a simple toy model [14, 15]. We construct a star network composed of N_b branches, where each branch is composed of n nodes. We then add a loop at distance ℓ from the center (see Fig. 5.5 for a sketch of this graph). We also consider here a more general case where the links are weighted and in this simplified model we assume that links on the branches have a weight equal to one and the loop segments between two consecutive branches have a weight given by w (the purely topological case then corresponds to the case $w = 1$). We then compute the BC using weighted shortest paths and this generalization allows us to discuss for example the impact of different velocities on a street network. In this case, w can be seen as the time spent on the segment and the weighted shortest path is then the quickest path.

   Within this simple toy model, we discuss under which conditions the loop is more central than the 'origin' at the center. Intuitively, for very large w, it is always less costly to avoid the loop, while for $w \to 0$, loops are very advantageous. The two main quantities of interest are therefore the centrality at the center denoted by $g_0(\ell, n, w)$ and the centrality at the intersection C of the branch and the loop, denoted by $g_C(\ell, n, w)$. We then compute the difference $\delta g = g_0 - g_C$ and will study under which condition it can be negative.

5.3.3.1   A Line and One Ring

This is the simplest case[3] which corresponds to $n_b = 2$ in Fig. 5.5. More precisely, we consider $2N + 1$ points along a line, from $-N$ to N. Each link between neighboring points is of length 1, and we introduce a shortcut from point $-l$ to point l. For the sake of simplicity, we give this shortcut a length 1.

---

[3] I thank Erwan Taillanter for sharing with me his notes on this calculation.

Fig. 5.5 Representation of the toy model discussed here. The number of branches is here $N_b = 5$, the number of nodes on each branch is $n = 11$ and the loop is located at a distance $\ell = 6$ from the center 0. The node C is at the intersection of a branch and the loop and T is the terminal node of a branch. Figure taken from [14]

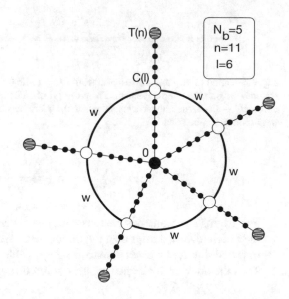

Let us determine the impact of this shortcut on the betweenness centrality of each point of the line. For everything that follows, we will consider paths from i to j such that $i < j$. If one wants to consider all possible paths, one should simply multiply all results by 2. We will call "direct path" a path ignoring the shortcut, and "alternative path" a path using the shortcut.

In the absence of a shortcut, a point in v is placed on all shortest paths from each point $i < v$ to each point $j > v$. The resulting BC is $BC_0(v) = (N - v)(N + v)$.

Points $v < -l$ or $v > l$ are not affected by the presence of the shortcut. Going from $i < v$ to $j > v$ requires to travel through v regardless of the presence of the shortcut, thus the BC remains unchanged. The number of this kind of paths going through v is

$$N_0(v) = (N - v)(N + v)$$

Now consider a point such that $-l < v < 0$.

1. For origins $i < -l$, and destinations $j > v$, one can note that the direct path is shorter than the alternative if and only if $j < 0$. As a result, the number of this kind of paths going through v reads

$$N_1(v) = (N - l)|v|$$

with $N - l$ the number of possible origins and $|v|$ the possible destinations

2. For origins $-l < i < v$, and destinations $j > v$, the shortest path goes through v if the direct path is shorter than the alternative path. This condition can be written as $j - i < (i - (-l)) + (l - j)$ or $j < l + i$.

We imposed $j > v$, thus for each value of i, there are $l + i - v$ possible values of j. The number of corresponding paths going through v writes

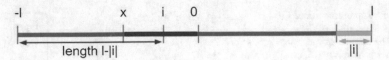

Fig. 5.6 Detail of the case where a path from i to j (with $0 > i > v > -l$ and $v < j < l$) will go through v using the alternative path. There are $|v|$ choices for i (red zone), and for a given i, to enforce $j - i > l$, the only acceptable values for j are between $l - |i|$ and l (green zone), i.e. $|i|$ possibilities. Figure from [15]

$$N_2(v) = \sum_{i=-l}^{v}(l - v + i) = \frac{(l + v)(l - v)}{2}$$

3. For origins $i > v$ and destinations $v < j < l$, the shortest path goes through v if the alternative is shorter than the direct path. This requires $j - i > l$. Since we imposed $j < l$, at a given i, there are $|i|$ possible values for j (cf. Fig. 5.6). Thus the number of alternative paths going through v is

$$N_3 = \sum_{i=v}^{0}|i| = \frac{v^2}{2}$$

4. For origins $i > v$ and destinations $j > l$, we have a case symmetric to case 1. Indeed, for the alternative path to be shorter than the direct path (and thus going through v), we need $i < 0$, leading to

$$N_4 = |v|(N - l)$$

We now consider the different contributions of these terms depending on the value of v. We will discuss all cases $v \le 0$, and the rest is obviously symmetrical.

- For $v < -l$ (resp. $v > l$), as mentioned, the BC is straightforward. The only contribution is N_0 and

$$g(v) = \mathcal{N}(N - v)(N + v) \tag{5.49}$$

- For $-l < v \le 0$ (resp. $0 \le v < l$), the BC is the sum of contributions from direct paths from $i < v$ to $j > v$ (i.e. $N_1 + N_2$) and from alternative paths from $i > v$ to $j > v$ (i.e. $N_3 + N_4$). This leads to

$$g(v) = \mathcal{N}(2(N - l)|v| + \frac{l^2}{2}) \tag{5.50}$$

- For $v = -l$ (resp. $v = l$), the BC is the sum of the direct or alternative paths from $i < -l$ to $j > -l$ (N_0), and of the alternative paths from $i > -l$ to $j > -l$ going through $-l$ ($N_1 + N_3$). We then obtain

$$g(-l) = \mathcal{N}(N^2 + Nl - \frac{3l^2}{2})\tag{5.51}$$

5.3.3.2   Approximate Formulas

The interest of this toy model lies in the fact that we can estimate analytically the BC for the center $g_0(\ell, w)$ and for the intersection nodes on the loop $g_C(\ell, w)$. We discuss in detail the derivation for these quantities in Chap. 17, and we give here the main results.

The exact expressions for the centralities g_0 and g_C are difficult to handle analytically, essentially because they are expressed as sums of complicated arguments. In order to derive analytical predictions we use a simple approximation scheme that allows to obtain the correct scaling for the most important quantities. In this derivation, we have to introduce the following quantity

$$\chi \equiv \min\left(\frac{N_b - 1}{2}, \left[\frac{2\ell}{w}\right]\right)\tag{5.52}$$

which basically compares the cost on the longest path along the loop (of order $(N_b - 1)/2 \times w$) to the cost of the path going through the center (2ℓ). After a lengthy calculation, we obtain the following expression for the centrality at the origin

$$g_0(w) \approx N_b\left\{\left(\frac{N_b - 1}{2} - \chi\right)n^2 + \chi\frac{(\ell - 1)(\ell - 2)}{2}\right\}\tag{5.53}$$

We note that this approximation recovers both exact limits

$$g_0 \simeq \begin{cases} N_b\frac{(N_b-1)}{2}\frac{(\ell-1)(\ell-2)}{2} & \text{for } w \to 0 \\ n^2 N_b\frac{(N_b-1)}{2} & \text{for } w \to \infty \end{cases}\tag{5.54}$$

In the following it will also be useful to consider the limit $\ell, n \to \infty$ with $x = \ell/n$ fixed which gives for $g_0(x, \chi) = g_0(\ell, n, w)/n^2$ (up to terms of order $1/n$)

$$g_0(x, \chi) \approx N_b\left\{\left(\frac{N_b - 1}{2} - \chi\right) + \frac{1}{2}\chi x^2\right\}\tag{5.55}$$

where the only dependence on w is now encoded in χ.

The same level of approximation can be used for computing the BC on the loop, and similarly to the case of the origin, it will be convenient for analyzing these expressions to consider the limit $n, \ell \to \infty$ such that $\ell/n = x$. Up to terms of order $1/n$ we then obtain for $g_C(x, \chi) = g_C(\ell, n, w)/n^2$

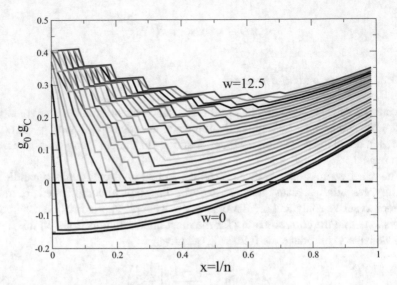

Fig. 5.7 $\delta g(\ell)$ versus ℓ for N_b and n fixed and for different values of w in the range $[0, 12.5]$. For values less than a threshold ($w_c \approx 4$ here) there is a minimum that is negative

$$g_C(x, \chi) = (1 - x)(x + N_b - 1)$$
$$+ 2\chi x(1 - \frac{x}{2})$$
$$+ \frac{\chi(\chi - 1)}{2}(1 - \frac{x^2}{2}) \tag{5.56}$$

In Chap. 17, we compare the exact result with these approximations, and show that they allow to understand and to predict the correct scaling for the important quantities.

5.3.3.3 Threshold Value of w and Optimal ℓ

We want to understand if the origin has always a larger BC than the loop, or if for some values of the parameters, this order can be inverted. The relevant quantity to understand is therefore the difference

$$\delta g(x, \chi) = g_0(x, \chi) - g_C(x, \chi) \tag{5.57}$$

given by Eqs. (5.55), (5.56). We first plot this quantity versus ℓ for different values of w and we observe the result shown in Fig. 5.7 This result shows that for w sufficiently small, δg can be negative. This demonstrates the existence of a threshold value w_c such that at $w = w_c$ the minimum is $\min_\ell \delta g = 0$. For $w < w_c$, the minimum of δg is negative and we can define an optimal value ℓ_{opt} which corresponds to this smallest value of δg. The quantity ℓ_{opt} thus gives the position of loop that maximizes the difference between the BC of the loop and the center.

In Chap. 17, we give the derivations for the quantities w_c and ℓ_{opt} and here we restrict ourselves to hand-waving arguments. In the case of ℓ_{opt}, when this quantity is small, most paths connecting nodes from different branches will go through 0 and we expect $\delta g > 0$. When ℓ is increasing more paths will go through the loop and will increase the value of g_C. However, when ℓ is too large, paths connecting the (large) fraction of nodes located on the lower branches will go through 0 again. In order to get a sufficient condition on ℓ_{opt}, we consider the path between the node C on a given branch and the corresponding node C' on the furthest branch $(N_b - 1)/2$. The optimal value for ℓ_{opt} is then such that the cost of the path from C to C' through 0 and which is 2ℓ is equal to the cost on the loop which is given by $w(N_b - 1)/2$. This immediately gives the result

$$\ell_{opt} \approx \frac{w(N_b - 1)}{4} \tag{5.58}$$

The threshold quantity w_c is obtained by imposing that the minimum of $\delta g(\ell = \ell_{opt})$ is equal to zero, but we can understand the scaling for w_c with the simple following argument. Indeed, a necessary condition on w is that ℓ_{opt} must be less than n. This gives the condition

$$w < w_c \sim 4\frac{n}{N_b} \tag{5.59}$$

If we come back to the 'topological' case where all weights are equal to 1, these results on this simple toy model show that the loop can be more central than the origin if $w_c > 1$ which implies that $n \gg N_b$. It thus suggests that the number and the spatial extension of radial branches are crucial ingredients that control the existence of central loops. If the extension of the network is large compared to the number of radial branches, w_c can be larger than one and central loops for $w = 1$ can be observed. In ordered systems—such as lattices—the effective number of branches is too large leading to a very small w_c and therefore prohibits the appearance of central loops in the 'topological' case ($w = 1$). In real-world planar graphs where randomness is present, the absence of some links can lead to a small number of 'effective' radial branches which in the framework of the toy model implies a large value of w_c and therefore a large probability to observe central loops.

5.3.4 Grid-Tree Network

In this part, we consider the grid-tree network discussed in [16]. This graph depicted in Fig. 5.8 consists of a central grid of size $w \times w$ and tree graphs representing the peripheral areas. The tree graphs have depth h and branching ratio r. The total number of nodes in the central grid is $N_G = w^2$ and the total number of nodes in a

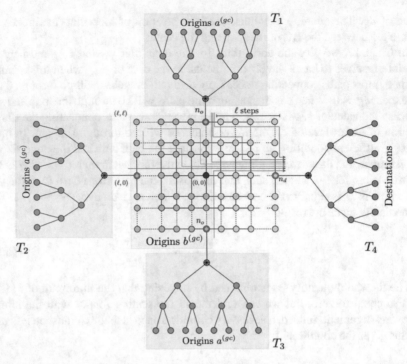

Fig. 5.8 Grid-tree model discussed in [16]. The grid has $N_G = w \times w$ nodes, the trees have a branching ratio r and depth h $(N_T = (r^{h+1} - 1)/(r - 1))$. The total number of nodes is then $N_{GT} = n_t N_T + N_G = w^2 + n_t N_T$. Figure from [16] and distributed under the terms of the Creative Commons Attribution 4.0 International license

tree is $N_T = \sum_{l=0}^{h} r^l = (r^{h+1} - 1)/(r - 1)$. The total number of nodes in this graph is therefore $N_{GT} = N_G + n_t N_T$ where n_t is the number of trees ($n_t = 4$ in Fig. 5.8).

It is interesting to note that this type of very simplified structure captures relatively well some aspects in real cities. In particular, one can compare the empirical distribution of the BC such as the monotonic decrease of the average BC with the distance to the city center (see Fig. 5.9) and the decrease of deviations from the average (large deviations for nodes close to the city center and small deviations for peripheral nodes).

Some nodes in this structure are more important than others and have a specific role. In particular, the central grid node located at $(0, 0)$ will in general experience a very large BC. Connector nodes are the entry points of the paths coming from trees and are directly connected to the roots of trees (Fig. 5.8). The idea for computing the BC for nodes in this graph and developped in [16] is the following one. First, we note that there are 3 types of paths: (i) from a tree to another one, (ii) from a tree to the grid, and (iii) within the grid. The number of paths of type (i) scales as N_T^2, of type (ii) as N_T and is independent from N_T in the case (iii). The BC of any node i can thus be written as a polynomial in N_T

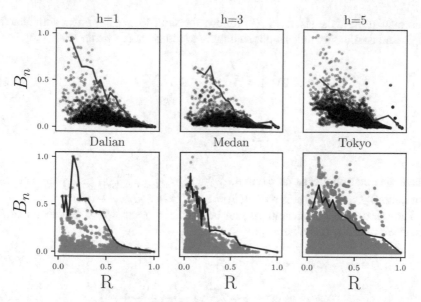

Fig. 5.9 (Top) BC versus the distance to the center for grid-tree models with parameters $w = 51$, $r = 2$, and $h = 1, 3, 5$ from left to right. (Bottom) Empirical result for cities of increasing size from left to right: Dalian (China), Medan (Indonesia), and Tokyo (Japan). The lines denote the average. Figure taken from [16] and distributed under the terms of the Creative Commons Attribution 4.0 International license

$$g(i) = a(i)N_T^2 + b(i)N_T + c(i) \tag{5.60}$$

where the different coefficients depend on the parameters defining the grid-tree graph: w, r, and h.

The analytical calculation of these coefficients can be tedious and for the sake of simplicity, we follow [16] and illustrate the calculation on the example of the central grid node BC $g(0)$. For this node, there are two paths from a tree to its opposite (T_2 to T_4 or T_1 to T_3). More complex are paths from a tree to its neighbor (typically from T_1 to T_4 for example). From the connector node n_o to the connector n_d there are $\pi_{\ell,\ell} = (2\ell)!/\ell!^2$ shortest paths of length 2ℓ and only one which goes through 0. This gives thus a contribution equal to $1/\pi_{\ell,\ell}$ to the BC and we therefore obtain

$$a(0) = 2 + 4\frac{\ell!^2}{(2\ell)!} \tag{5.61}$$

where the factor 4 comes from all possible neighboring trees. Similarly, Lampo et al. obtain for $b(0)$ the

$$b(0) = 4\ell + 8\sum_{x=0}^{\ell}\sum_{y=1}^{\ell}\frac{\pi_{x,y}}{\pi_{x+\ell,y}} \tag{5.62}$$

The calculation of the last term $c(0)$ follows the same line of argument with now the origin and destination nodes both belonging to the grid. The result reads [16]

$$c(0) = 2\ell^2 + 4 \sum_{\alpha,y=1}^{\ell} \frac{1}{\pi_{\alpha,y}} + 8 \sum_{\alpha,x,y=1}^{\ell} \frac{\pi_{x,y}}{\pi_{x+\alpha,y}} \tag{5.63}$$

$$+ 2 \sum_{a,b,x,y=1}^{\ell} \frac{\pi_{x,y}\pi_{a,b}}{\pi_{x+a,y+b}} \tag{5.64}$$

where the notation means for example $\sum_{\alpha,y=1} = \sum_{\alpha=1}^{\ell} \sum_{y=1}^{\ell}$. Using these expressions, the BC at the origin is then $g(0) = a(0)N_T^2 + b(0)N_T + c(0)$.

The same type of calculations can be used for other nodes. For example for connector nodes c, one obtains [16]

$$a(c) = 3 \tag{5.65}$$

$$b(c) = w^2 - 1 + 2 \sum_{y=1}^{\ell} \left(\frac{1}{\pi_{2\ell,y}} + \frac{\pi_{\ell,\ell}}{\pi_{\ell,\ell+y}} \right) \tag{5.66}$$

$$c(c) = \ell^2 + 2 \sum_{a=1}^{2\ell} \sum_{b=0}^{\ell} \sum_{y=1}^{\ell} \frac{\pi_{a,b}}{\pi_{a,b+y}} \tag{5.67}$$

The first result for $a(c)$ just counts the number of other trees, while the first term of $b(c)$ counts all the number of nodes in the grid (except c itself). The sum in $b(c)$ corresponds to paths going from the grid to the other trees than the one connected to c. The first term in the sum corresponds to an origin o starting below y of the connector and going to a node in the opposite tree. There is then one path from this node to c and to the opposite tree (T_2) among $\pi_{2\ell,y}$ possible shortest paths. The second term in the sum corresponds to all possible shortest path starting from this node o to the tree T_1 adjacent of the connector's one.

Finally, it is also possible to compute the BC of a node in the trees. The symmetry of the problem implies that it depends on the height v of the node in the tree $v \in \{0, 1, \ldots, h\}$ and will therefore be denoted by $g(v)$. Since we have a tree (and therefore the absence of loops) the number of paths is always equal to one. There are two types of paths: paths coming from one children of the node and going to another child and paths going through the node and exploring the upper levels of the tree. A simple calculation then gives [16]

$$g(v) = \frac{r(r^{h-v} - 1)}{2(r - 1)^2} (2r^{h+1} - r^{h-v+1} - r^{h-v} - r + 1) \tag{5.68}$$

Beyond their own theoretical interest, these calculations also allowed the authors of [16] to discuss how congestion can spread in a road network. In particular, they could show that for increasing size of the trees, there are multiple transitions corresponding to the increase of BC from the center to the periphery of the graph, mimicking the spatial spread of congestion in urban roads.

5.4 The BC in a Disk: The Continuous Limit

5.4.1 The Infinite Density Limit

We will follow here the derivation given in [17] for a random spatial network in the dense limit where $\rho \to \infty$. In this limit the use of the continuous approximation allows to compute analytically the BC. We will consider that the domain is a disk which implies that the centrality depends on the distance κ to the center only, and following [17], we write for the BC $g(\kappa)$ of a node at distance κ from the center

$$g(\kappa) = \frac{1}{2} \int \frac{dr_i}{V} \frac{dr_j}{V} \chi_{ij}(\kappa) \tag{5.69}$$

where $\chi_{ij}(\kappa)$ is equal to one if the geodesic path from i to j goes through the node κ (we identify here the name of the node with its distance to the center) and zero otherwise. In this continuous limit the geodesic are straight lines and the number of shortest paths is equal to one which justifies this expression Eq. (5.69). The prefactor $1/2$ here accounts for the double counting of paths. Using the projection $\kappa_\perp$ of κ on the line $i - j$ (see Fig. 5.10), we can rewrite the integral as

$$\int dr_i dr_j \chi_{ij} = \int dr_i dr_j \delta(\kappa_\perp) \tag{5.70}$$

which simply means that the straight line from i to j goes exactly through κ.

We choose a polar coordinate system centered at κ and we denote the coordinates of the nodes i and j by (r_i, θ_i) and (r_j, θ_j). Elementary trigonometry implies that

$$\kappa_\perp = r_i \sin \phi_i \tag{5.71}$$
$$\kappa_\perp = r_j \sin \phi_j \tag{5.72}$$

and

$$\phi_i + \phi_j = \pi - \theta_j + \theta_i \tag{5.73}$$

which implies in the limit $\kappa_\perp \simeq 0$ (which is indeed the interesting domain imposed by the delta function) the following expression

Fig. 5.10 Representation of the different quantities used for the calculation of the BC at κ in the continuous limit. Figure redrawn from [17]

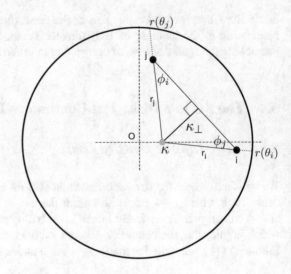

$$\kappa_\perp = \frac{\theta_i - \theta_j + \pi}{\frac{1}{r_i} + \frac{1}{r_j}} \tag{5.74}$$

Inserting this expression in Eq. (5.69) we then obtain

$$g(\kappa) = \frac{1}{2V^2} \int_0^{2\pi} d\theta_i \int_0^{2\pi} d\theta_j \delta(\theta_i - \theta_j + \pi)$$
$$\times \int_0^{r(\theta_i)} dr_j r_j \int_0^{r(\theta_j)} dr_i r_i \left(\frac{1}{r_i} + \frac{1}{r_j} \right) \tag{5.75}$$

where

$$r(\theta) = \sqrt{R^2 - \kappa^2 \sin^2 \theta} - \kappa \cos \theta \tag{5.76}$$

denotes the polar equation of the circle bounding the domain. Performing some of the integrals leads to

$$g(\kappa) = \frac{2(R^2 - \kappa^2)}{\pi^2 R^3} E\left(\frac{\kappa}{R} \right) \tag{5.77}$$

where

$$E(x) = \int_0^{\pi/2} d\theta \sqrt{1 - x^2 \sin^2 \theta} \tag{5.78}$$

is the complete elliptic integral of the second kind. Once normalized with the maximum value $g^*(\kappa) = g(\kappa)/g(0)$ we obtain [17]

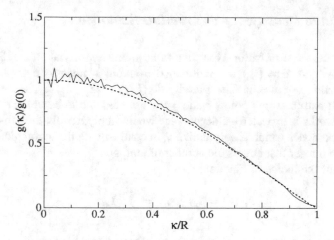

Fig. 5.11 Comparison between the numerical result obtained for $\rho = 200$ in a disk $R = 1$ (averaged over 1000 realizations) and the theoretical result Eq. (5.79) shown by a dotted line

$$g^*(\kappa) = \frac{2}{\pi}\left(1 - \frac{\kappa^2}{R^2}\right) E\left(\frac{\kappa}{R}\right) \tag{5.79}$$

This quantity is a monotonically decreasing function and the expansions around the origin and the boundary are given by

$$g^*(\kappa \ll 1) = 1 - \frac{5\kappa^2}{4R^2} + \mathcal{O}(\kappa^4) \tag{5.80}$$

$$g^*(\kappa \simeq R) = \frac{4}{\pi}\left(1 - \frac{\kappa}{R}\right) + \mathcal{O}((R - \kappa)^2) \tag{5.81}$$

We can note that for practical applications we can consider that $2/\pi E(x) \approx 1$ for $x \in [0, 1]$ which leads to a quadratic form $g^*(x) \approx 1 - (x/R)^2$. More generally, as expected from the 1d case, this function is decreasing from the center in a parabolic fashion and linearly close to the boundary. Equation (5.81) is the solution for $\rho \to \infty$ and the rate of convergence to this solution was quickly discussed in [17] but is not yet clarified. We show in Fig. 5.11 the comparison between the exact result and the simulation for a random geometric graph. We note that fluctuations are the largest close to zero which is probably due to the fact that we have a smaller number of points in this zone.

This calculation actually corresponds to a continuous limit and the universality of the result shows that the important ingredient is that shortest paths are straight lines, independently from the structure of the graph. It is then be interesting to understand the convergence towards this limit or equivalently to find the next term in a large density expansion and which corresponds to geodesics that are not perfect straight lines. This is what we will do in the next section.

5.4.2 Finite Density: A Perturbation Expansion

We present here a perturbation expansion at the lowest non-trivial order [18] for large densities. We denote by (i, j, κ) random nodes (among N nodes of a graph G) inside a disk domain $\mathscr{D}$ of area V. The quantity $SP(i, j) = \{x_{j_1}, ..., x_{j_m}, ..., x_{j_n}\}$ denotes the shortest path between points i and j (that we assume to be unique, which for spatial networks is expected - a degeneracy would imply exactly the same distance between two nodes which is very unlikely, in contrast with the topological distance which is an integer that counts the number of jumps).

We define the indicator function

$$\sigma_{ij}(\kappa) = \mathbf{1}_{\kappa \in SP(i,j)} = \sum_m \mathbf{1}_{\kappa = x_{jm}} \tag{5.82}$$

This indicator function $\sigma_{ij}(\kappa)$ is equal to unity if κ is in $SP(i, j)$ and zero otherwise. The betweenness centrality for the node κ is then

$$g(\kappa) = \frac{1}{2} \sum_i \sum_j \frac{\sigma_{ij}(\kappa)}{\sigma_{ij}} \tag{5.83}$$

where i and j are nodes of the graph. For large ρ, we use a continuous approximation and write

$$\sigma_{ij}(\kappa) = \int_{t_i}^{t_j} dt\, \delta(\kappa - x(t)) \tag{5.84}$$

where the shortest path $\{x(t) \in SP(i, j)\}$ is parametrized by $t \in [t_1, t_2]$ where t_1 and t_2 correspond to the endpoints (δ is the Dirac delta function). The BC for κ is then

$$g(\kappa) = \frac{1}{2V^2} \int_{\mathscr{D}} d\mathbf{r}_i \int_{\mathscr{D}} d\mathbf{r}_j\, \sigma_{ij}(\kappa) \tag{5.85}$$

In the continuous limit [17] $\rho \to \infty$, the shortest paths are straight segments and the indicator reads as $\sigma_{ij} = \delta(x \cos(\phi) + y \sin(\phi) - p)$ where $\kappa = (x, y)$ and where the segment (i, j) is parametrized by p and ϕ (for this type of parametrization, see for example [19]). This means that κ is in $SP(i, j)$ if and only if κ is on the line between i and j. In particular, it is easy to check that $\int_{\mathscr{D}} d\kappa\, \sigma_{ij}(\kappa) = |t_2 - t_1|$ as expected.

In the quasi-dense limit ($1 \ll \rho < \infty$), we define the average betweenness centrality for κ as the expectation the BC for κ

$$\overline{g(\kappa)} = \mathbb{E}_G(g(\kappa)) \tag{5.86}$$

where $\mathbb{E}_G$ denotes the average over all the graphs (for a given connection rule) constructed over an ensemble of points realizations. We then obtain

Fig. 5.12 Sketch of the system considered and notations. The origin of the polar system is κ and the nodes i and j have the coordinates (r_i, θ_i) and (r_j, θ_j) in this system. The deviation of the two segments $(i\,\kappa)$ and (κj) from the straight line is characterized by the angle $\varepsilon_{ij} = \theta_i - \theta_j + \pi$

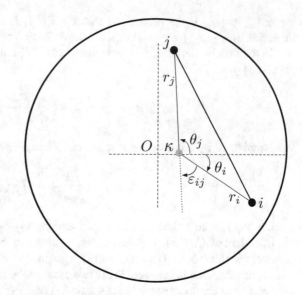

$$\overline{g(\kappa)} = \frac{1}{2V^2} \int_{\mathscr{D}} d\mathbf{r}_i \int_{\mathscr{D}} d\mathbf{r}_j \, \mathbb{E}_G(\sigma_{ij}(\kappa)) \tag{5.87}$$

The quantity $\sigma_{ij}(\kappa)$ is an indicator function and its average is then a probability

$$\mathbb{E}_G(\sigma_{ij}(\kappa)) = \text{Prob}\,(\kappa \in SP(i, j)) \tag{5.88}$$

which we will denote by $\chi_{ij}(\kappa) = \text{Prob}\,(\kappa \in SP(i, j))$. In the dense limit, shortest paths are straight lines and we have

$$\chi_{ij}(\kappa) = \delta(x \cos(\phi) + y \sin(\phi) - p) \tag{5.89}$$

When the density is finite, the shortest paths deviate from the straight line and we define the angular deviation ε_{ij} from the straight line (i, j) in the frame of origin κ: $\varepsilon_{ij} = \theta_i - \theta_j + \pi$ (see Fig. 5.12). Due to the statistical isotropy of the problem, it is enough to consider the node at distance κ from the center and at polar angle $\theta = 0$ and here and in the following we will work with the polar coordinate centered on this point.

We now express the probability $\chi_{ij}(\kappa)$ that κ is in $SP(i, j)$ for a given value of ε_{ij}, and the average BC can then formally be rewritten as the 5d integral

$$\overline{g(\kappa)} = \frac{1}{2V^2} \int_{\mathscr{D}^2} d\mathbf{r}_i d\mathbf{r}_j$$

$$\int_0^\infty d\varepsilon_{ij} \delta(\theta_i - \theta_j + \pi - \varepsilon_{ij}) \chi_{ij}(\kappa | \varepsilon_{ij}) \tag{5.90}$$

where $\chi_{ij}(\kappa|\varepsilon_{ij})$ is the probability that $\kappa \in SP(i, j)$ conditioned by ε_{ij}. The delta function $\delta(\theta_i - \theta_j + \pi - \varepsilon_{ij})$ ensures the definition of the angle ε_{ij}.

In the infinite density limit, we know from [17] that this conditional probability is given by

$$\chi_{ij}(\kappa|\varepsilon_{ij}, \rho = \infty) \propto \left(\frac{1}{r_i} + \frac{1}{r_j}\right) \delta(\varepsilon_{ij}) \tag{5.91}$$

and motivated by this case we assume the following generalization

$$\chi_{ij}(\kappa|\varepsilon_{ij}) = \left(\frac{1}{r_i} + \frac{1}{r_j}\right) \chi(\varepsilon_{ij}) \tag{5.92}$$

where $\chi(\varepsilon)$ is an unknown function. Denoting ε_{ij} by ε, we assume that $\chi(\varepsilon)$ is independent of (i, j, κ). This is a strong assumption that we empirically show to be correct for the k-NN, RGG, DT and GG graphs (as we will further discuss below, this approximation is incorrect in the two cases of the MST and the RNG graphs and suggests that for these graphs the assumption that $\chi(\varepsilon)$ is independent of (i, j, κ) is not correct). Indeed, we show empirically that for these graphs, the function

$$\chi(\varepsilon) = \frac{\chi_{ij}(\kappa|\varepsilon)}{\left(\frac{1}{r_i} + \frac{1}{r_j}\right)} \tag{5.93}$$

can be fitted by a decreasing exponential function of ε with parameter ε_0 ($R^2 > 0.8$ for all graphs except for the MST and RNG)

$$\chi(\varepsilon) = \varepsilon_0(\rho) \mathcal{N} e^{-\varepsilon/\varepsilon_0(\rho)} \tag{5.94}$$

where $\varepsilon_0(\rho)$ is a smooth decreasing function of ρ (see Fig. 5.13) and $\mathcal{N}$ a normalization.

Using this form Eq. (5.93), we obtain

$$\overline{g(\kappa)} = \frac{1}{2V^2} \int_0^\infty d\varepsilon \, \chi(\varepsilon) \int_{\mathscr{D}} d\mathbf{r}_i \int_{\mathscr{D}} d\mathbf{r}_j$$
$$\left(\frac{1}{r_i} + \frac{1}{r_j}\right) \delta\left(\theta_i - \theta_j + \pi - \varepsilon\right) \tag{5.95}$$

In the dense limit, we have $\chi(\varepsilon) = \mathcal{N}\delta(\varepsilon)$ and we recover the known result of [17]. In order to go beyond this infinite density result, we expand this function $\chi(\varepsilon)$ around 0 to the second order in ε_o

$$\chi(\varepsilon) = \varepsilon_0(\rho) \mathcal{N} e^{-\varepsilon/\varepsilon_0(\rho)}$$
$$\simeq \mathcal{N} \left(\delta(\varepsilon) - \varepsilon_0(\rho)\delta'(\varepsilon) + \varepsilon_0^2(\rho)\delta''(\varepsilon)\right) \tag{5.96}$$

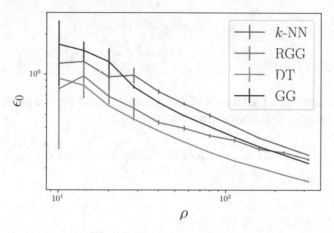

Fig. 5.13 Writing $\chi(\varepsilon) = \varepsilon_0(\rho)\mathcal{N}e^{-\varepsilon/\varepsilon_0(\rho)}$, we show that ε_0 is a smooth decreasing function of the density ρ, validating the shape of $\chi(\varepsilon)$ for k-NN, RGG, DT and GG graphs (the vertical error bars correspond to the dispersion). The graph suggests a power law relation of the form $\varepsilon_0(\rho) \simeq \rho^{-\beta}$ with $\beta \simeq 0.4 \pm 0.1$. We note that the exponent is not the exact same for all graphs, nor is the prefactor, thus making some types of graphs converging faster towards the dense regime than others. Figure taken from [18]

Here, we used the distributional derivative of the Dirac delta function, which is defined so that for any compactly supported smooth test function ϕ, we have

$$\int \mathrm{d}x \phi(x)\delta'(x) = -\int \mathrm{d}x \phi'(x)\delta(x) \tag{5.97}$$

Inserting the expansion of Eq. (5.96) into the expression Eq. (5.95), we get

$$\overline{g(\kappa)} = \frac{\mathcal{N}}{2V^2} \int_0^\infty \mathrm{d}\varepsilon \left(\delta(\varepsilon) - \varepsilon_0(\rho)\delta'(\varepsilon) + \varepsilon_0^2(\rho)\delta''(\varepsilon)\right)$$

$$\int_{\mathscr{D}} \mathrm{d}\mathbf{r}_i \int_{\mathscr{D}} \mathrm{d}\mathbf{r}_j \left(\frac{1}{r_i} + \frac{1}{r_j}\right) \delta\left(\theta_i - \theta_j + \pi - \varepsilon\right)$$

$$= \frac{\mathcal{N}}{4V^2} \int_0^{2\pi} \mathrm{d}\theta_i \, r(\theta_i)r(\theta_i + \pi)\left(r(\theta_i) + r(\theta_i + \pi)\right)$$

$$- \varepsilon_0(\rho) \int_0^{2\pi} \mathrm{d}\theta_i \, r(\theta_i)r'(\theta_i + \pi)\left(r(\theta_i) + 2r(\theta_i + \pi)\right)$$

$$- \varepsilon_0^2(\rho) \int_0^{2\pi} \mathrm{d}\theta_i \, r(\theta_i)r''(\theta_i + \pi)\left(r(\theta_i) + 2r(\theta_i + \pi)\right)$$

In the polar coordinate centered at κ, the frontier of the disc is given by $r(\theta) = \sqrt{R^2 - \kappa^2 \sin^2(\theta)} - \kappa \cos(\theta)$ and the previous expressions can be now rewritten as

$$\overline{g(\kappa)} = \frac{\mathcal{N}}{4V^2}\left[2\int_0^{2\pi} d\theta \ (R^2 - \kappa^2)\sqrt{R^2 - \kappa^2\sin^2(\theta)}\right.$$
$$\left. - \varepsilon_0^2(\rho)\int_0^{2\pi} d\theta \ \frac{\kappa^2(R^2-\kappa^2)(R^2(6\sin^2\theta - 1) - 5\kappa^2\sin^4\theta)}{(R^2 - \kappa^2\sin^2\theta)^{3/2}}\right] \quad (5.98)$$

The first order term (coefficient of ε_0) is equal to 0 and the first non-trivial term is of second order. We introduce the functions

$$I_0(\kappa, R) = 2\int_0^{2\pi} d\theta \ (R^2 - \kappa^2)\sqrt{R^2 - \kappa^2\sin^2(\theta)} \quad (5.99)$$

and

$$I_2(\kappa, R) =$$
$$\int_0^{2\pi} d\theta \ \frac{\kappa^2\left(R^2 - \kappa^2\right)\left(R^2\left(6\sin^2(\theta) - 1\right) - 5\kappa^2\sin^4(\theta)\right)}{\left(R^2 - \kappa^2\sin^2(\theta)\right)^{3/2}} \quad (5.100)$$

and the BC can be rewritten as

$$\overline{g(\kappa)} = \frac{\mathcal{N}}{4V^2}[I_0(\kappa) - \varepsilon_0^2(\rho)I_2(\kappa, R)] \quad (5.101)$$

In the result of Eq. (5.101), the infinite density limit which corresponds to the first term is universal, i.e. independent from the graph structure. In contrast, the second term (term in $\varepsilon_0(\rho)$) does depend on the graph and encodes the deviation of shortest paths from the straight line which varies from a graph to another. This implies that in general (and as expected) the BC of a graph at finite density is not universal and depends on the graph considered. In particular, the numerical result of Fig. 5.13 suggests a power law relation of the form $\varepsilon_0(\rho) \simeq \rho^{-\beta}$ with $\beta \simeq 0.4 \pm 0.1$. We observe that the value of the exponent β and the prefactor are not exactly the same for all graphs, implying different rates of convergence towards the dense regime.

We can express the integrals appearing in Eq. (5.101) using special functions and we get

$$I_0(R, \kappa) = 2R(R^2 - \kappa^2)E\left(\frac{\kappa}{R}\right) \quad (5.102)$$

and

$$I_2(R, \kappa) = 8R^3\left[3\left(\frac{\kappa}{R}\right)^2 K\left(\frac{\kappa}{R}\right) + 2K\left(\frac{\kappa}{R}\right)\right.$$
$$\left. +2\left(\frac{\kappa}{R}\right)^2 E\left(\frac{\kappa}{R}\right) - 2E\left(\frac{\kappa}{R}\right)\right] \quad (5.103)$$

where $K(x)$ and $E(x)$ are respectively the elliptic integrals of the first and second kind.

If $\kappa \ll R$, we can show that

$$I_2(R, \kappa) = 4\pi R^3 \left(\frac{\kappa}{R}\right)^2 + o\left(\left(\frac{\kappa}{R}\right)^2\right) \tag{5.104}$$

while if $R - \kappa \ll R$

$$I_2(R, \kappa) \simeq 16R^3 \left(1 - \frac{\kappa}{R}\right) \log\left(1 - \frac{\kappa}{R}\right) \tag{5.105}$$

Normalizing the BC by $\overline{g(0)} = \frac{\mathcal{N}}{4V^2} \times 8\pi R^3$, we obtain

$$g^*(\kappa) = \frac{\overline{g(\kappa)}}{\overline{g(0)}} = \frac{1}{8\pi R^3}[I_0(\kappa) - \varepsilon_0^2(\rho)I_2(\kappa, R)] \tag{5.106}$$

If $\kappa \ll R$, it gives

$$g^*(\kappa) \simeq 1 - \left(5 + \frac{1}{2}\varepsilon_0^2(\rho)\right)\left(\frac{\kappa}{R}\right)^2 \tag{5.107}$$

while if $R - \kappa \ll R$

$$g^*(\kappa) \simeq \left(\frac{4}{\pi} - 2\varepsilon_0^2(\rho) \log\left(1 - \frac{\kappa}{R}\right)\right)\left(1 - \frac{\kappa}{R}\right) \tag{5.108}$$

We note that under such a normalization we have always $g^*(0) = 1$ and $g^*(R) = 0$ (which can be easily proven). In order to use this result and to compare it with the simulations, we need to specify the deviation characterized by $\varepsilon_0(\rho)$. We test numerically this analytical result Eq. (5.106) on various graphs: the DT, the k-NN, the RGG, the GG and the MST. For the GG, DT, k-NN and the RGG, we observe an excellent agreement between the analytical result and our numerical simulations (averaged over 5 000 runs) for $1 \ll \rho$ (we show here results for the DT on Fig. 5.14, for the other graphs see [18]). The discrepancies between the quasi-dense and the dense regime are larger around $\frac{\kappa}{R} = 0.8$. We observe that for these different graphs the speed of convergence to the infinite density limit is not the same. The convergence for the GG, RGG and the DT is fast while it is slower for the k-NN. For the GG, it is so fast that the infinite regime approximation is a good approximation for densities as low as 3 points per square unit (graphs of 10 points). In general, however, it is hard to predict or to understand why one type of graph would converge faster towards the infinite limit than the others. For k-NN and RGG graphs, the approximation is good from densities as low as 6 points per square unit (which corresponds to less than 20 points in the disc) while the approximation is valid for smaller densities (about 10 points in the disc) for DT and the Gabriel graphs (not shown here). We

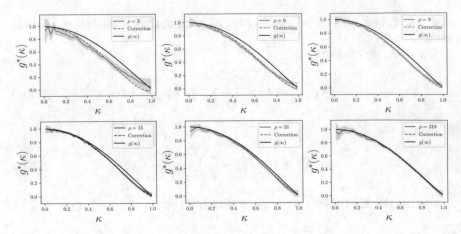

Fig. 5.14 Comparison of the expansion of Eq. (5.106) with the numerical result for the DT. The quality of the approximation increases with the density but is insightful at surprisingly low densities (6 points per square unit is less than 20 points in the disc). The number of points is $N = \rho\pi$. Figure taken from [18]

also note that Gabriel graphs (GG) being subgraphs of DT, we would naïvely expect that GG converge slower towards the dense regime limit than DT. This would result from the fact that the shortest paths are closer to straigth lines (and hence the dense regime) when more points are added in the network. This is not true however since the normalized expected BC of κ depends on both the average BC in κ and the maximal BC in the graph (in 0). Adding more points to the system can both decrease the BC on average but increase the maximal BC, leading to non-trivial behaviors of convergence towards the dense regime.

The expansion Eq. (5.106) however does not work for the MST and RNG graphs since the assumption stating that $\chi(\varepsilon)$ is independent of (i, j, κ) seems not to be valid for these graphs. The BC converges towards the infinitely dense limit result of [17] (see Fig. 5.15 for the MST) but not at the rate predicted by the result Eq. (5.106).

This expansion is a step towards the study of the BC in spatial networks but many questions are left open. As mentioned, it is unclear why the behavior of the MST and the RNG is so different from the other graphs. More work is certainly needed in order to understand how shortest paths in these systems become always more straight when the density increases. Also, an open question concerns the spatial patterns of the BC in disordered spatial networks and it would interesting to understand from a theoretical point of view the effect of disorder.

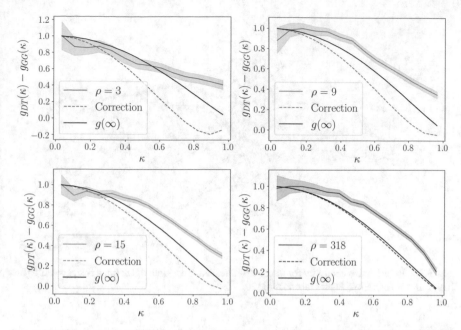

Fig. 5.15 Comparison of the expansion of Eq. (5.106) with the numerical result for the MST. The average BC converges towards the dense regime limit but the second-order approximation we used is not valid in this specific case. Figure taken from [18]

5.5 Empirical Measures on Street Networks

We will discuss here some empirical results for the BC on street networks, its spatial pattern, its distribution and the impact of one-ways.

5.5.1 Spatial Patterns of Large BC Nodes

The Fig. 5.16 displays the spatial distribution of the betweenness centrality for the city of Dresden and as expected, zones which are central from a geographical point of view also have a large betweenness centrality. We however see that other roads or zones can have a large betweenness centrality pointing to a complex pattern of flow distribution in cities.

Many other studies [20–34] considered different aspects of the street network and observed non-trivial structures in the BC spatial distribution. In particular, in [20] it has been observed that the distribution of the BC can display non-trivial spatial patterns and in [33] the authors showed that during the evolution of the street network of Paris (France) most 'standard' measures were unable to detect the

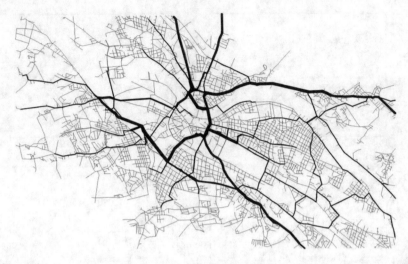

Fig. 5.16 Betweenness centrality for the city of Dresden. The width of the links corresponds to the betweenness centrality. Figure taken from [20]

important structural changes that occurred in the 19th century, while in contrast, the spatial distribution of the BC displayed dramatic changes (see Chap. 10).

Using the road network obtained from city extracts (the data has been obtained from the Mapzen website [35]), the BC can be computed for the different cities shown in Fig. 5.17. For all these real-world cases, we observe that indeed non-trivial structures appear and in particular the appearance of loops made of central links and of different sizes. The stability of these loops can be tested by filtering these networks for different values of the BC larger than a threshold g^* and computing the perimeter of the main loop. The results for Dresden, Los Angeles, and Paris are shown in Fig. 5.18. We observe on this plot the presence of various plateaus at intermediate values of g^* suggesting that these loops are indeed very central and stable.

We note that in general, boundary effects can be important and can affect the measures done on spatial networks [36, 37] and we test this in the case of the street network of Paris [14]. The area enclosed in the largest loop was measured on the same network but at different scales (i.e. with different boundaries, going from central Paris to almost the whole Ile-de-France region). We observe that at least the area of the largest loop remains relatively stable but further systematic studies are certainly needed in order to understand which patterns are stable and which ones are not, and what are the conditions on the boundaries in order to ensure stability of the main spatial patterns.

We now discuss the existence of very central nodes far away form the center with a simple model of a random planar graph built with a percolation cluster. In order to do this, we consider the percolation on a regular lattice and we assume that each link has a probabilty p to be present (and $1 - p$ to be removed). Above the percolation threshold $p \geq p_c$, the system displays a giant component which connects a non-zero

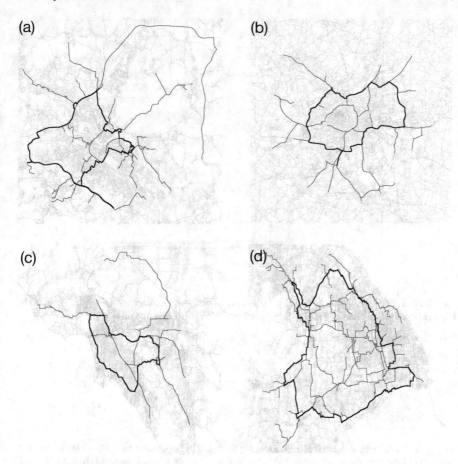

Fig. 5.17 Real-world networks: **a** Dresden, Germany, **b** Paris, France, **c** Los Angeles, USA, **d** Shanghai, China. We show here the links e with a BC larger than a certain threshold $g(e) > g^*$ and we highlight the largest loop. Figure taken from [14]

fraction of the nodes. We study the BC on this giant component and filter the nodes for different threshold g^*. We show in Fig. 5.19 the set of links that belong to the giant component and with a BC larger than g^*. For different values of this threshold we observe that the set of most central links forms a non-trivial pattern, and in general displays many loops (we show the largest one in Fig. 5.19). We can go further in the analysis of the structure of the percolating cluster by analyzing the ratio

$$\eta(r, \theta) = \frac{g(r, \theta)}{\max_{r' < r, \theta' \in [0, 2\pi]} g(r, \theta')} \tag{5.109}$$

which compares the BC at one point with the maximum BC of nodes in the region closer to the origin. For a percolating cluster obtained at $p = 0.8$ (well above the

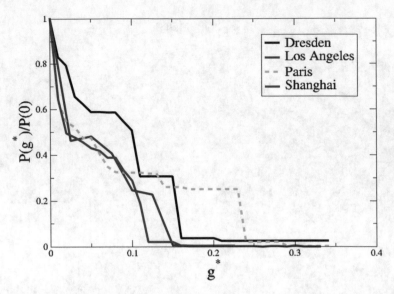

Fig. 5.18 Perimeter $P(g^*)$ of the main loop (normalized by perimeter of the loop at $g^* = 0$) for the road network of Dresden, Los Angeles, and Paris. Figure taken from [14]

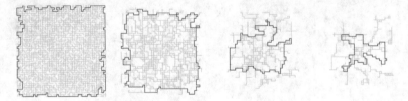

Fig. 5.19 Links belonging to the giant component obtained for percolation with $p = 0.15$ and with normalized BC larger than g^*. From left to right, the threshold g^* increases and the largest loop is here highlighted with a thick line. Figure taken from [14]

percolation threshold) and on a square lattice 100×100, we observe a very broad distribution of η. For values larger than one we obtain on average $\overline{\eta} \simeq 3$ and a very large dispersion of order 10^3. We can identify the points for which we have a ratio $\eta > 1$ and plot (Fig. 5.20) the distribution of the distance to the center for these points (normalized by the maximum distance d_{max}). This Fig. 5.20 shows that the location of nodes with a very large BC (at least larger than the BC of the nodes closer to the center) can be of order the system size. This shows that—depending on the disorder—the 'central' area composed of the geometrical center and its surroundings are composed of nodes with a relatively small BC. This reinforces the need to understand in which cases the monotoneous decrease of the BC with the distance to center can be strongly modified by fluctuations.

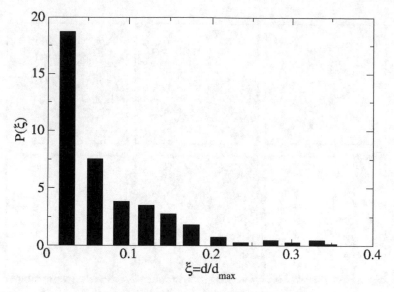

Fig. 5.20 Distribution of the distance to the center for nodes with a ratio $\eta > 1$ (the distance is here normalized by the maximum distance on the lattice). These results were obtained for $p = 0.8$ on a 100×100 lattice and averaged over 30 configurations. Figure taken from [14]

5.5.2   The BC and Socio-Economic Indicators

In addition to have a relation with the traffic and possibly the congestion, a study [38] proposed an interesting direction in the general context of connecting topological measures of networks to socio-economical indices. In particular, these authors analyzed the distribution of commerce and service activities in the city of Bologna and compared it to the spatial distribution of the BC. They observed that there is indeed a clear correlation between the betweenness centrality and the presence of commercial activities, a statement that can be quantified more precisely (see [38] for details).

5.5.3   The BC Probability Distribution for Street Networks

Lämmer et al. [20] studied the German road network and obtained very broad distributions of betweenness centrality with a power law exponent in the range [1.3, 1.5] (for Dresden ≈ 1.36). The fact that the BC is broadly distributed signals the existence of a strong heterogeneity of the network in terms of traffic, with the existence of a few very central roads which probably experience congestion traffic problems. Also the absence of a scale in a power law distribution suggests that the importance of roads is organized in a hierarchical way, a property expected for many transporta-

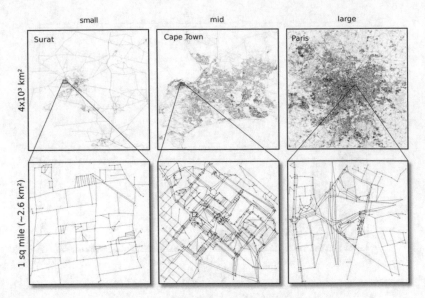

Fig. 5.21 The street networks are split into three categories based on their number of nodes (inter-sections) in the sampled street networks: small ($N \sim 10^3$), medium ($N \sim 10^4$) and large ($N \sim 10^5$). The upper panel shows the networks at the full sampled range of 10^3 square-kilometers, whereas the lower panel shows selected smaller samples (on the order of one-square-mile). The street networks used in this analysis were constructed from the OpenStreetMaps (OSM) database [42]. Figure taken from [41] and distributed under a Creative Commons Attribution 4.0 International License

tion networks [39]. The broadness of the betweenness centrality distribution does not seem however to be universal. Indeed, in [25, 40], the betweenness centrality distribution is peaked (depending on the city either exponentially or according to a Gaussian) which signals the existence of a scale and therefore of a finite number of congested points in the city.

This problem was thoroughly discussed in [41] where the authors analyzed the BC distribution for street networks from 97 of the most populous cities sampled from all the six inhabited continents. The different street networks can be grouped in three main categories according to their size (Fig. 5.21), from small ($N \sim 10^3$ nodes), medium ($N \sim 10^4$) to large road networks ($N \sim 10^5$). The aggregate statistics for these networks are shown in Table 5.1.

We use here the conventional definition of the BC $g(i)$ of the node i (Eq. (5.1)) but in contrast with many other studies and in order to understand the scaling with the system size, we use here the unnormalized version with $\mathcal{N} = 1$.

A range of behaviors in the tail of the BC distribution is observed for these different cities, ranging from peaked to broad depending on the value of N (Fig. 5.22a). In general the BC distributions for cities within each size category are virtually identical, and displays a bimodal feature with two regimes separated by a bump roughly located at $g \sim N$. For larger values of the BC we observe a slow decay signaling a broad

Table 5.1 Aggregate statistics for the 97 street networks. Shown are the average, standard deviation, minimum, maximum and various percentile values for the area A, number of intersections (nodes) in the network N, number of roads (edges) e, total length of streets L_{tot} and the density $\rho = N/A$ of intersections. Table from [41]

	Nodes N	Edges e	Length L_{tot} (km)	Area A (km^2)	Density ρ
Mean	83528.87	130253.05	17461.68	4600.08	18.02
Stdev	90335.10	143060.21	15052.83	1926.00	15.43
Min	3349.00	5020.00	1793.45	777.07	1.00
25%	18925.00	28518.00	5789.36	3184.32	5.35
50%	62451.00	95797.00	12812.46	4411.81	14.98
75%	118712.00	178773.00	23751.22	5873.67	26.59
Max	612418.00	976040.00	82586.30	11562.73	93.47

distribution. One then sees a dramatic difference when the scale is of order $3,000km^2$ where the distribution in all cities start to look similar.

By rescaling the betweenness of each node by the number of vertices in the network $g \to \tilde{g} = g/N$, we see that the distributions collapse on a single curve with a unique bump separating two clear regimes as seen in Fig. 5.22b (although some variability exists resulting from differences in the number of edges). Fitting the distribution of $\tilde{g} = g/N$ with the function

$$p(\tilde{g}) \sim \tilde{g}^{-\alpha} e^{-\tilde{g}/\beta}, \tag{5.110}$$

results in a tightly bound range for $\alpha \approx 1$ and a broad size-dependent distribution for β. Rescaling the tail with respect to β results in a collapse of the curves for all cities as seen in Fig. 5.22c.

A clue for the bimodal behavior stems from the fact that it is peaked at N, a feature reminiscent of nodes adjacent to the leaves of the MST (see Chap. 18). The MST, as a tree, has a BC of order $\mathcal{O}(N)$, specifically $N - 2$ for degree two nodes adjacent to leaves (which for street networks are the dead-ends). Indeed all paths from the leaf to $N - 2$ other nodes have to go through this node. While deriving an exact analytical expression for the BC distribution of the MST is challenging, one can make progress by approximating it as a k-ary tree (where each node has a branching ratio bounded by k). The following approximate result can be derived (see section above)

$$P(g) \approx \frac{k^{\log_k \left(\frac{AN}{gB} \right)}}{N} = Ag^{-1}, \tag{5.111}$$

indicating that the node betweenness of a Cayley tree scales with exponent $\alpha = 1$, consistent with previous calculations of the link betweenness [43]. This provides a possible explanation for both the scaling with N as well as the form of the tail found in the empirical measurements of the BC of city streets (Eq. (5.110)). Indeed, this

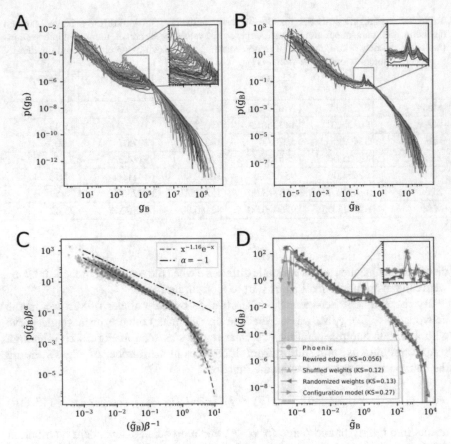

Fig. 5.22 Betweenness invariance in urban streets. **a** The betweenness distribution for all the 97 cities considered in [41]. The peak of the distribution for each city is shown as inset. **b** The version of the distribution after rescaling by the number of nodes N showing the alignment of the peaks across all cities (also shown as inset). **c** The collapse of the tails after rescaling with respect to β. The dashed line shows the analytically computed asymptotic scaling for a Cayley-tree Eq. (5.111). **d** The BC distribution of various random graph models considered in [41] compared to the baseline distribution of Phoenix as a representative example. Shaded area reflects fluctuations around the average over hundred realizations of each model. Figure from [41] and distributed under a Creative Commons Attribution 4.0 International License

implies an underlying tree structure on which the high BC nodes of all cities lie, indicating that the majority of flow is concentrated around *a spanning tree* of the street network [44].

The Cayley approximation neglects loops which are however present in street networks. Starting from a tree, addition of edges will necessarily produce loops leading to alternate paths: a large fraction of the previously high BC nodes on the MST are then bypassed. This induces the emergence of a low BC regime as well as increasingly sharp cutoffs in the tail, in agreement with empirical observations

(Fig. 5.22). In order to test quantitatively the effect of increasing edges on the BC distribution, a simple model of random planar graphs can be used [41]. The control parameter for this model is the edge density defined by

$$\rho_e = \frac{E}{E_{DT}} \tag{5.112}$$

defined as the fraction of edges E compared to the maximal number of possible edges E_{DT} determined by the Delaunay triangulation, and varies between $\rho_e \approx 1/3$ for the MST to $\rho_e \approx 1$ for the DT. Since we have $E_{DT} \approx 3N$, this parameter ρ_e is proportional—in the context of street networks—to the average degree $\langle k \rangle$ of street intersections.

We start from a uniform distribution of N nodes in the 2d plane, and in order to be able to vary the edge density, we generate the Delaunay triangulation on the set of nodes and remove randomly edges until we reach the desired value of ρ_e. We show in Fig. 5.23 the BC distribution for increasing values of ρ_e from the MST to the Delaunay triangulation.

We observe that the distribution for the MST has indeed a peak at N and is bounded by N^2. In this interval $[N, N^2]$, the distribution follows a form close to the Cayley tree calculation $P(g) \sim 1/g$ Eq. (5.111). When edges are added and loops are created, we observe the emergence of a bimodal form with a low betweenness regime due to alternate paths that reduce the BC of some nodes (Fig. 5.23b). For larger ρ_e, the distribution gets more homogeneous but still peaked around N even in the limiting case of the DT (Fig. 5.23d). On these figures, we can thus separate the 'tree contribution' (in the interval $[N, N^2]$) from the 'loop contribution' (interval $[0, N]$) separared by $g \sim N$ and visualize these different regions with shades.

This simple model thus confirms that the observed bimodality in street networks results from a combination of a large BC backbone belonging to the MST and a low BC region induced by loops. The frontier between these two regimes is determined by the lowest (non-zero) betweenness value for the MST. Progressively decorating the tree with loops leads to arbitrarily low betweenness values due to the creation of many alternate paths which smoothens out the distribution.

5.5.4 The Effect of One-Way Streets on the BC

Cars have to follow the direction of links and consequently one-way streets govern the spatial structure of traffic. The theoretical question is then to understand what happens to the patterns of shortest paths when we turn an undirected link into a one-way street. This can for instance be measured by comparing the betweenness centrality (BC) of nodes [45]. We denote by $g_G(i)$ the BC of node i on the graph G (normalized by $\mathcal{N} = (N-1)(N-2)$) and by $g_{\vec{G}}(i)$ the BC of node i when we include one-ways. We analyze the relative variation

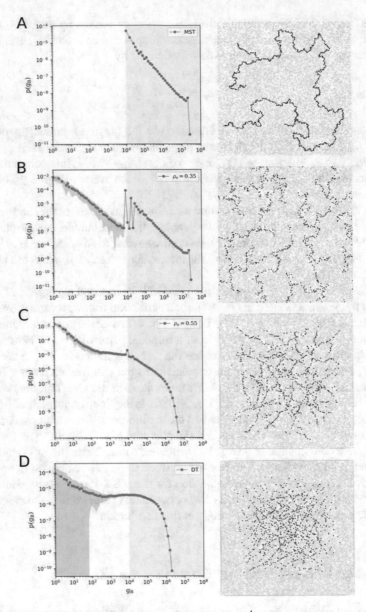

Fig. 5.23 Effect of the edge-density ρ_e on the BC. $N = 10^4$ nodes were randomly distributed on the 2D plane and their DT was generated. Edges are removed until the desired edge-density ρ_e is reached. The left panel shows the averaging over a hundred realizations of the resulting BC distribution ranging from the MST constructed over all nodes **a** to the DT **d** with increasing ρ_e. The shaded area corresponds to fluctuations around the average of the realizations, while the silver and white shades separate the 'tree-like' region from the 'loop-region' respectively. The right panel shows a single instance of the actual generated network corresponding to each ρ_e. Shown in red are the nodes in the 90'th percentile and above in terms of their BC value. Theses results are obtained for hundred realizations and $N = 104$. Figure taken from [41] and distributed under a Creative Commons Attribution 4.0 International License

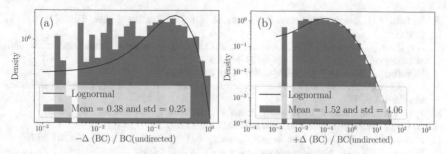

Fig. 5.24 Distribution of the relative variation Δ of the betweenness centrality (BC) due to one-way streets in the Parisian network for negative (the BC decreases) and positive values of Δ (the BC increases with one-ways). Both distributions can be fitted by a lognormal and parameters are: $\mu = 0.38$, $\sigma = 0.25$ (negative values) and $\mu = 1.52$ and $\sigma = 4.06$. Figure taken from [45]

$$\Delta = \frac{g_{\vec{G}}(i) - g_G(i)}{g_G(i)} \tag{5.113}$$

In the case of Paris for example, we find that 53% of the nodes have a smaller BC ($\Delta < 0$) due to one-way streets with 27% of them having less than half the undirected BC and 3% less than 10%. For the other 47% with $\Delta > 0$ the BC is increased, more than doubled for 31% of them and the BC is ten times higher in 3% of cases. We thus observe here the dual effect of one-way streets: certain nodes are preserved and experience a reduced traffic while this simultaneously create bottlenecks where the BC can be very large. More generally, we observe (see Fig. 5.24) that the distribution of Δ is not symmetric (with a global average of ~ 0.59) and skewed towards positive values indicating that the bottlenecks due to the deviated traffic can be extremely busy.

References

1. L.C. Freeman, Sociometry, pp. 35–41 (1977)
2. M. Newman, Phys. Rev. E **64**, 016132 (2001)
3. K.I. Goh, B. Kahng, D. Kim, Phys. Rev. Lett. **87**, 278701 (2001)
4. M. Barthelemy, Phys. Rev. Lett. **91**, 189803 (2003)
5. M. Barthelemy, Eur. Phys. J. B **38**, 163 (2004)
6. S. Gago, J.C. Hurajová, T. Madaras, Quantitative Graph Theory: Mathematical Foundations and Applications, p. 233 (2014)
7. S. Gago, J. Hurajová, T. Madaras, Mathematica Slovaca **62**(1), 1 (2012)
8. F. Comellas, S. Gago, Linear Algebr. Appl. **423**(1), 74 (2007)
9. U. Brandes, J. Math. Sociol. **25**, 163 (2001)
10. L. Benguigui, I. Porat, Physica A: Stat. Mech. Appl. **505**, 243 (2018)
11. O. Narayan, I. Saniee, Phys. Rev. E **84**(6), 066108 (2011)
12. E. Jonckheere, M. Lou, F. Bonahon, Y. Baryshnikov, Internet Math. **7**(1), 1 (2011)
13. M. Barthelemy, Eur. Phys. J. B-Condens. Matter Complex Syst. **38**(2), 163 (2004)
14. B. Lion, M. Barthelemy, Phys. Rev. E **95**(4), 042310 (2017)
15. E. Taillanter, M. Barthelemy, In preparation

16. A. Lampo, J. Borge-Holthoefer, S. Gómez, A. Solé-Ribalta, Phys. Rev. Res. **3**(1), 013267 (2021)
17. A.P. Giles, O. Georgiou, C.P. Dettmann, in *2015 IEEE International Conference on Communications (ICC)* (IEEE, 2015), pp. 6450–6455
18. V. Verbavatz, M. Barthelemy, Phys. Rev. E, to appear (2022)
19. L.A.S. Sors, L.A. Santaló, *Integral Geometry and Geometric Probability* (Cambridge University Press, 2004)
20. S. Lämmer, B. Gehlsen, D. Helbing, Physica A: Stat. Mech. Appl. **363**(1), 89 (2006)
21. B. Jiang, C. Claramunt, Environ. Plan. B **31**(1), 151 (2004)
22. M. Rosvall, A. Trusina, P. Minnhagen, K. Sneppen, Phys. Rev. Lett. **94**(2), 028701 (2005)
23. S. Porta, P. Crucitti, V. Latora, Environ. Plann. B **33**, 705 (2006)
24. S. Porta, P. Crucitti, V. Latora, Physica A: Stat. Mech. Appl. **369**(2), 853 (2006)
25. P. Crucitti, V. Latora, S. Porta, Phys. Rev. E **73**(3), 036125 (2006)
26. A. Cardillo, S. Scellato, V. Latora, S. Porta, Phys. Rev. E **73**(6), 066107 (2006)
27. F. Xie, D. Levinson, Geograp. Anal. **39**(3), 336 (2007)
28. B. Jiang, Physica A: Stat. Mech. Appl. **384**(2), 647 (2007)
29. A. Masucci, D. Smith, A. Crooks, M. Batty, Eur. Phys. J. B-Condens. Matter Complex Syst. **71**(2), 259 (2009)
30. S.H. Chan, R.V. Donner, S. Lämmer, Eur. Phys. J. B-Condens. Matter Complex Syst. **84**(4), 563 (2011)
31. T. Courtat, C. Gloaguen, S. Douady, Phys. Rev. E **83**(3), 036106 (2011)
32. E. Strano, V. Nicosia, V. Latora, S. Porta, M. Barthelemy, Sci. Rep. **2** (2012)
33. M. Barthelemy, P. Bordin, H. Berestycki, M. Gribaudi, Sci. Rep. **3** (2013)
34. S. Porta, O. Romice, J.A. Maxwell, P. Russell, D. Baird, Urban Studies, p. 0042098013519833 (2014)
35. Mapzen website. http://www.mapzen.com
36. A. Rheinwalt, N. Marwan, J. Kurths, P. Werner, F.W. Gerstengarbe, EPL (Europhysics Letters) **100**(2), 28002 (2012)
37. J. Gil, Environ. Plan. B: Plan. Des. 0265813516650678 (2016)
38. E. Strano, A. Cardillo, V. Iacoviello, V. Latora, R. Messora, S. Porta, S. Scellato, Env. Plann. B: Plann. Des. **36**, 450 (2009)
39. B.M. Yerra, D.M. Levinson, Ann. Reg. Sci. **39**(3), 541 (2005)
40. S. Scellato, A. Cardillo, V. Latora, S. Porta, Eur. Phys. J. B **50**, 221 (2006)
41. A. Kirkley, H. Barbosa, M. Barthelemy, G. Ghoshal, Nat. Commun. **9**(1), 1 (2018)
42. Open street map project. http://www.openstreetmap.org
43. G. Szabó, M. Alava, J. Kertész, Phys. Rev. E **66**, 026101 (2002)
44. Z. Wu, L. Braunstein, S. Havlin, H. Stanley, Phys. Rev. Lett. **96**, 148702 (2006)
45. V. Verbavatz, M. Barthelemy, Phys. Rev. E **103**(4), 042313 (2021)

Chapter 6
The Shape of Shortest Paths

Shortest paths constitute an important aspect of the study of networks and inform us about many processes such as the appearance of bottlenecks in traffic flow, weak points of the system, navigability properties, or structural changes during the evolution of networks. In addition, in spatial networks, we can compute different types of shortest paths between a pair of nodes. In particular, if we weight the links by a time or a distance, we can define the quickest or shortest distance path. In this chapter, we will consider graphs where we weight the edges with their Euclidean length (see Fig. 6.1). These shortest paths in spatial networks are geometrical objects and it is natural to ask what are their typical shapes. There are many ways to characterize such a shape and we will mainly discuss here the exponents that describe transversal fluctuations and the variance of the shortest path length. This problem is related to the first passage percolation in statistical physics that we will describe here. We will see that for spatial networks, we have two classes characterized by different exponent values.

6.1 (Euclidean) First Passage Percolation

6.1.1 Models and Definitions

First passage percolation (FPP) [1] attempts to capture the features of shortest paths with a probabilistic model. In FPP one usually considers a deterministic lattice such as $\mathbb{Z}^d$ with independent, identically distributed weights, known as *local passage times*, on the edges. With a fluid flowing outward from a point, the question is:

What is the minimum passage time over all possible routes between the source and another distant point, where routing is quicker along lower weighted edges?

© The Author(s), under exclusive license to Springer Nature Switzerland AG 2022
M. Barthelemy, *Spatial Networks*,
https://doi.org/10.1007/978-3-030-94106-2_6

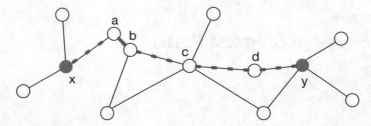

Fig. 6.1 Illustration of the weighted shortest path on a small spatial network. The network is constructed over a set of points denoted by circles here and the edges are denoted by lines. For a pair of nodes (x, y) we look for the shortest path (shown here by a dotted line) where the length of the path is given by the sum of all edges length: $d(x, y) = ||x - a|| + ||a - b|| + ||b - c|| + ||c - d|| + ||d - y||$

More than 50 years of intensive study of FPP has been carried out [2] and has lead to many results such as the subadditive ergodic theorem, a key tool in probability theory, but also a number of insights in crystal and interface growth [3], the statistical physics of traffic jams [4], and key ideas of universality and scale invariance in the shape of shortest paths [5]. In particular, the lengths of paths between pairs of points and the corresponding transversal deviations of the geodesics, have been the focus of in-depth research [2]. They exist as minima over collections of correlated random variables and the travel times are conjectured in the independent identically distributed (i.i.d.) case to converge to the Tracy-Widom distribution (TW), found throughout various models of statistical physics, see e.g. [6, Sect. 1]. This links the model to random matrix theory, where β-TW appears as the limiting distribution of the largest eigenvalue of a random matrix in the β-Hermite ensemble, where the parameter β is 1, 2 or 4 [7].

A particular case of FPP is the Euclidean first passage percolation (EFPP). This is a probabilistic model of fluid flow between points of a d-dimensional Euclidean space where one studies optimal routes from a source node to each possible destination node in a spatial network built either randomly or deterministically on the points. Introduced by Howard and Newman in 1997 [8], this model is defined on the complete graph constructed on a point process. Long paths are discouraged by taking powers of interpoint distances as edge weights. Here, we will focus on the case where edge weights are given by the Euclidean distances between the spatial points themselves, in sharp contrast with the usual FPP problem, where weights are i.i.d. random variables. This is a crucial difference that distinguishes the FPP and the EFPP. Indeed, in the usual FPP, given i.i.d weights, paths are sums of independent identically distributed random variables, while in the EFFP, weights are correlated.

The precise definition of the model is the following. We will consider a Poisson point process in an unbounded region with links connecting pairs of points according to a given rule [9, 10], rather than the totality of the weighted complete graph. Here, we will consider a variety of networks such as the random geometric graph with unit disk and Rayleigh fading connection functions, the k-nearest neighbour graph, the

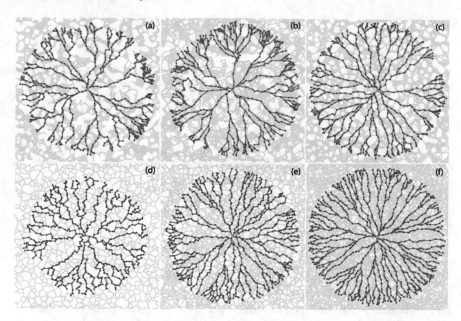

Fig. 6.2 Spatial networks built on a realization of a simple, stationary Poisson point processes of expected $\rho = 1000$ points in the unit square $\mathscr{V} = [-1/2, 1/2]^2$, but with different connection rules. The boundary points at time $t = 1/2$ of the first passage process are shown in red, while their respective geodesics are shown in blue. The networks considered here are: **a** Hard RGG with unit disk connectivity. **b** Soft RGG with Rayleigh fading connection function $H(r) = \exp(-\beta r^2)$, **c** 7-NNG, **d** Relative neighbourhood graph, which is the lune-based β-skeleton for $\beta = 2$, **e** Gabriel graph, which is the lune-based β-skeleton for $\beta = 1$, and **f** the Delaunay triangulation. Figure taken from [11]

Delaunay triangulation, the relative neighbourhood graph, the Gabriel graph, and the complete graph with (in this case only) the edge weights raised to the power $\alpha > 1$. More generally, we will distinguish proximity graphs (see Chap. 15) which are determined by a proximity rule such as the RGG, and excluded region graphs (see Chap. 16) based on the absence of points in a given region between two nodes. We show in the Fig. 6.2 examples of networks considered here.

Once the graph is defined over the set of points, we study the random length and transversal deviation of the shortest paths between two nodes in the network, denoted x and y, conditioned to lie at mutual Euclidean separation $||x - y||$, as a function of the point process density and other parameters of the model used (here and in the following $||x||$ denotes the usual norm in euclidean space). The study of the scaling with $||x - y||$ of the length and the deviation allow to define the fluctuation and wandering exponents. We denote by $\tau(e)$ the weight of edge e and the 'passage time' $T(x, y)$ from vertices x to y which corresponds to the length of the shortest path is a random variable given as the minimum over all possible paths $\mathscr{P}(x, y)$ on the lattice connecting these points, of the sum of the τ's over a given path

$$T(x, y) = \min_{P \in \mathscr{P}} \sum_{e \in P} \tau(e) \qquad (6.1)$$

This minimum path is a geodesic, and it is almost surely unique when the edge weights are continuous. The average travel time is proportional to the distance

$$\overline{T(x, y)} \sim ||x - y|| \qquad (6.2)$$

where here and in the following we denote the average of a quantity by $\overline{\cdot}$. More generally, if the ratio of the geodesic length and the Euclidean distance is less than a finite number t (the maximum value of this ratio is called the stretch), the network is a t-*spanner* [12] (see Chap. 18). For example, many important networks are t-spanners, including the Delaunay triangulation of a Poisson point process, which has $\pi/2 < t < 1.998$ [13, 14].

The variance of the passage time is also important, and scales as

$$\mathrm{Var}\,(T\,(x, y)) \sim ||x - y||^{2\chi}, \qquad (6.3)$$

This exponent, χ, informs us about the variance of the random length of shortest paths. Another way to see this exponent is to consider a ball of radius R around any point. For large R, the ball has an average radius proportional to R and the fluctuations around this average grow as R^χ [6]. With $\chi < 1$ the fluctuations die away $R \to \infty$, leading to the *shape theorem*, see e.g. [2, Sect. 1].

The maximum deviation $D(x, y)$ of the geodesic from the straight line from x to y is characterized by the wandering exponent ξ

$$\overline{D(x, y)} \sim ||x - y||^\xi \qquad (6.4)$$

for large $|x - y|$. Knowing ξ informs us about the geometry of geodesics between two distant points. See for example Fig. 6.3 for an illustration of wandering on different networks.

6.1.2   Known Results About Exponents

A well-known result in the 2d lattice case [15] is that $\chi = 1/3$, $\xi = 2/3$. Also, another belief is that $\chi = 0$ for dimensions d large enough. Many physicists, see for example [15–21], also conjecture that independently from the dimension, one should have the so-called KPZ relation between these exponents

$$\chi = 2\xi - 1 \qquad (6.5)$$

This is connected with the KPZ universality class of random geometry, apparent in many physical situations, including traffic and data flows, and their respective models,

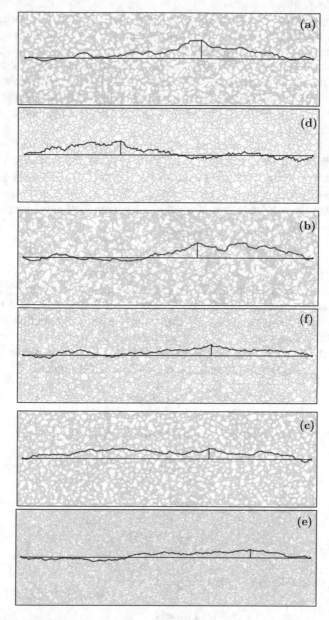

Fig. 6.3 Examples of Euclidean geodesics (blue) running between two end nodes of a simple, stationary Poisson point process (red). The maximal transversal deviation is shown (vertical black line). The Euclidean distance between the endpoints is the horizontal black line (the point density is equal for each model). **a** Hard RGG, **b** Soft RGG with connectivity probability $H(r) = \exp(-r^2)$, **c** 7-NNG, **d** RNG, **e** GG, **f** DT. Figure from [11]

including the corner growth model, ASEP, TASEP, etc. [4, 22, 23]. In particular, FPP is in direct correspondence with important problems in statistical physics [24] such as the directed polymer in random media (DPRM) and the KPZ equation, in which case the dynamical exponent z corresponds to the wandering exponent ξ defined for the FPP [6, 25]. Also, the 'rotational invariance' of the Poisson point process implies the KPZ relation Eq. (6.5) is satisfied in each spatial network considered here (see for example [2] for a discussion of the generality of the relation, and the notion of rotational invariance).

The situation regarding exact results is more complex [2, 26]. The majority of results are based on the model on $\mathbb{Z}^d$. Kesten [27] proved that $\chi \leq 1/2$ in any dimension, and for the wandering exponent ξ, Licea et al. [28] gave some hints that possibly $\xi \geq 1/2$ in any dimension and $\xi \geq 3/5$ for $d = 2$.

Concerning the KPZ relation, Wehr and Aizenman [29] and Licea et al. [28] proved the inequality

$$\chi \geq (1 - d\xi)/2 \tag{6.6}$$

in d dimensions. Newman and Piza [30] gave some hints that possibly $\chi \geq 2\xi - 1$. Finally, Chatterjee [26] proved Eq. (6.5) for $\mathbb{Z}^d$ with independent and identically distributed random edge weights, with some restrictions on distributional properties of the weights. These were lifted by independent work of Auffinger and Damron [2].

Licea et al. [28] showed that for the standard first-passage percolation on $\mathbb{Z}^d$ with $d \geq 2$, the wandering exponent satisfies $\xi(d) \geq 1/2$ and specifically

$$\xi(2) \geq 3/5 \tag{6.7}$$

In Euclidean FPP, however, these bounds do *not* hold, and we have [31, 32]

$$\frac{1}{d+1} \leq \xi \leq 3/4 \tag{6.8}$$

and, for the wandering exponent,

$$\chi \geq \frac{1 - (d-1)\xi}{2}. \tag{6.9}$$

Combining these different results then yields, for $d = 2$

$$\begin{cases} 1/8 \leq \chi \\ 1/3 \leq \xi \leq 3/4 \end{cases} \tag{6.10}$$

Since the KPZ relation of Eq. (6.5) apparently holds in the setting considered here, the lower bound for χ implies then a better bound for ξ, namely

$$\xi \geq \frac{3}{3+d} \tag{6.11}$$

which in the two dimensional case leads to $\xi \geq 3/5$, the same result as in the standard FPP.

Finally, we can find some bounds for the various networks we are considering here. In particular, we consider four excluded region graphs and some of them obey the inclusion relation (see for example [33])

$$RNG \subset GG \subset DT \tag{6.12}$$

where RNG stands for the relative neighborhood graph, GG for the Gabriel graph, and DT for the Delaunay triangulation. This nested relation trivially implies the following inequality

$$\xi_{RNG} \geq \xi_{GG} \geq \xi_{DT} \tag{6.13}$$

as adding links can only decrease the wandering exponent. We are not aware of a similar relation for χ. We will also consider here three distinct proximity graphs (Chap. 15) such as the hard and soft RGG, and the k-nearest neighbour graph.

6.1.3 Numerical Results

The scaling exponents χ and ξ and the distribution of the travel time fluctuations can be evaluated numerically for various networks [11] and the results are shown in Fig. 6.4. These plots, shown in loglog, reveal a power law behaviour of T and D, and the linear growth of typical travel time with Euclidean span. Power law fits on these numerical results suggest the existence of two classes of spatial networks characterized by different values of the exponents. The proximity graphs (hard and soft RGG, and k-NNG) are in one class, with exponents

$$\chi_{RGG,NNG} = 0.20 \pm 0.01 \tag{6.14}$$
$$\xi_{RGG,NNG} = 0.60 \pm 0.01 \tag{6.15}$$

The excluded region graphs (he β-skeletons and Delaunay triangulation), and Howard's EFPP model with $\alpha > 1$ belong in another class with

$$\chi_{DT,\beta\text{-skel},EFPP} = 0.40 \pm 0.01 \tag{6.16}$$
$$\xi_{DT,\beta\text{-skel},EFPP} = 0.70 \pm 0.01 \tag{6.17}$$

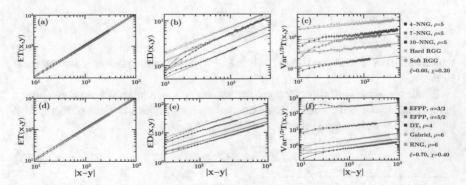

Fig. 6.4 Numerical results for the expected travel time (**a**) and (**d**), expected wandering (**b**) and (**e**), and standard deviation of the travel time (**c**) and (**f**). The power law exponents are indicated in the legend. Error bars of one standard deviation are shown for each point. The top plots show the results from the models in first universality class, while the lower plots show the second class. The RGG and NNG are distinguished with different colours (green and blue), as are EFPP on the complete graph, the DT, and the two beta skeletons (Gabriel graph, and relative neighbourhood graph). The point process density ρ points per unit area is given for each model. Figure from [11]

Clearly, the KPZ relation of Eq. (6.5) is satisfied up to the numerical accuracy of these simulations. These results led the authors of [11] to conjecture that for the proximity graphs (the DT and the β-skeletons for all β)

$$\text{Var} \left(T \left(x, y \right) \right) \sim ||x - y||^{4/5}, \tag{6.18}$$

$$\mathbb{E} \left(D \left(x, y \right) \right) \sim ||x - y||^{7/10} \tag{6.19}$$

and, for the RGGs and the k-NNG,

$$\text{Var} \left(T \left(x, y \right) \right) \sim ||x - y||^{2/5}, \tag{6.20}$$

$$\mathbb{E} \left(D \left(x, y \right) \right) \sim ||x - y||^{3/5}, \tag{6.21}$$

These results are summarized in the Table 6.1. These two classes thus correspond to the following two broad classes of networks: firstly, based on joining vertices according to critical proximity, such as in the RGG and the NNG, and secondly, based on graphs which connect vertices based on excluded regions, as in the lune-based β-skeletons, or the DT. Intuitively, the most efficient way to connect two points uses the nearest neighbour, which suggests that ξ for proximity graphs should on the whole be smaller than for exclusion-based graphs, which is in agreement with the results reported in [11].

We note that it is surprising that these exponents are apparently rational numbers and further mathematical studies are needed in improve our understanding of the euclidean first passage process.

6.1.4 Travel Time and Transversal Fluctuations

The travel time distribution seems to be normal in most cases (see Fig. 6.5) and the results are summarized in the Table 6.1 (the skewness and the kurtosis for the travel time fluctuations were also computed providing additional information about the distribution, as the skewness is 0 and the kurtosis 3 for a Gaussian distribution, while the Tracy-Widom distribution displays other values, see [11] for more details and discussions.) The distribution of T is compared against four test distributions, the Gaussian orthogonal, unitary and symplectic Tracy Widom distributions, and also the standard normal distribution (Fig. 6.5).

There are only a few cases where gaussian fluctuations occur and it thus seems that the EFPP on spatial networks belong to this family. We note that the Tracy-Widom distribution is thought to occur in problems where matrices represent collections of totally uncorrelated random variables [34]. However, in the EFPP case, the important quantity is the shortest path distance between nodes which leads to spatial correlations in the adjacency matrix. The fact that we don't have i.i.d. values in this matrix is potentially the reason why we do not observe Tracy-Widom distributions.

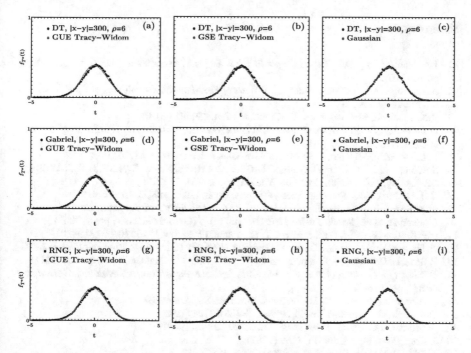

Fig. 6.5 Numerical results for the travel time distributions for the DT **(a)**–**(c)**, RNG **(d)**–**(f)**, and Gabriel **(g)**–(**i**) graphs, compared with the GUE and GSE Tracy-Widom ensembles, and the Gaussian distribution. The point process density ρ points per unit area is given for each model. In all cases, the Gaussian distribution provides a better fit (and consistent values for the skewness and kurtosis). Figure from [11]

Table 6.1 Exponents ξ and χ, and passage time distribution for the various networks considered in [11]

Network	ξ	χ	Distribution of T
Proximity graphs			
Hard RGG	3/5	1/5	Normal (Conj.)
Soft RGG with Rayleigh fading	3/5	1/5	Normal (Conj.)
k-NNG	3/5	1/5	Normal
Excluded region graphs			
DT	7/10	2/5	Normal
GG	7/10	2/5	Normal
β−skeletons	7/10	2/5	Normal
RNG	7/10	2/5	Normal
Euclidean FPP			
With $\alpha = 3/2$	7/10	2/5	Normal
With $\alpha = 5/2$	7/10	2/5	Normal

References

1. J.M. Hammersley, D.J. Welsh, *Bernoulli 1713, Bayes 1763, Laplace 1813* (Springer, 1965), pp. 61–110
2. A. Auffinger, M. Damron, J. Hanson, *50 Years of First-Passage Percolation*, vol. 68 (American Mathematical Soc., 2017)
3. K.A. Takeuchi, M. Sano, Phys. Rev. Lett. **104**(23), 230601 (2010)
4. G. Grimmett, *Probability on Graphs: Random Processes on Graphs and Lattices*, vol. 8 (Cambridge University Press, 2018)
5. K.A. Takeuchi, M. Sano, T. Sasamoto, H. Spohn, Sci. Rep. **1**(1), 1 (2011)
6. S.N. Santalla, J. Rodríguez-Laguna, T. LaGatta, R. Cuerno, New J. Phys. **17**(3), 033018 (2015)
7. J.P. Keating, N.C. Snaith, Commun. Math. Phys. **214**(1), 57 (2000)
8. C.D. Howard, C.M. Newman, Probab. Theory Relat. Fields **108**(2), 153 (1997)
9. B. Bollobás, B. Béla, *Random Graphs*, vol. 73 (Cambridge university press, 2001)
10. M. Penrose, et al., *Random Geometric Graphs*, vol. 5 (Oxford university press, 2003)
11. A.P. Kartun-Giles, M. Barthelemy, C.P. Dettmann, Phys. Rev. E **100**(3), 032315 (2019)
12. G. Narasimhan, M. Smid, *Geometric Spanner Networks* (Cambridge University Press, 2007)
13. P. Bose, L. Devroye, M. Löffler, J. Snoeyink, V. Verma, Comput. Geom. **44**(2), 121 (2011)
14. G. Xia, in *Proceedings of the Twenty-Seventh Annual Symposium on Computational Geometry* (2011), pp. 264–273
15. M. Kardar, G. Parisi, Y.C. Zhang, Phys. Rev. Lett. **56**(9), 889 (1986)
16. D.A. Huse, C.L. Henley, Phys. Rev. Lett. **54**(25), 2708 (1985)
17. M. Kardar, Y.C. Zhang, Phys. Rev. Lett. **58**(20), 2087 (1987)
18. J. Krug, Phys. Rev. A **36**(11), 5465 (1987)
19. J. Krug, P. Meakin, Phys. Rev. A **40**(4), 2064 (1989)
20. E. Medina, T. Hwa, M. Kardar, Y.C. Zhang, Phys. Rev. A **39**(6), 3053 (1989)
21. J. Krug, H. Spohn. Kinetic roughening of growing surfaces, in *Solids Far from Equilibrium: Growth, Morphology and Defects*, c. godr eche ed (1991)
22. P. Deift, arXiv preprint arXiv:math-ph/0603038 (2006)
23. B. Derrida, Phys. Rep. **301**(1–3), 65 (1998)

24. T. Halpin-Healy, Y.C. Zhang, Phys. Rep. **254**(4–6), 215 (1995)
25. P. Calabrese, P. Le Doussal, Phys. Rev. Lett. **106**(25), 250603 (2011)
26. S. Chatterjee, Ann. Math. pp. 663–697 (2013)
27. H. Kesten, Ann. Appl. Probab. pp. 296–338 (1993)
28. C. Licea, C.M. Newman, M.S. Piza, Probab. Theory Relat. Fields **106**(4), 559 (1996)
29. J. Wehr, M. Aizenman, J. Stat. Phys. **60**(3), 287 (1990)
30. C.M. Newman, M.S. Piza, Ann. Probab. pp. 977–1005 (1995)
31. C.D. Howard, J. Appl. Probab. pp. 1061–1073 (2000)
32. C.D. Howard, C.M. Newman, Ann. Probab. pp. 577–623 (2001)
33. D.J. Aldous, J. Shun, Stat. Sci. **25**(3), 275 (2010)
34. When should we expect Tracy Widom? https://mathoverflow.net/questions/71306/when-should-we-expect-tracy-widom

Chapter 7
Simplicity and Entropy

In this chapter, we will discuss important aspects of planar graphs that are intimately connected to their geometrical organization. The statistical comparison of the lengths of the shortest and simplest paths (with the smallest number of turns) provides a non-trivial and non-local information about the spatial organization of these graphs. We define the simplicity index as the average ratio of these lengths and the simplicity profile characterizes the simplicity at different scales. We present results of this measure on artificial (roads, highways, railways) and natural networks (leaves, slime mould, insect wings). This enable us to show that there are fundamental differences in the organization of urban and biological systems, related to their function, navigation or distribution: straight lines are organized hierarchically in biological cases, and have random lengths and locations in urban systems. In the case of time evolving networks, the simplicity is able to reveal important structural changes during their evolution (see Chap. 10).

In a second part of this chapter we characterize the complexity of paths with the help of an entropy. This allows us to estimate the level of complexity of graphs and to compare them with each other. In order to illustrate this measure, we apply it on road networks and on subways. In particular we will see that complexity of multilayer systems largely exceed human cognitive capacities.

7.1 Simplicity

7.1.1 Simplest Paths

Generally speaking, we can define different types of paths on a graph for a given pair of nodes (i, j). A usual quantity is the shortest euclidean path of length $\ell(i, j)$ which minimizes the distance travelled to go from i to j. We can however ask for another path which minimizes the number of turns – the simplest path – of length $\ell^*(i, j)$ (if there are more than one such path we choose the shortest one). Figure 7.1a displays an example of the shortest and simplest path for a given pair of nodes on the Oxford (UK) street network.

To identify the simplest path, we first convert the graph from the primal to the dual representation, where each node corresponds to a straight line in the primal graph. Edges in dual space, in turn, represent the intersection of straight lines in the primal graph.

The straight lines in the graph are determined by a continuity negotiation-like algorithm [1]. In this algorithm, given an edge (i, j), we search among the adjacent edges attached to j, (j, k), that one that is most aligned to (i, j). If the angle $\theta_{i,j,k}$

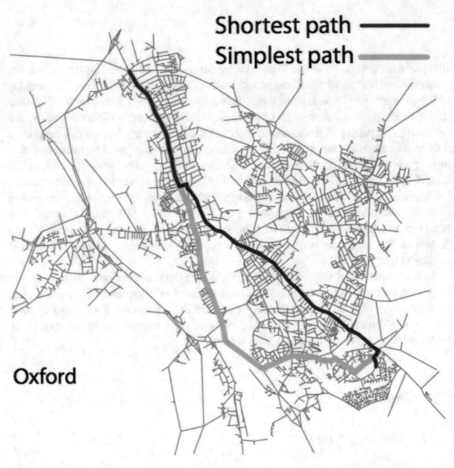

Fig. 7.1 Example of shortest (black line) and simplest (gray line) paths illustration on the Oxford (UK) street network. The simplest path has less turns at the expense of being longer than the shortest path. Figure taken from [2]

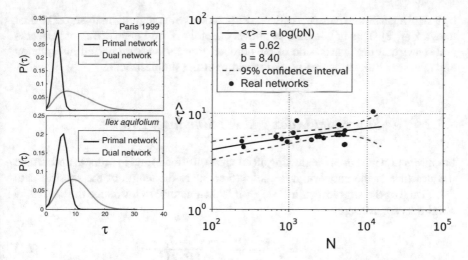

Fig. 7.2 **a** Probability distribution of number of turns for Paris (1999), a leaf (Ilex aquifolium). **b** Average number of turns $\langle \tau \rangle$ versus the size of the network N for all the networks studied here. The fit here is logarithmic showing a very slow dependence of τ versus N, a behavior typical of small-world networks

between (i, j) and (j, k) is smaller or equal to $\theta_c = 30°$, these two edges are assumed to belong to the same straight line. This procedure continues until no more edges are assigned to the same straight line. Then, the procedure is repeated in opposite direction starting from the adjacent edges attached to node i. Once assigned to a straight line, an edge is removed from the network. As it is, this algorithm produces different networks depending on the choice for the initial edge. To overcome this ambiguity, the algorithm always starts with the edge that gives the longest straight line for a given network. After this straight line is fully detected and its edges deleted, the next edge is chosen so that it will give us the second longest straight line and so on. The algorithm ends when there are no more edges left in the network.

We define the number of turns τ of a given path as the number of switches from one straight line to another when walking along this path. As we will see in the next section, this quantity is intimately related to the amount of information required to move along the path. The probability distribution $P(\tau)$ has been computed for all shortest and simplest paths for the networks of Paris (1999) and of the leaf *Ilex aquifolium* and the results are shown in Fig. 7.2a. These results confirm that the number of turns along simplest paths is smaller than for shortest paths, as expected. In addition, it can be seen that $P(\tau)$ for the simplest paths is well fitted by a normal distribution centered in $\tau_c \approx 6$ turns, which means that on average, any pair of nodes is separated by a simplest path made of 6 turns, regardless the nature of the network (Paris or Ilex).

We also show in Fig. 7.2 (right panel) the average number of turns $\langle \tau \rangle$ as a function of the number of nodes N which displays a small-world type behavior characterized by a slow logarithmic increase with N, consistently with previous analysis of the dual network [1, 3]. This feature is thus not very useful to distinguish different networks and shows that the distribution of the number of turns is a very partial information and tells very little about the spatial structure of the simplest paths.

7.1.2   The Simplicity Index and the Simplicity Profile

For navigation purposes (neglecting all congestion effects) and in order to understand the structure of the network, it is useful to compare the length of the shortest path $\ell(i, j)$ and of the simplest path $\ell^*(i, j)$. It is then natural to introduce the *simplicity index S* as the average

$$S = \frac{1}{N(N-1)} \sum_{i \neq j} \frac{\ell^*(i, j)}{\ell(i, j)}. \tag{7.1}$$

The simplicity index is larger than one and exactly equal to one for a regular square lattice and any tree-like network for example. Large values of S indicate that the simplest paths are on average much longer than the shortest ones, and that the network is not easily navigable. This metric is a first indication about the spatial structure of simplest paths but mixes various scales, and in order to obtain a more detailed information, we define the *simplicity profile*

$$S(d) = \frac{1}{N(d)} \sum_{i, j/d_E(i,j)=d} \frac{\ell^*(i, j)}{\ell(i, j)}, \tag{7.2}$$

where $d_E(i, j)$ is the euclidean distance between i and j and where $N(d)$ is the number of pairs of nodes at euclidean distance d. This quantity $S(d)$ is larger than one and its variation with d informs us about the large scale structure of these graphs. We can draw a generic shape of this profile: for small d, we are at the scale of nearest neighbors and there is a large probability that the simplest and shortest paths have the same length, yielding $S(d \to 0) \sim 1$, and increasing for small d. For very large d, it is almost always beneficial to take long straight lines when they exist, thus reducing the difference between the simplest and the shortest paths. As a result we expect $S(d)$ to decrease when $d \to d_{max}$ (note that a similar behavior is observed for another quantity, the route-length efficiency, see Chap. 4). The simplicity profile will then display in general at least one maximum at an intermediate scale d^* for which the length differences between the shortest and the simplest path is maximum. The length d^* thus represents the typical size of domains not crossed by long straight lines. At this intermediate scale, the detour needed to find long straight lines for the simplest paths is very large.

We finally note here that these indices are not limited to planar networks but can be used for all networks in which the notion of straight lines has a sense and can be computed. This would be the case for example for spatial networks which are not perfectly planar.

7.1.3 A Null Model

In order to provide a simple benchmark to further analyze the results obtained with the simplicity, we introduce a null model [2]. The goal is to compare empirical results with a very simple model based on a minimal number of assumptions. This model is constructed as follows (see Fig. 7.3). We generate N points randomly distributed in a square of unit area and we construct the Voronoi graph (see Chap. 14) on these points. We then add a tunable fraction of straight lines of length distributed according to

$$P(\ell) \sim \ell^{-\alpha} \tag{7.3}$$

We show in Fig. 7.3 the results for $\alpha = 0$. The density of straight lines ρ is defined as the ratio $\rho = L_{SL}/L$ where L_{SL} is the total length of straight lines. The total network length L is given by $L = L_{SL} + L_{se}$, where L_{se} corresponds to the total length of single edges (not belonging to any straight line) shown in gray in Fig. 7.3.

The calculation of the simplicity could be computationally costly and small samples of pairs of nodes are enough to compute accurately the simplicity index. In Fig. 7.4 we represent the convergence versus the fraction of pairs of nodes used to compute the index. We see that for a small fraction such as 2.5% we already have an accuracy of 10^{-3} (for the real networks presented here, the simplicity index was however computed exactly). This sampling method proved to be very useful and efficient for this quantity and obviously could be used for other metrics in large networks.

We show in Fig. 7.5, the simplicity profile obtained for the model for $\alpha = 0$ and $\alpha = 2$ for different values of the density of straight lines. In both cases, we see a non-monotonous behavior when the density is increased: for low density, we have a small simplicity and for a large density there are enough straight lines to provide a simple path not too different from the shortest one. The value of the peak d^* decreases with the density, consistently with the picture that d^* represents the size of domains 'free' of long straight lines.

We finally note that the effect of straight lines is more pronounced in the case $\alpha = 0$ where their length displays a wider variety than in the case $\alpha = 2$.

7.1.4 Measures on Real-World Networks

We first show results on static networks (see Fig. 7.6) such as the streets of cities (Bologna, Italy; Oxford, UK; Nantes, France), the national highway network of Australia, the national UK railway system, and the water supply network of central

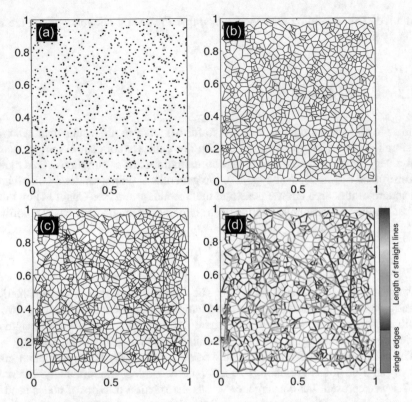

Fig. 7.3 Illustration of the null model: We start from a random set of points (**a**), construct the Voronoi lattice over this set (**b**) and add straight lines of random locations and random lengths distributed according to Eq. (7.3) (**c**). In **d** we indicate the straight lines with a color code according to their length, and single edges are represented in gray (we considered here the case $\alpha = 0$). Figure taken from [2]

Nantes (France). In the case of biological networks, we study the veination patterns of leaves (*Ilex aquifolium* and *Hymenanthera chatamica*), and of a dragonfly wing (for details see [2]). We observe on the Fig. 7.6 that basically for most of these systems, the simplicity profile displays the generic shape with a maximum at an intermediate scale. In urban cases, such as Bologna and central Nantes, we have a typical monocentric system with a dense center and a few important radial straight lines, leading to a simple profile $S(d)$.

In order to characterize macroscopically the different networks, the authors of [2] considered

- the density of straight lines ρ, defined as the ratio of the total length of straight lines, over the total length of the network
- the Gini coefficient G of lines lengths which quantifies here the inequalities of the lengths of straight lines, and is defined as (see for example [4])

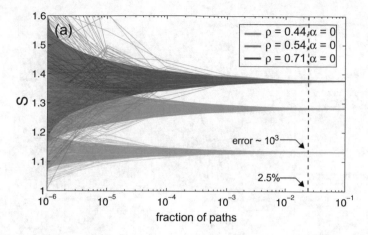

Fig. 7.4 a Illustration of the convergence of the simplicity calculation versus the fraction of pairs of nodes used for different densities (here in the case $\alpha = 0$). Figure taken from [2]

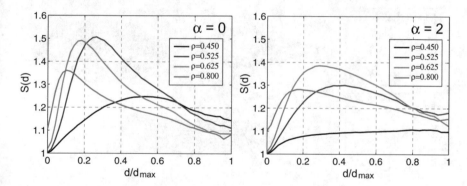

Fig. 7.5 Simplicity profiles for the null model for different values of α: **a** $\alpha = 0$ and **b** $\alpha = 2$ and for different values of the density of straight lines. Figure taken from [2]

$$G = \frac{1}{2E^2\bar{\ell}} \sum_{i,j=1}^{E} |\ell_i - \ell_j| \qquad (7.4)$$

where $\bar{\ell}$ is the average length of straight lines and E is the number of straight lines. The Gini coefficient lies in the range [0, 1] and $G = 0$ when all lengths are equal. On the other hand, if all lengths but one are very small, the Gini coefficient will be close to 1.

In the case of Oxford and the Australian highway network, the polycentric organization leads to multiple peaks in the simplicity profile (Fig. 7.6). Interestingly, we observe that the profiles for australian highways and railways in the UK are very different, despite their similar scale, density ρ, and Gini coefficient G. In particular,

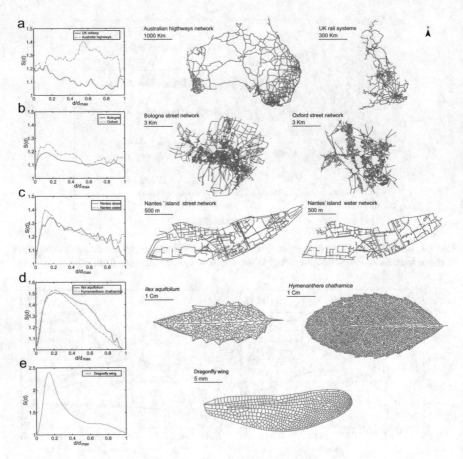

Fig. 7.6 Simplicity profiles for various real-world networks ranging from large scale networks ($10^6 m^2$) to small scales of order $10^{-3} m$. We see on these different examples the effect of the presence of long straight lines and of a polycentric structure. In particular for cases (d, e), we can clearly see that the peak at $d^* \sim 0.2 d_{max}$ corresponds to the size of domains not crossed by long straight lines. Figure taken from [2]

the UK railway displays small values of the simplicity (less than 1.2) while for the Australian highway network there are many pairs of nodes for which the simplest path is much longer than the shortest one. We also observe that the profile for both street and water systems of Nantes have a very similar shape, pointing to the fact that these networks are strongly correlated. In addition, the position and the height of the peak (≈ 1.4) observed for the Nantes water system suggests that this distribution system has similar features compared to biological systems such as vein networks in leaves (see below) whose function is also distribution.

Compared to urban systems, the simplicity profile of biological networks have a single well-defined, and much more pronounced peak. We observe values of order $S_{max} \approx 1.5$ and 2.5 for $d^*/d_{max} \approx 0.2$, meaning that for this range of distance, the

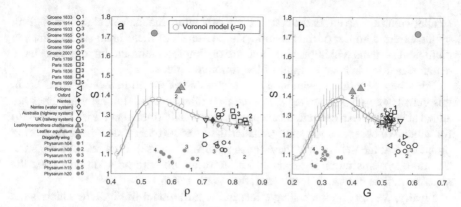

Fig. 7.7 Simplicity index versus **a** the density of straight line ρ and **b** the Gini coefficient for the length of straight lines. In both plots, the symbols correspond to the different networks studied in [2]. We also represented the result for the null model (for $\alpha = 0$) and its cubic spline interpolation (continuous line). From **a** we see that biological networks are limited to the region $\rho \le 0.7$ and have a large simplicity index, and from **b** we see that urban networks have simultaneously higher values of G and relatively small values of S. Figure taken from [2]

detour made by the simplest path is very large. This peak is related to the existence of domains of typical size d^* not crossed by large veins. We see here a clear effect of the existence of the spatial organization of long straight lines in these systems, probably optimized for the distribution (of water for leaves). The decay for large d is also much faster in the biological case compared to urban systems: this shows that in biological systems there are long straight lines allowing to connect far away nodes. This is particularly evident on the leaves shown in Fig. 7.6 where we can see the first levels (primary and secondary) veins, the rest forming a network. For streets, the organization is much less rigid and the hierarchy less strict: we have a more uniform spatial distribution of straight lines, leading to a smoother decrease of $S(d)$.

We now show results for the simplicity index S for the various datasets and for the null model as well. The results are shown in Fig. 7.7 as a function of the density of straight lines ρ and the Gini coefficient G. The first observation from Fig. 7.7 is that the simplicity index encodes information which is neither contained in the density ρ nor in the Gini coefficient G, and reveals how the straight lines are distributed in space and participate in the flows on the network. In Fig. 7.7a, we observe that the density of straight lines is always larger for urban systems. More precisely, in the biological systems the density lies in the range $\rho \in [0.55, 0.7]$, while we observe $\rho > 0.7$ for artificial systems. Except for the Physarum, which appears to be close to a regular lattice with a small simplicity and small Gini coefficient, the simplicity index for the wing and the leaves is larger than the values obtained for the null model. These results indicate that the organization of straight lines in biological systems is very different from artificial systems, that have very similar values of ρ, G, and S. In particular, we observe a hierarchy of straight lines in biological systems (see Fig. 7.6): a main artery (the midrib for leaves) connects to veins which in turn are connect to smaller veins and so on. In the case of dragonfly wing, the main straight line is given by the external border of the network. The existence of these main straight lines in biological systems will impact the structure of simplest paths and impose some large detour, resulting in a larger value of the simplicity index.

For urban systems, the simplicity is very close to the null model (of order 1.3 in this density range), suggesting that in dense urban systems, long straight lines are added at random (An exception concerns, the pre-Haussmannian Paris (1789–1836) for which we observe a simplicity smaller than for the null model, the reason being probably that the networks at these times were very sparse). As a result, navigation on urban systems requires relatively less information with no additional cost: the simplest path is not too different from the shortest path.

Finally, we note an interesting effect in the null model in Fig. 7.7a which is the existence of a maximum of the simplicity at densities of order $\rho \sim 0.55$. In this density regime, using straight lines implies having to make large detours. However, when the density exceeds 0.6, there are enough straight lines to enable a simplest path which differs not too much from the shortest one.

These results show how the simplicity can inform us about the structure of a spatial network. Here, we could observe the structural differences between biological and artificial networks. In the former, we have a clear spatial organization of straight lines, with a clear hierarchy of lines (midrib, veins, etc.), leading to simplest paths that require a very small number of turns but at the cost of large detours. In contrast, there is no such strong spatial organization in urban systems, where the simplicity is usually smaller and comparable to a null model with straight lines of random length and location. These differences between biological and urban systems might be related to the different functions of these networks: biological networks are mainly distribution networks serving the purpose of providing important fluids and materials. In contrast, the role of road networks is not only to distribute goods but to enable individuals to move from one point of the city to another. In addition, while biological networks are usually the result of a single process, urban systems are the product of a more complex evolution corresponding to different needs and technologies.

7.2 Information Perspective

The number of 'megacities' – urban areas whose human population is larger than 10 million – has tripled since 1990 [5]. New York City, one of the first megacities, reached that level in the 1950s, and the world now has almost 30 megacities. The growth of such large urban areas usually also includes the development of transportation infrastructure and an increase in the number and the use of different transportation modes [6]. For example, about 80% of cities with populations larger than 5 millions have a subway system [7]. This leads to a natural question:

Is navigating transportation systems in very large cities too difficult for humans? Additionally, how does one quantitatively characterize this difficulty?

When navigating for the first time between two unfamiliar places and having a transportation map as one's only support, a traveler has to compare different path options to find an optimal route. Here, the traveler does not need to simultaneously visualize the whole route; it is sufficient to identify and keep track of the position of the connecting stations on the map. Therefore, a first important point to consider is that humans can track information on a maximum of about four objects in their visual working memory [8]. This implies that a person can easily keep in mind the key locations (origin, destination, and connection points) for trips with no more than two connections (which corresponds exactly to four different points). In addition, recent studies on visual search strategies [9, 10] show a transition in search strategies between the simple cases of the Stuttgart and Hong Kong metropolitan networks and the case of Paris, which has one of the most complicated transportation networks in the world. The time needed to find a route in a transportation network grows with the complexity of its map, and the pattern of eye fixations also changes from following metro lines to a random scattering of eye focus all over the map [9]. A similar transition from directional to isotropic random search has been observed for visual search of hidden objects when increasing the number of distractors [11]. The ability to manage complex "mental maps" is thus limited, and only extensive training on spatial navigation can push this limit with morphological changes in the Hippocampus [12]. Human-constructed environments have far exceeded these limits, and it is interesting to ask whether there is a navigation analog of the Dunbar number and a cognitive limit to human navigation ability, such that it becomes necessary to rely on artificial systems to navigate in transportation systems in large cities.

7.2.1 Entropy and Simplest Paths

We discuss here a measure of 'information search' associated to a trip that goes from one route to another [3, 6]. In most networks, many different paths connect a pair of nodes, and one generally seeks the fastest path that minimizes the total time to reach a destination. However, it tends to be more natural for most individuals to instead consider the simplest path (see above), which has the minimum number of connections [2]. As above, we will compute the simplest path as the shortest path in the dual space. In Fig. 7.8, we give a simple example of a subway network with 8 different lines and its dual.

Once we have the dual, we can then easily compute the simplest path and compare it with the shortest path. In Fig. 7.9, we give such an example for the New York City subway.

Rosvall et al. [3] proposed a measure – in dual space – for the information that is needed to encode a shortest path from a route s to another route t. The amount of necessary information can however depend strongly on the initial and final nodes, and we consider a trip from an origin node i in route s to a destination node j in route t. This trip is embedded in real space and among all possible simplest paths [3] (which need not to be unique), we pick the fastest one $p(i, s; j, t)$, which can differ from an

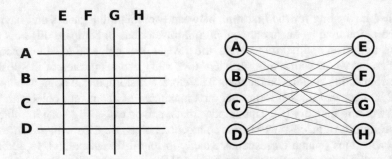

Fig. 7.8 (Left) Primal and (right) dual networks for a square lattice. In this example, the lattice has $N = 8$ routes. Each route has $k = N/2 = 4$ connections, so the total number of connections is $K_{tot} = k^2 = 16$. In the dual network, the four East-West routes (A, B, C, D) and the four North-South routes (E, F, G, H) yield a graph with a diameter of 2. Figure taken from [13]

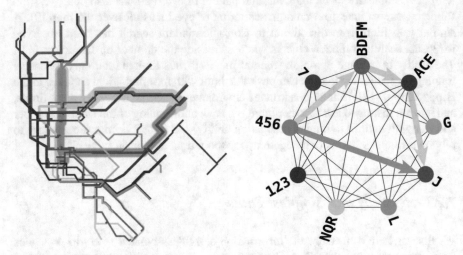

Fig. 7.9 Fastest and Simplest paths in Primal and Dual networks. (Left) In the primal network of the New York City metropolitan system, the simplest path (highlighted in light blue) from 125th St. on line 5 (dark green) to 121st St. on line J (brown) differs significantly from the fastest path (highlighted in grey). There is only one connection for the simplest path (*Brooklyn Bridge – City Hall/Chambers Street*) in Lower Manhattan. In contrast, the fastest path needs three connections ($5 \rightarrow F \rightarrow E \rightarrow J$). The fastest path using travel times from the Metropolitan Transportation Authority (MTA) Data Feeds was computed, neglecting walking and waiting times. (Right) In the dual space, nodes represent routes and edges represent connections. A simplest path in the primal space is defined as a shortest path with the minimal number of edges in the dual space (light blue arrow). It has a length of $C = 1$ and occurs along the direct connection between line 5 (dark green node) and line J (brown node). The fastest path in the primal space has a length of $C = 3$ (grey arrows) in the dual space, as one has to change lines three times). Figure taken from [6]

actual fastest path between i and j (see the left panel of Fig. 7.9). For computing the travel time of a trip, we neglect the contribution of walking and waiting times [6]. Extending the approach proposed in [3], we then propose that the total information for knowing the fastest simplest path is given by

$$S(i, s; j, t) = -\log_2 \left(\frac{1}{k_s} \prod_{n \in p(i,s;j,t)} \frac{1}{k_n - 1} \right), \tag{7.5}$$

where $p(i, s; j, t)$ is the sequence of routes needed for connecting i in route s to j in route t. The term k_s is the number of routes connected to s (it is the degree of route s in the dual space). At each node along a path, $n \in p(i, s; j; t)$ with $n \neq t$, one has a choice between $k_n - 1$ routes.

The idea behind Eq. (7.5) is that when tracking a trip along a map (with the eyes or a finger), the connections that one has to exclude represent – similarly to the number of distractors in visual search tasks [11] – the information that has to be processed and thus temporarily stored into working memory [14]. One can therefore construct the measure of entropy (7.5) as a proxy for the accumulated cognitive load that is associated to the trip, and it is analogous to the total amount of load experienced during a task [15]. For this reason, this measure of entropy seems to be appropriate for estimating the information limit associated to the observed transition in the visual search strategy [9, 11]. From a map user's perspective, the existence of several alternative simplest paths is not necessarily a significant factor, as one only needs a single simplest path for successful transportation from origin to destination. Consequently, we use the entropy in Eq. (7.5) rather than the one proposed in Ref. [3].

In order to produce a single summary statistic for a path, we average $S(i, s; j, t)$ over all nodes $i \in s$ and $j \in t$ (we denote this mean using brackets $\langle \cdot \rangle$) to obtain

$$\bar{S}(s, t) = \langle S(i, s; j, t) \rangle, \tag{7.6}$$

which is the main quantity that we use to describe the complexity of a trip and which will allow us to extract an empirical upper limit to the information that a human is able to process for navigating.

7.2.2 Quantifying the Complexity

We use the measure $\bar{S}(s, t)$ given by Eq. (7.6) to characterize the complexity of the 15 largest urban metropolitan systems in the world. The values of $\bar{S}(s, t)$ in a network tend to grow with the number C of connections (i.e. line changes) that appear in a simplest path as well as with the mean degree $\langle k \rangle$ of the nodes in the dual space (which is related to the total number of connections in a network). Adding new routes can thus have a negative impact from the information perspective. Although new routes can be useful for shortening the simplest paths for some (s, t) pairs, new connections simultaneously increase the mean degree of a network and can make it more difficult to navigate in a network. We thus want to estimate the maximum possible information that an individual can reasonably process to navigate in a transportation system. For that purpose, we consider the world's 15 metro networks with the largest number of stations. The characteristics of metro networks were examined in previous papers [7, 16–21], and navigation strategies have been considered in transportation networks (see [22] and references therein). For each network, we consider the shortest simplest paths with $C = 2$ connections. This corresponds to paths that use 3 different lines: such a path starts from a source route s, connects to an intermediate route r, and then connects to a destination route t. There are two distinct reasons for this choice: (i) the limit of four objects in the visual working memory [8]; (ii) in most of the 15 cities, two connections correspond to the diameter of the dual network. From a map user's perspective, after having checked that each pair of consecutive stations is connected by a direct line, the locations to keep in mind are the origin, the destination, and connecting stations. These nodes correspond to the places that one has to "highlight" on the map to record the trajectory. The capacity of visual working memory thus allows one to easily keep in mind only trajectories with 2 connections, leading to a total of 4 stations. Interestingly, the value 2 is also the diameter of the dual network in most of the metropolitan systems. In such situations, all pairs of nodes can be connected by paths with at most 2 connections, staying below the working memory limit of humans.

In Fig. 7.10, we show the cumulative distribution of entropies $\bar{S}(s, t|C = 2)$ for these 2-connections paths. We find that the New York City metro system is the largest and most complex metropolitan system in the world; it has a maximal value of $S_{\max} \approx 8.1 \approx \log_2(274)$ bits. Paris' transportation system reaches a similar value if one takes into account the light rail and tram system in the multilayer Metro-Rail-Tramway (MRT) network displayed in the official metro map.

Navigation in such large networks is already non-trivial [10], and it has been observed that there is an eye-movement behavioral transition when the system becomes too large (i.e., when there are too many connections) [9]. The value $S_{\max}$ for trips with two connections thus provides a natural limit, above which human cognitive capabilities are challenged and for which it becomes extremely difficult to find a simplest path. We thus make the reasonable choice to take $S_{\max}$ as the cognitive limit for public transportation: a human needs an information entropy of $\bar{S}(s, t) \leq S_{\max}$ to be able to navigate in a network successfully without assistance from information technology tools.

In order to understand the cognitive limit $S_{\max}$, we estimate $S(s, t|C = 2)$ for a regular lattice (like the one in Fig. 7.8) with N lines that are each connected with $N/2$ other lines (i.e., $k_r = N/2$ for all r). This choice of a lattice is justified by the results in [7] that most large metropolitan transportation networks consist of a core set of nodes with branches that radiate from it. The core is rather dense and has a peaked degree distribution, so it is reasonable to use a regular lattice for comparison. In the dual space of the regular lattice, the degree k_s of route s is equal to $N/2$, and we thus obtain

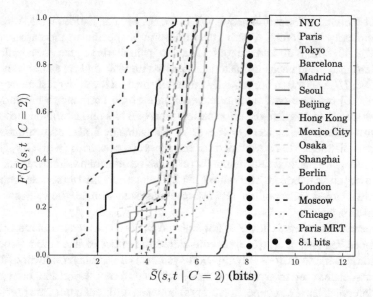

Fig. 7.10 Information threshold. Cumulative distribution of the information needed to encode trips with two connections in the 15 largest metro networks. The largest value occurs for the New York City metro system (solid red curve), which has trips with a maximum of $S_{max} \approx 8.1$ bits. Among the 15 networks, the Hong Kong (dashed, red) and Beijing (solid, black) metro networks have the smallest number of total connections and need the smallest amount of information for navigation. The Paris MRT (Metro, Light Rail, and Tramway) network (orange, dash-dotted) from the official metro map includes three transportation modes and reaches values that are similar to those in the larger NYC Metro. Figure taken from [6]

$$\bar{S}(s, t | C = 2) = \log_2[k_s(k_r - 1)]$$

$$\approx \log_2(\langle k \rangle^2) = \log_2\left(\sum_{i=1}^{N} k_i / 2\right), \tag{7.7}$$

where $\langle k \rangle$ denotes the mean degree. The last equality in Eq. (7.7) comes from the relation for the total degree of a regular lattice

$$\sum_{i=1}^{N} k_i = \langle k \rangle N = 2\langle k \rangle^2 \tag{7.8}$$

The key quantity for understanding S_{max} is therefore the total number of undirected connections $K_{tot} = \sum_{i=1}^{N} k_i / 2$ in the dual space. As indicated in Eq. (7.8), this is identical to the square of the mean degree $\langle k \rangle^2$ in a lattice. For Paris, for example, we obtain $\langle k \rangle \approx 9.75$, which leads to $9.75^2 \approx 95$ connections for the corresponding lattice. The actual Paris metropolitan network has a total of 78 connections, and the difference comes from the fact that the real network is not a perfectly regular lattice. At this stage, it is important to make two remarks. First, the apparently paradoxical fact that the total information grows with size for regular lattices while intuitively the complexity for finding a path stays constant is specific to the case in which there is an 'algorithm' to find the route. Indeed, in perfectly regular rectangular (i.e., Cartesian) grids, one needs to make at most two turns to find a desired route. This result Eq. (7.7) was tested numerically and confirmed on the 15 largest metropolitan networks [6] and also on the temporal evolution of the Paris metro network. This equation allows to translate the information limit of 8 bits into a limit on the number T of intra-route connections ($S = \log_2(T)$). The value of T also corresponds to the number of distractors to be excluded for the most complex trips (with $C(s, t) = 2$). This process of exclusion thus demands progressive information integration, which causes a cognitive overload. The value $T \approx 250$ represents the worst-case scenario in the world's largest metropolitan network and thus overestimates the values at which the transition occurs.

The subway network is however not isolated and is connected to other transportation modes. Gallotti et al. [6] thus considered the effect of inter-model coupling. The natural framework for studying such systems is a multilayer network [23–25], which associates each transportation mode with a different 'layer' in a network and where interchanges (i.e., connection points) between different modes are represented by inter-layer edges. they considered three large cities on three different continents: New York City, Paris, and Tokyo (in this work one layer is associated to bus routes and another to metro lines). The distribution of $\bar{S}(s, t)$ appears as a superimposition of peaks associated to different values of C. By comparing the distribution of $\bar{S}(s, t)$ for the bus monolayer network and the full multilayer transportation network, we can distinguish two competing effects of multimodality: (i) it tends to reduce the number C of connections and thereby reduces $\bar{S}$; and (ii) it increases $\bar{S}$, because new routes increase the node degrees in the dual space. However, these two contributions do not compensate each other. In Fig. 7.11, we show the cumulative distribution $F(\bar{S}(s, t))$ of information entropy values for the New York, Paris, and Tokyo multimodal (i.e., multilayer) transportation systems. We observe that most of the trips require more information than the cognitive limit $S_{\max} \approx 8.1$. The fraction of trips under this threshold are 15.6% for NYC, 10.7% for Paris, and 16.6% for Tokyo. The dashed curves are associated to all possible paths in a metro layer; in this case, the amount of information is always under the threshold, except for Paris, which includes trips with $C = 3$. We also note that the threshold value lies in a relatively stable part of all three cumulative distributions, which suggests that our results are robust with respect to small variations of the threshold.

These results imply that more than 80% of the trajectories in the complete public transportation networks of these major cities require more information than the most complicated trajectory in the largest metro networks. The other 20% correspond to pairs of nodes for which the trip has essentially one connection (for NYC) or at most two connections (for Paris and Tokyo), as the simplest paths that carry a small amount of information are those that avoid using too many major hubs. The number of connections acting as distractor for the case of the Paris MRT is already so large that it has a crucial impact in the route search (it takes roughly 30 seconds on average for such as search [10]), and the complexity of the bus layer (and therefore of the coupled

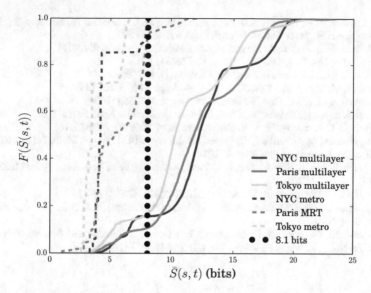

Fig. 7.11 Information entropy of multilayer networks. The solid curves represent the cumulative distributions of $\bar{S}(s, t)$ for multilayer networks that include a metro layer for New York City, Paris, and Tokyo. Figure taken from [6]

metro and bus system) will therefore exceed the human capacity. Consequently, traditional maps that represent all existing bus routes have a very limited utility. This result thus calls for the need of thinking about a user-friendly way to present and to use bus routes. For example, unwiring some bus-bus connections lowers the information, and leads to the idea that a design centered around the metro layer could be efficient. Further work is however needed to reach an efficient, 'optimal' design from a user perspective.

References

1. S. Porta, P. Crucitti, V. Latora, Phys. A: Stat. Mech. Appl. **369**(2), 853 (2006)
2. M.P. Viana, E. Strano, P. Bordin, M. Barthelemy, Sci. Rep. **3** (2013)
3. M. Rosvall, A. Trusina, P. Minnhagen, K. Sneppen, Phys. Rev. Lett. **94**(2), 028701 (2005)
4. P.M. Dixon, J. Weiner, T. Mitchell-Olds, R. Woodley, Ecology, pp. 1548–1551 (1987)
5. Un population division. http://www.unpopulation.org
6. R. Gallotti, M. Barthelemy, Sci. Rep. **4** (2014)
7. C. Roth, S.M. Kang, M. Batty, M. Barthelemy, J. R. Soc. Interface **9**(75), 2540–2550 (2012)
8. S.J. Luck, E.K. Vogel, Nature **390**(6657), 279 (1997)
9. M. Burch, K. Kurzhals, M. Raschke, T. Blascheck, D. Weiskopf, in *Proceedings of First Workshop on Schematic Mapping* (2014). https://sites.google.com/site/schematicmapping/BurchEyeTracking.pdf
10. M. Burch, K. Kurzhals, D. Weiskopf, in *Proceedings of the 2nd International Workshop on Eye Tracking for Spatial Research*, pp. 32–36 (2014)

11. H.F. Credidio, E.N. Teixeira, S.D. Reis, A.A. Moreira, J.S. Andrade Jr., Sci. Rep. **2** (2012)
12. E. Maguire et al., Proc. Natl. Acad. Sci. USA **97**, 4398 (2000)
13. R. Gallotti, M.A. Porter, M. Barthelemy, Sci. Adv. **2**(2), e1500445 (2016)
14. A. Baddeley, Nat. Rev. Neurosci. **4**(10), 829 (2003)
15. F. Paas, A. Renkl, J. Sweller, Educ. Psychol. **38**(1), 1 (2003)
16. V. Latora, M. Marchiori, Phys. A: Stat. Mech. Appl. **314**(1), 109 (2002)
17. P. Angeloudis, D. Fisk, Phys. A: Stat. Mech. Appl. **367**, 553 (2006)
18. K. Lee, W.S. Jung, J.S. Park, M. Choi, Phys. A: Stat. Mech. Appl. **387**(24), 6231 (2008)
19. S. Derrible, C. Kennedy, Phys. A: Stat. Mech. Appl. **389**(17), 3678 (2010)
20. M.P. Rombach, M.A. Porter, J.H. Fowler, P.J. Mucha, SIAM J. Appl. Math. **74**(1), 167 (2014)
21. R. Louf, M. Barthelemy, Sci. Rep. **4** (2014)
22. M. De Domenico, A. Solé-Ribalta, S. Gómez, A. Arenas, Proc. Natl. Acad. Sci. **111**(23), 8351 (2014)
23. R.G. Morris, M. Barthelemy, Phys. Rev. Lett. **109**(12), 128703 (2012)
24. M. Kivelä, A. Arenas, M. Barthelemy, J.P. Gleeson, Y. Moreno, M.A. Porter, J. Complex Netw. **2**(3), 203 (2014)
25. S. Boccaletti, G. Bianconi, R. Criado, C. Del Genio, J. Gómez-Gardeñes, M. Romance, I. Sendina-Nadal, Z. Wang, M. Zanin, Phys. Rep. **544**(1), 1 (2014)

Chapter 8
Large-Scale Tools

It is clear that the difficulty for spatial systems – including spatial networks – is to go beyond local measures and characterization. This is a very general problem which finds its root in the existence of spatial correlations between the elements of the system. Simple correlation measures can be helpful, but it is desirable to be able to have a global characterization of a spatial setting. In this chapter, we will present different approaches that can help for characterizing a system of points in the plane, that are not only based on spatial networks but that could also be applied to spatial networks. We will start with the problem of points that carry an information. In the case where nodes represent cities for example, this extra information could be the population. The location of a city is of primary importance but we understand here that it makes a difference if the city is very large or not. We will thus start this chapter by approaches that deal with these marked systems (the mark being an information attached to the node). More precisely, we will discuss the notion of spatial dominance introduced by Okabe and Sadihiro [1] based on a recursive Voronoi tessellation constructed on the set of points. We will also show that this notion allows us to introduce a new tool for spatial statistics. More generally, we believe that secondary structures such as spatial networks constructed from a set of points could be of great help for identifying relevant tools in spatial statistics that go beyond deviation from uniformity.

We will then present another example of such a new tool and that is based on first passage times of a random walk on a graph constructed over a set of points. We will then discuss community detection for spatial networks with marks and the conditions under which the information gained by the detection of communities is non-trivial. In particular, when using methods based on the optimization of the modularity, one compares the structure of the graph with a random case. The resulting communities will therefore depend strongly on the null model chosen and we will discuss the effects of null models that incorporate spatial information.

Finally, we will discuss new tools that emerged from topological data analysis. Historically, algebraic topology has provided a framework for describing rigorously and quantitatively the global structure of a space, and recent advances in topological data analysis have produced tools for analyzing data. More precisely, we will discuss tools such as persistent homology for characterizing a set of points and discuss also an extension of this method to the analysis of spatial networks. We will introduce these tools in a simple way that doesn't necessitate all the knowledge about these topological notions and will discuss applications on street networks in cities and spider webs.

8.1   Spatial Dominance

The origin of spatial dominance and the discussion brought by Okabe and Sadihiro [1] lies in the study of the organization of a system of cities and empirical tests of the famous 'central place theory' proposed by Christaller [2]. The point here is not to discuss the validity of Christaller's theory (see for example [3]) but to present new tools [4] for characterizing spatial networks with attributes and we will follow the approach proposed by Okabe and Sadihiro [1]. We believe that the tools developed in this paper – in particular for characterizing a spatial hierarchy – can be useful for studying spatial networks in general and could also trigger interesting research in spatial statistics where an important goal is to characterize a distribution of points.

8.1.1   Constructing the Dominance Tree

Okabe and Sadahiro consider the general problem of points randomly distributed in the 2d space. These points represent activity places and can be towns or regions for example. We start from N points that are distributed in the plane with coordinate x_i ($i = 1, \ldots, N$) and we assume that they are also described by an attribute a_i that can be the population or any other quantity that allows to rank these nodes.

The first step is to construct a Voronoi tesselation on the points (see Chap. 8). The Voronoi cell V_i of a node i is the set of points that are closer to i than to any other node:

$$V_i = \{ x \mid d(x, x_i) < d(x, x_j), \forall j \neq i \} \tag{8.1}$$

We show in Fig. 8.1 an example of a Voronoi tesselation computed for 100 nodes distributed uniformly in the plane.

The nodes can also be sorted according to their attribute but the main difficulty – as pointed in [1] – is to characterize mathematically the notion of spatial dominance and of a spatial hierarchy. They proposed to proceed in three steps:

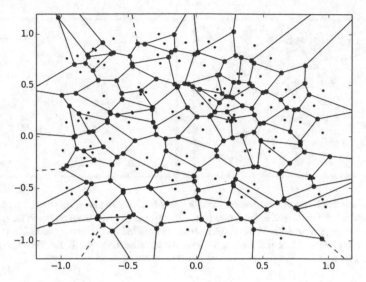

Fig. 8.1 Voronoi tesselation for 100 nodes in $[-1, 1]^2$

1. Find the local centers
2. Rank the local centers
3. Determine the spatial relations among local centers.

For the first task, local centers are defined accordingly to the very intuitive idea of a local maxima: for any point i we can define with the help of the Voronoi tesselation the set of neighbors $\Gamma(i)$ that are the nodes whose Voronoi cell is adjacent to V_i. A node i is then a local center when its attribute a_i is larger than the attributes of its neighbors

$$i \text{ is a local center} \iff a_i > a_j \ \forall \ j \in \Gamma(i) \tag{8.2}$$

Starting from an initial configuration of nodes, denoted by $P^{(0)}$, we can then construct the set of local centers $P^{(1)}$, and continue this process recursively until there is only one node left at a certain level m (see a simple illustration of one step of this process in Fig. 8.2). We thus have a series of sets which satisfies $P^{(m)} \subset P^{(m-1)} \subset \cdots \subset P^{(1)} \subset P^{(0)}$. The number of local centers at each level naturally decreases and if the number of neighbors is roughly constant $|\Gamma|$ we expect an exponential decrease

$$|P^{(k)}| \sim N e^{-k \log |\Gamma|} \tag{8.3}$$

which implies that for N initial nodes, there is about $m \sim \log N$ levels in order to reach one single local center. We can now define the rank of local centers. The first rank center is the node left at the last m-th level. The second rank centers are those

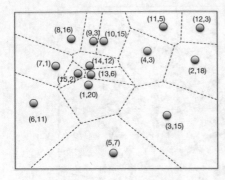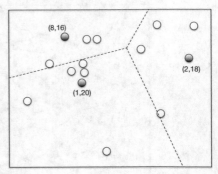

Fig. 8.2 Different levels of local centers. In the figure left, all nodes are represented by (i, a_i) where i is the index of the node, and a_i the corresponding attribute (such as the population for example). The dotted lines represent the Voronoi tesselation constructed on the nodes present at that level. (Right) We represent the local centers at the first level (nodes 1, 2, and 8) and the corresponding Voronoi tesselation

that are present at level $m - 1$ but not at level m. In general, the centers of rank j are present until level $m - j + 1$ but not at level $m - j + 2$.

We now have all the tools for defining the spatial dominance according to [1]. The starting idea is that in the Voronoi tesselation all points inside a given Voronoi cell i are 'dominated' by this node i, in the sense that it is the nearest point in $P^{(0)}$ to all these points. From a 'marketing' point of view, the consumers in V_i minimize their distance to the facility located in i. Following this idea, a local center of level $k - 1$ is spatially dominated by the local center i if it is included in the Voronoi cell of i at level k. The spatial dominance allows to construct a tree for all the nodes (see Fig. 8.3) which represents the various level of centers and their relations. We note that since it is a tree, there is a spatial dominance relation only between certain nodes but not among all of them. It is interesting to note that this process bear some resemblance to the coarse-graining obtained by means of real-space renormalization [5].

Although the method described here can be applied to any marked point process where points have a specific location in the plane and are characterized by a mark (that is a scalar), we will illustrate it on the case of cities in a country. We assume that a given country has N cities and each city i is characterized by its location x_i in some coordinate system, and its population $P_i(t)$ which can vary over time. In order to characterize this system we have to define a data structure that encodes both the spatial information (the location of cities) and their importance (their population).

We thus represent this iterative process by the dominance tree where we keep at each level the remaining points and where the links denote the spatial dominance relation. At height zero (the leaves of the tree), we then have all cities of the system that are locally dominated by a larger neighbor city. After one iteration of the process, we have the local centers, etc. until we reach the root of the tree which is the city with the largest population. Okabe and Sadahiro proposed this construction [1] and used it

Fig. 8.3 Tree graph that
represents the spatial
hierarchy of nodes in the
example of Fig. 8.2

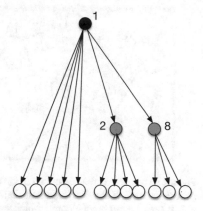

to prove that the most important quantitative statements in Christaller's central place theory were actually observed also for a completely random spatial point process. We can then represent a system of nodes (cities) thanks to a dominance tree, where the height of a city is the largest iteration before it is absorbed by a larger local center. Once we have constructed the dominance tree, each city i has its height h_i (see Fig. 8.4). We recall here that, by definition, the height of a node in a tree is the number of edges between this node and the furthest leaf going down in the tree. In other words, it is the length of the longest path (i.e. its number of edges) from the node to the deepest leaf (in contrast, the depth of a node in a tree is the distance from a node to the root). The height of a leaf is then $h = 0$ while the root has the largest height (denoted by H in the following). Due to the statistical properties of Voronoi tesselations built from spatial Poisson point processes, $H \approx log_6(N)$ where N is the number of cities in the system, and H is in general of order 5 or 6 in empirical measures. We can then introduce $n_f(h)$ which is the number of cities such that their final height is h. In contrast, the total number of cities at a given height h is denoted by $n_i(h)$ and we have the following relation $n_i(h) = \sum_{h' \geq h} n_f(h')$.

Once we have constructed the dominance tree, we have to characterize it. There is a large number of possible measures, but we will focus here on the height of a city, and we will discuss it for toy models and for empirical data. A city i at time t is then characterized by its population $P_i(t)$ (or equivalently by its rank $r_i(t)$ when population are sorted by decreasing order), and its height $h_i(t)$ in the dominance tree. This height characterizes the position of the city in the hierarchical organization of the urban system. While in the following we analyze urban systems at the national scale, the method could be applied at different levels as well, such as the regional or continental scale. As we will show on toy models and then on empirical data, the study of a city's height in the tree will allow us to draw some conclusions about the importance of this city in a country, and its evolution in time.

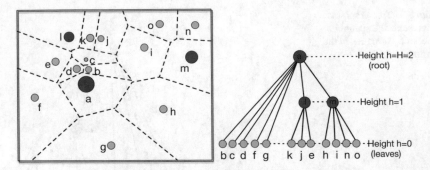

Fig. 8.4 The construction described in the text applied to the distribution shown in the left figure (where the size of the disk represent their importance) leads to a dominance tree of the form shown here. Leaves are the points which are not local maxima, and the children of a node are the points belonging to its attraction basin. By construction, the largest city will always be the root of the dominance tree, but for the other cities their height in this tree will not only depend on their size/rank in the urban hierarchy, but also on their location. Figures taken from [4]

8.2   Class First-Passage Times

The difficulty with characterizing spatial networks is that heterogeneity and correlations can appear simultaneously at multiple scales. We saw that persistent homology could detect some of the correlations between motifs and we discuss here another method proposed in [11]. This non-parametric framework is based on a normalized mean first passage time between classes of nodes for a uniform random walk on the graph of interconnections among the units of the system. In the case of a social system, the graph will be constructed from the relations between individuals and for an urban system, the graph can be obtained from the network of adjacency between census tracts or the connections among census tracts due to human mobility flows. The normalization of the class mean first passage time will be done with respect to a null model where classes are reassigned to nodes uniformly at random. In this way, it becomes possible to quantify and compare the heterogeneity of class distributions in systems of different nature, size and shape.

This method allows to extract from the diffusion structure the organization and functioning of a networked complex system. It is important to note that here we are concerned about correlations between locations and classes and not about the structure of the network only. Indeed, random walks have been widely used to identify some properties of graphs such as the existence of communities, navigability, distribution of node roles, etc., but here the focus is on systems whose nodes have preassigned classes that do not necessarily coincide with their location in the graph. More precisely, we consider a graph make E edges and N nodes, and described by an adjacency matric $A = \{a_{ij}\}$. Here, we are interested in nodes belonging to different classes and we model this by a function which associate to each node i a discrete label $f(i) = c_i$. The idea is then to consider a uniform random walk on this graph and which is defined by the transition matrix $\Pi = \{\pi_{ij}\}$ where π_{ij} is the probability

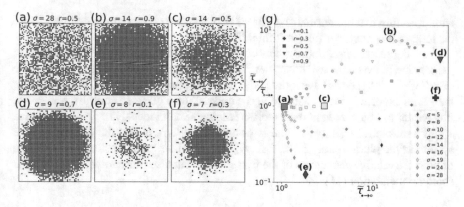

Fig. 8.5 Distribution of colors in a 2d gaussian cluster model obtained for **a–f** different values of σ and r (the lattice has here 60×60 nodes and two colours). Figure from [11]

that a walker at node i will jump to node j. The proposed method focuses on the distribution of the 'class mean first passage time' (CMFPT) $\tau_{alpha\beta}$ defined as the expected number of steps needed for a random walker to visit the first time a node of a certain class β when its starts from a node of class α. In this way, spatial correlations between classes are encoded in the distribution of CMFPT.

This method has been applied on various systems and we will present here the case of synthetic colourings in two-dimensional lattices with $N = L \times L$ nodes [11] (we refer the interested reader to the original paper for other applications of their method). We associate to each node one of two possible classes denoted by the subscript $\bullet$ or $\circ$. The relative abundance of $\bullet$ nodes is denoted by $r = \frac{N_\bullet}{N}$. $N_\bullet$ nodes are then assigned to the class $\bullet$ and their coordinates (x, y) are drawn from a symmetric two-dimensional Gaussian distribution centered in $(L/2, L/2)$ with standard deviation equal to σ. By tuning these two parameters (r, σ), we can then explore a variety of models between a uniform distribution of colors and patterns and strong segregation cases. Various cases are shown in Fig. 8.5a–f: for large values of σ, the distribution is mostly uniform, regardless of the relative abundance of the two colors. In contrast, for smaller values of σ, the nodes of the class $\bullet$ are more clustered around the center. The role of the relative abundance of colors is evident from the comparison of panel (b) and panel (c), which are two configurations with the same σ but with different values $r = 0.9$ and $r = 0.5$, respectively.

The authors of [11] studied the ratio of normalized mean first passage times from one color to the other and vice-versa

$$\widetilde{\tau}_{\bullet \to \circ} / \widetilde{\tau}_{\circ \to \bullet} \tag{8.4}$$

versus the normalized MFPT $\widetilde{\tau}_{\bullet \to \circ}$. For large values of σ, the colors are more homogeneously distributed and we expect $\widetilde{\tau}_{\bullet \to \circ} \simeq 1$ showing that the model is not very different from the null one. We also observe in this case that $\widetilde{\tau}_{\bullet \to \circ} / \widetilde{\tau}_{\circ \to \bullet}$ is close to 1 as well, meaning that the normalised CMFPTs of the two classes are indistinguishable (as long as r is not too different from 0.5).

As σ decreases, the $\bullet$ nodes are more concentrated in the center of the system and the relation between the two quantities $\widetilde{\tau}_{\bullet \to \circ}$ and $\widetilde{\tau}_{\bullet \to \circ}/\widetilde{\tau}_{\circ \to \bullet}$ will depend on the value of r. For $r > 0.5$ (i.e. we have more nodes of the $\bullet$ class), the ratio $\widetilde{\tau}_{\bullet \to \circ}/\widetilde{\tau}_{\circ \to \bullet}$ is larger than 1, as expected as it is easier to find a node in the $\bullet$ class. In contrast, if $r < 0.5$, it is harder to find a node in the e $\bullet$ class leading simultaneously to $\widetilde{\tau}_{\circ \to \bullet}$ larger than 1 and a ratio $\widetilde{\tau}_{\bullet \to \circ}/\widetilde{\tau}_{\circ \to \bullet} < 1$. The important point here is that the behavior of the system in the phase space $(\widetilde{\tau}_{\bullet \to \circ}, \widetilde{\tau}_{\bullet \to \circ}/\widetilde{\tau}_{\circ \to \bullet})$ priovides an interesting visualization of both the relative abundance of the two species, but also more importantly of their spatial organization [11].

8.3 Community Detection in Spatial Networks

Community detection is an important tool for exploring and classifying the properties of large complex networks and can be of great help for understanding the structure of spatial networks. A community is a mesoscale object in a network and is loosely defined as a set of nodes that have dense connections among themselves but sparse links to other communities [12]. Finding communities has many potential applications in various cases such as social or biological networks for example where the goal in many instances is to extract useful high-level information from very large networks. There are many methods available (see the review [12]), but after almost a decade of efforts, there is no definitive method of identification of communities, but instead many different methods with their respective advantage and drawbacks.

Community detection can also have several purposes in spatial networks [13, 14], but probably the main one is to disentangle various aspects such as spatial correlations. In most cases [13, 14] communities are determined by geography only, which results from the simple fact that the most important flows are among nodes in the same geographical regions. In this sense, community detection in spatial networks offers a visual representation of dominant exchange zones. This suggests that community detection might be an important tool in geography and in the determination of new administrative or economical boundaries [15]. More generally, community detection should incorporate metadata or temporal information about nodes or edges [16].

In the case of existing metadata for the network, each node has a mark that can be a real variable or a category. A community can then be naturally defined as a group of network elements having the same mark value such as the language or the age for social networks, or the internet domain name for web pages. In spatial networks, each node is described by its coordinates but has in general other attributes. For individuals, it can be any cultural or socio-economical parameter. For infrastructure networks such as power grids, it can be the voltage at the electric substations. In general, this mark depends on space and the resulting network displays entangled layers of parameters. An important goal in the analysis of these networks is to disentangle these different levels and to extract mesoscopic information from the spatial network structure. If one is interested in studying effects beyond space [17], one should have a straightforward way to 'subtract' it from the network, or in other words, to disentangle space and the other marks.

Community detection procedures consist in finding these groups of nodes in the network and various methods were proposed so far (we refer the interested reader to the review [12]). In particular, the Newman-Girvan method [18] which relies on the optimization of a quantity called modularity is frequently used (or the so-called Louvain method [19]) and despite its intrinsic limits shown in [20], it possesses the advantage of being simple and relatively easy to implement. When optimizing this modularity, one compares the existing links with the one obtained from a null model. This null model was chosen in the original work [18] to be the random graph (obtained by preserving the degrees of nodes but reshuffling the links), and the communities are then obtained by identifying the set of nodes that are more densely connected than what could be observed by chance. Obviously, the communities obtained with this method depend crucially on the choice of the null model. It is therefore important to analyze this dependence and to propose a set of null models that are adapted to the problem at hand. In this chapter we will discuss such models, based on correlations between attributes [21] and based on various mobility models [16].

In the following we will therefore focus on two main points. First, we recall basic definitions about community detection and we will define the modularity. We then present spatial benchmarks used for testing various community detection methods. These benchmarks are embedded in space and can include attributes. It is important to develop synthetic spatial benchmark networks with a small number of control parameters in order to test various community detection methods and in this chapter, we will discuss some examples that can be found in the literature [16, 21]. Second, we will discuss modifications of the modularity optimization method adapted to spatial networks. For more details on the application of these various methods on the synthetic benchmarks and on data we refer the interested reader to [16, 21].

8.3.1 Modularity

Loosely speaking, a community (or a 'module') is a set of nodes which has more connections among themselves than with the rest of nodes. One of the first and simplest method to detect these modules is the modularity optimization and consists in maximizing the quantity called modularity defined as [18]

$$Q = \sum_{s=1}^{n_M} \frac{\ell_s}{E} - \left(\frac{d_s}{2E} \right)^2 \qquad (8.5)$$

where the sum is over the n_M modules of the partition, ℓ_s is the number of links inside module s, E is the total number of links in the network, and d_s is the total degree of the nodes in module s. The first term of the summand in this equation is the fraction of links inside module s and the second term represents the expected fraction of links in that module if links were located at random in the network (and by keeping the same degree distribution). The number of modules n_M is also a variable whose value is obtained by the maximization. If for a subgraph S of a network the first term is much larger than the second, it means that there are many more links inside S than what one could expect by random chance, and S is indeed a module. The comparison with the null model represented by the randomized network is however misleading and small modules connected by at least a link might be seen as one single module. This resolution limit was demonstrated in [20] where it is shown that modules of size $\sqrt{E}$ or smaller might not be detected by this method. Modularity detection was however applied in many different domains and is still used (usually with the Louvain algorithm). In the case of spatial networks, it is the only method which was used so far but it is clear that community detection in spatial networks is a very interesting problem which might receive a specific answer. In particular, it would be interesting to see how the classification of nodes proposed in [22] applies to spatial networks.

In [13], Guimera et al. used modularity optimization with simulated annealing in order to identify communities defined by the worldwide air network. They observed that most communities are actually determined by geographical factors and are therefore not very informative: the most important flows are among nodes in the same geographical regions. More interesting are the spatial 'anomalies' which belong to a community from the modularity point of view but which are in another geographical region. For example, the community of western Europe also contains airports from Asian Russia. More generally, it is clear that in the case of spatial networks, community detection offers a visual representation of large exchange zones. It also allows to identify the inter-communities links which probably play a very important role in many processes such as disease spread for example.

8.3.2   A Null Model for Spatial Networks with Marks

In order to test how community detection acts on spatial networks, we define a simple model of spatial networks with marks [21]. These marks could be anything and we will restrict ourselves – without loss of generality – to the simple binary case where they have two possible values at each node (for example either french or flemish as in the Brussels case studied in [23]). We assume that nodes and their marks are randomly distributed in space. In general, the marks can be delocalized in space or, on the contrary, be localized in some well-defined region. In some cases, some mark community could emerge in space, but our target community structure will always be the partition of the network in the two subgraphs composed of nodes with the same mark and we will test how various methods can recover these two communities. In this respect, the main focus in this case is to disentangle the sole attribute network features beyond spatial node arrangements.

The construction of the first simple benchmark follows these steps:

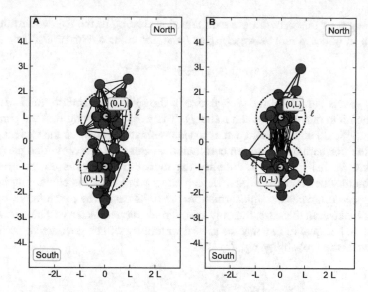

Fig. 8.6 The two spatial communities North and South are well separated having their average size $\ell = L$. In the A panel we present the case $\varepsilon = 0$ where there is a perfect correlation between the space and the marks (green and red colors). In the B panel, the uncorrelated case $\varepsilon = 0.5$ is presented where the mark colors are randomly distributed between the two segregated spatial communities (for the sake of clarity, only 40 out of the 100 nodes used in our simulations are shown here, and $\beta = 1.0$). Figure taken from [21]

1. We first generate points/nodes in the $2d$ space $(x - z)$ in two spatial communities, say the North and the South, around the two centers $(x, z) = (0, +L)$ and $(x, z) = (0, -L)$ (see Fig. 8.6). A simple way to do that is to generate points i around the two centers according to the probability

$$p(x_i, z_i) \propto e^{-d_{ci}/\ell} \tag{8.6}$$

where d_{ci} is the euclidean distance between one of the centers c and the node i of coordinates (x_i, z_i).

2. We then assign a mark S_i to each node i. As discussed above, we will focus on the simplest case where this mark can take only two values $S_i = \pm 1$ (which we will represent by red and green colors). A simple way to control correlations between mark and space is to choose $S_i = +1$ with probability q for $z > 0$ and $S_i = -1$ with probability $1 - q$. In order to tune the various cases we introduce the parameter ε, with $q = 1 - \varepsilon$, that determines the mixing between space and marks, ranging from 0.0 to 0.5. In the case $\varepsilon = 0.0$ space and marks are strongly correlated, while for $\varepsilon = 0.5$ they are totally uncorrelated.

The relevant parameters for the generation of network nodes are therefore ℓ and ε.

The edges of this network are then added according to the following rules. For each pair of nodes, a link between nodes i and j is created with probability

$$p_{link}(i, j) \propto e^{\beta S_i S_j - d_{ij}/\ell_0} \tag{8.7}$$

where ℓ_0 plays the role of the typical size of the spatial community (and where d_{ij} is the euclidian distance between i and j). It is worth observing that the parameter ℓ_0 is the typical length of links when space dominates while ℓ is the typical spatial size of the northern and southern communities. Here the relevant edge parameters are β and ℓ_0, but in order to simplify the model and to focus on the efficiency of community detection methods, we can choose $\ell = \ell_0$. This choice implies that when space dominates the link formation, the links cannot be much larger than the community size. In this case, the only spatial relevant parameter will be ℓ/L and we can fix L to be equal to 1 so that the spatial variability will be governed by ℓ. We can then rewrite the probability $p_{link}(i, j)$ as

$$p_{link}(i, j) = \frac{1}{\mathcal{N}} e^{\beta(S_i S_j - d_{ij}/\ell\beta)} \tag{8.8}$$

where $\mathcal{N} = \sum_{i<j} \exp(\beta S_i S_j - d_{ij}/\ell)$ is the normalization constant. As in the Erdos-Renyi random graph, the number of edges is a random variable with small fluctuations around its average. The number of nodes is thus fixed in each network but not the number of edges or the average degree, and this implies that we will have to average the observables over different realizations of the network.

When $\beta\ell$ is large, links are essentially between nodes with the same attribute (irrespective of their distance) and if $\beta\ell$ is small then space is the governing factor and links are essentially between neighboring nodes.

This set of rules allow to construct a benchmark network where the probability associated to a link depends on both space and mark, and the correlation between marks and space can be controlled. If the mark is the same between two nodes the probability to have a link will be reinforced, otherwise it will be weakened, the interplay being controlled by the parameter β. Concerning the spatial factor, the closer the nodes and the larger the probability to have an edge between them.

The generation of marks is also an important point. We have two values of the attribute only so that we need to generate marks for only half ($N/2$) of the nodes. We will focus here on the specific case of a mark community structure of equal size communities: half of the nodes has mark $S_i = +1$ and the other half has $S_i = -1$. We can tune the model in order to recover two extreme cases:

- Marks and space uncorrelated: this case is obtained by choosing $\varepsilon = 1/2$.
- Marks and space are strongly correlated. For this, we choose ε small. In this case, the spatial communities are also mark communities.

Furthermore we can distinguish two different spatial arrangements for the northern and southern communities. The first case corresponds to a situation where the two communities are well separated with their average size $\ell \leq L$ and the spatial effects dominate the community structure (see Fig. 8.6). The second situation corresponds to a larger value of the average community size ℓ where the two communities start mixing up while ℓ approaches L (see Fig. 8.7).

This model therefore takes into account in a simple way the correlation between space and node marks. The interplay between space and marks can lead to various situations that need to be understood within the framework of community detection. Indeed we have two main regimes $\beta\ell \gg 1$ and $\beta\ell \ll 1$ (see also Table 8.1):

1. $\beta\ell \gg 1$. In this case, the spatial component of the links becomes irrelevant (see Eq. (8.8)) and for a given value of β the community structure due to the node marks will emerge, independently from the correlation between space and marks. In this regime any community detection method should work.
2. $\beta\ell \ll 1$. Here we have two subcases depending on the correlation between space and marks:

   - ($\varepsilon = 0.0$) Space and marks are correlated: any regular community detection will work and moreover if you carefully remove the spatial effect the mark community structure will be recovered.
   - ($\varepsilon = 0.5$) Space and marks are uncorrelated: in this case the links are between neighboring nodes but the attributes are anywhere in space. Standard commu-

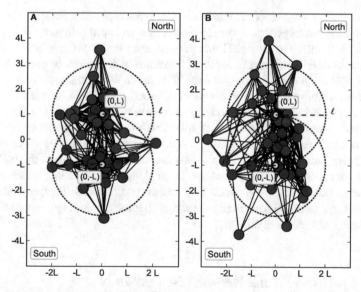

Fig. 8.7 The two communities North and South are mixing up each other with their average size ℓ approaching the value of L (in this case $\ell = 2L$). In the A panel, we display the case $\varepsilon = 0.0$. Even if the spatial correlation is fading away the space-mark correlation is still strong enough to display a mark community. In the B panel, we show the extreme case $\varepsilon = 0.5$ where the marks are not correlated with space. In this case spatial mixing destroys the mark community structure (for the sake of clarity, only 40 out of the 100 nodes used in the simulations are shown here, and $\beta = 1.0$). Figure taken from [21]

Table 8.1 Behavior of the spatial null model defined in the text for the two regimes $\beta\ell \ll 1$ and $\beta\ell \ll 1$ both in the correlated ($\varepsilon = 0.0$) and uncorrelated ($\varepsilon = 0.5$) case

Spatial correlation ε	$\beta\ell \ll 1$: Space is the governing factor	$\beta\ell \gg 1$: The spatial component of the links is irrelevant
Spatially correlated: ($\varepsilon = 0.0$)	• Links are between neighboring nodes but spatial communities correspond to the mark ones • Any regular community detection will work	• Links are between nodes with the same mark • Any community detection method should work
Spatially uncorrelated: ($\varepsilon = 0.5$)	• Links are between neighboring nodes but the marks are anywhere in space • It is necessary to 'remove' space in order to uncover the mark communities	• Links are between nodes with the same mark • Any community detection method should work

nity detection methods won't work and it is then necessary to 'remove' space in order to uncover the attribute communities.

We thus saw a simple model that allows to test community detection on spatial networks. This model generates simple graphs that mix both geographical properties and marks and allows to show [21] that the existence of correlations between marks and space drastically affects the result of community detection. In particular, community detection in spatial networks should be taken with great care, and including space in community detection methods could lead to results difficult to interpret. For weak correlations, most community detection methods work, but for stronger correlations, community detection methods which remove the spatial component of the network can lead to incorrect results. It is thus important to have some information on the correlations between space and marks in order to assess the validity of the results of community detection methods. In practical applications however, these mark-space correlations are generally not known and this calls for the need of new approaches, for example community detection methods including in a tunable form the existence of such correlations.

8.3.3 Synthetic Spatial Network Benchmarks

The authors of [16] proposed synthetic models with a planted community structure that incorporate spatial effects and temporal correlations. We will discuss the spatial effects here and for temporal aspects, we refer the interested reader to the original paper.

The authors of [16] assign N nodes uniformly at random to positions on the lattice $\{1, 2, \ldots, l\} \times \{1, 2, \ldots, l\}$ and a 'population' n_i to each node i which is either uniform or random. In the first model called 'distance benchmark', the probability p_{ij}^{dist} that an edge exists between nodes i and j depends only on the geographical distance d_{ij} between nodes and on their community assignments. After having assigned nodes uniformly at random to one of two communities, the probability is chosen to be

$$p_{ij}^{\text{dist}} = \frac{\lambda(c_i, c_j)}{Z_1 d_{ij}}, \tag{8.9}$$

where Z_1 is a normalization constant and where c_i is the community that contains node i and the function $\lambda(c_i, c_j) = 1$ if nodes i and j are in the same community and $\lambda(c_i, c_j) = \lambda_d$ otherwise. This last parameter λ_d controls the degree of mixing between communities: for $\lambda_d = 0$, only nodes in the same community are connected, while for $\lambda_d = 1$, there are no distinct communities. A number of $L = \mu N(N - 1)/2$ edges is constructed and the parameter μ determines the network density.

In the second model of 'flux benchmark' of [16], the goal is to mimic the spread of a disease on a network. The weight of an edge is thus given by the mobility flow on it and is assumed to be given by the radiation model [24]. The edge probability p_{ij}^{flux} is then proportional to the average flow predicted by the radiation model $\hat{T}_{ij}$

$$p_{ij}^{\text{flux}} = \frac{\lambda(c_i, c_j)\hat{T}_{ij}}{Z_2}, \tag{8.10}$$

where Z_2 is the normalization constant that ensures that $\sum_{i>j} p_{ij}^{\text{flux}} = 1$. We note here that in [16], they extend this single-layer construction to the multilayer case, and also incorporate temporal evolution. They are then able to evaluate the performance of the Newman-Girvan algorithms and the other models (see below) by comparing their result with the planted community structure (see [16] for more details).

8.3.4 Modifying the Modularity

In modularity optimization the choice of a null model has a crucial impact on the communities found and their interpretation. The best null model depends then on the question posed and on the available data. Here we will follow [16, 21] and discuss various choices for the null model in the context of spatial networks.

The modularity can be written for static, weighted network as [18]

$$Q = \frac{1}{2w} \sum_{ij} (W_{ij} - \gamma P_{ij})\delta(c_i, c_j) \tag{8.11}$$

where W_{ij} is the weight on edge between nodes i and j, $2w = \sum_{ij} W_{ij}$ is the total weight. The element P_{ij} is the matrix that characterizes the null model (and γ is the resolution parameter [25] that controls the size of communities). In the standard case of the random graph as the null model we have

$$P_{ij}^{NG} = \frac{k_i k_j}{2w} \tag{8.12}$$

where $k_i = \sum_j W_{ij}$ is the strength (i.e. weighted degree) of the node i. This simple form does not integrate any specific information such as space or other metadata, and in the following we will discuss examples that are designed for taking into account attributes or mobility features.

8.3.4.1 Including Mark Correlations

As discussed in [21], the goal is to subtract the spatial component and to recover the mark communities. Indeed, in the general case, we do not know to what extent the existence of a link between a pair of nodes is due to a specific factor or to space only. The link could exist because of a strong mark affinity between the nodes, or in the other extreme case, because they are close neighbors. In general, one could expect a combination of these two effects. If we are interested in recovering communities defined by a mark (such as language for example) from the network structure, we then have to consider various assumptions such as the correlation between link formation, mark values and space. In order to understand the effect of the underlying correlations, we can consider two extreme cases. When the links are purely spatial and independent from the marks, removing the spatial component leads to random communities (obtained for a random graph) which contain a random number of nodes with random marks. In this situation, community detection is unapplicable and there is no way to recover mark communities from the network structure. The other extreme case is when the formation of a link depends on the marks only. In this case, space is irrelevant and any standard community detection method should give sensible results, i.e. communities made of nodes with the same mark.

As discussed above, Eq. (8.11) gives the modularity function which needs to be optimized is defined as [18]. For an unweighted network, one can choose $P_{ij} = \frac{k_i k_j}{2m}$ which amounts to take as a null model a random network with the same degree sequence as the original network. In order to introduce explicitly space, the idea is to change the null model defined by P_{ij} and to compare the actual network with this null model. Such a proposal was recently made in [17] where the quantity P_{ij} is directly obtained from the data describing the network. More precisely, Expert et al. [17] used the following form

$$P_{ij}^{Data} = N_i N_j f(d_{ij}) \tag{8.13}$$

where N_i is related to the importance of the node i (such as the population for example). This form is reminiscent of the gravitational model for traffic flows (see for example [26]) where flows are proportional to the product of populations and decrease with distance. In [17], the authors proposed to estimate the unknown function f directly from the empirical data by

$$f(d) = \frac{\sum_{i,j|d_{ij}=d} A_{ij}}{\sum_{i,j|d_{ij}=d} N_i N_j} \tag{8.14}$$

which can be seen as the probability to have two nodes connected at a distance d. Note that there is a binning procedure implicit in Eq. (8.14) and the usual way to proceed in these cases consists in introducing a discretization of the space in bins that capture classes of distances. Expert et al. [17] applied this method to the specific case of the phone network in Belgium in order to reconstruct linguistic communities (Flemish and French) beyond individuals spatial location. This choice is probably the best one if there are no correlations between the mark under study (in their case the linguistic membership of the people calling each other) and space. In this specific case, extracting the node spatial dependencies from the actual link distribution present in the network data is the most effective way to subtract the spatial component. Otherwise if there are any correlations between space and node marks, the dataset contains in an unknown proportion the two informations (space and mark) and their method needs to be reformulated. One possible way to do this is to explicitly guess a spatial dependency of the link distribution and to put it as an independent factor in the optimization function definition. In order to be able to deal with the correlated case and to remove spatial effect only, the following explicit function of space for P_{ij} can be chosen

$$P_{ij}^{Spatial} = \frac{1}{Z} k_i k_j g(d_{ij}) \tag{8.15}$$

where Z is the normalization constant, k_i the degree of the node i, d_{ij} the euclidean distance between node i and node j. The function $g(d)$ is a decreasing function of distance and its role is to remove the spatial effect. A simple choice is

$$g(d) = e^{-d/\bar{\ell}} \tag{8.16}$$

where $\bar{\ell}$ is the average distance between nodes in the network. Of course $\bar{\ell}$ is a rough approximation of the real ℓ value, but as shown in [21] it is in fact enough to capture the essence of the spatial signature of the network.

We note here that other proposals can be found in the literature such as in [27] where the authors focused on finding geographically-disperse communities and used a distance modularity defined with

$$P_{dist} = \frac{k_i k_j f(d_{ij})}{\sum_q k_q f(d_{qi}} \tag{8.17}$$

where f is the deterrence function chosen as $f(d_{ij}) = e^{-d_{ij}/\sigma^2}$ where σ is a constant.

8.3.4.2   Spatial Null Models: Gravity Model

We discuss here the integration of the gravity model into community detection [16, 17]. The main idea is that in many spatial networks, proximity has a very strong effect and neighboring nodes are more likely to connect than far away ones. It is thus important to incorporate the expected influence of proximity in a null model for community detection in order to identify relevant structures.

In the classical gravity model (see for example [26]), the traffic T_{ij} between locations i and j with respective populations n_i and n_j and separated by a distance d_{ij} is written a

$$T_{ij} = K n_i^\alpha n_j^\beta f(d_{ij}), \tag{8.18}$$

where K is a constant. The 'deterrence function' $f(d)$ describes the effect of space and decreases with distance. Standard choices include the power law decay

$$f(d_{ij}) = d_{ij}^{-\kappa} \tag{8.19}$$

or the exponential decay

$$f(d_{ij}) = e^{-d_{ij}} \tag{8.20}$$

The parameters α, β, and κ are usually determined by regression on data.

The simplest form of Eq. (8.18), with $\alpha = \beta = 1$ and $\kappa = 1$, was incorporated into a 'gravity null model' [17] with

$$P_{ij}^{grav} = I_i I_j f(d_{ij}), \tag{8.21}$$

where I_i characterizes the importance of node i. The deterrence function is then estimated from data with

$$f(d) = \frac{\sum_{\{k,l|d_{kl}=d\}} W_{kl}}{\sum_{\{k,l|d_{kl}=d\}} (I_k I_l)}, \tag{8.22}$$

where bins for pairs of nodes at distance d are used. The gravity null model can then be written as

$$P_{ij}^{\text{grav}} = I_i I_j \frac{\sum_{\{k,l|d_{kl}=d_{ij}\}} W_{kl}}{\sum_{\{k,l|d_{kl}=d_{ij}\}} (I_k I_l)} . \tag{8.23}$$

which was used in [17] in the case of linguistic partition in Brussels. This gravity null model was then subsequently extended to the multilayer case in [16].

8.3.4.3 Spatial Null Models: Radiation Model

The gravity model has several drawbacks: it does not include congestion effects and does not have a clear theoretical foundation (see for example Chap. 5 in [3]). Simini et al. provided an alternative [24] that does not have all these problems and that does not rely heavily on data fitting. This 'radiation model' is designed to capture human mobility (more specifically commuting between homes and offices) and might be useful in many processes involving the displacement of individuals such as epidemic spread. For these reasons, Sarzynska et al. [16] constructed a community detection method based on the radiation model. In this model [24], the average commuting flow between locations i and j with populations n_i and n_j is given by [24]

$$T_{ij} = T_i \frac{n_i n_j}{(n_i + r_{ij})(n_i + n_j + s_{ij})}, \tag{8.24}$$

where s_{ij} is the population in the disk centered on i and of radius d_{ij} (minus the populations at i and j), and T_i is the number of commuters in location i. In order to avoid problems wit directions, we can symmetrize the flows $\hat{T}_{ij} = (T_{ij} + T_{ji})/2$ and the 'radiation null model' is then defined by [16]

$$P_{ij}^{\text{rad}} = \hat{T}_{ij} \frac{\sum_{\{k,l|d_{kl}=d_{ij}\}} W_{kl}}{\sum_{\{k,l|d_{kl}=d_{ij}\}} \hat{T}_{kl}} . \tag{8.25}$$

We note here that this radiation null model was also extended in [16] to the multilayer case.

8.4 Persistent Homology

8.4.1 Topological Data Analysis

In a set of a large number of points, we can sometimes distinguish some shape. The natural question is then if we can recover this shape using a finite number of points sampled from it [28]. Topology which is the study of shape has become increasingly quantitative and constitute the basis of what is now called topological data analysis (TDA) and which provide tools for answering this question. This method is increasingly applied to various datasets from physics, signal processing, computational sciences, biology or social sciences [29].

The central ingredient in TDA is persistent homology and is based on the study of homology at different scales. Without entering too many details, homology relates to classifying the topology of objects. In particular, homology focuses on properties such as the existence of connected components, one-dimensional or two-dimensional holes. This properties can be characterized quantitatively with the use of Betti numbers B_d which refer to the number of d-dimensional holes in a topological surface. A d-dimensional hole is a d-dimension cycle that is not a boundary of a $(d + 1)$ dimensional object. For example, a circle is not a disk and encloses a one-dimensional hole, while the sphere encloses a two-dimensional hole.

The first Betti numbers are then β_0 which counts to the number of connected components, β_1 to circular 'flat' holes and β_2 to trapped volumes (or 2d cavities). For example, a torus has one connected component implying $\beta_0 = 1$, two circular holes (equatorial and meridional) implying $\beta_1 = 2$, and a single cavity enclosed within the surface leading to $\beta_2 = 1$. We can also see the number β_d as the maximum number of d-dimensional curves that can be removed while keeping the object connected. If we remove two 1-dimensional curves (equatorial and meridional) to the torus, it remains connected, implying $\beta_1 = 2$ (higher dimension Betti number are however more difficult to apprehend intuitively).

In order to apply these tools to a set of points, we first need to construct a simplicial complex from the data. We note that constructing a graph would not be enough as we are interested in describing high order correlations in the data (and not counting all the small cycles for example). Instead of constructing a graph from the data points, we then construct a simplicial complex which allows extract large scale features. In addition, there are algorithms based on linear algebra that allow to compute persistent homology in simplicial complexes (in general with a complexity of order $\mathcal{O}(N^3)$.) A simple way to construct a graph and then a simplicial complex (there are many different methods and we will see an example of another one below) is to place a ball of diameter ε around each point and to connect nodes if their distance is less than this proximity parameter ε (see Fig. 8.8). Each point is a zero-dimensional simplex, each edge a 1-dimensional simplex and triangles are 2-dimensional simplexes and from the graph we then obtain a simplicial complex known as the Vietories-Rips complex. In the last figure of our example shown in Fig. 8.8, we observe a quadrilateral because the four nodes are connected in a cyclic manner and which corresponds to a flat hole.

The main idea of persistent homology is then to detect features that are 'persistent' and really characterize the data. We thus need a sequence of the simplicial complex and generally it is obtained by increasing the proximity parameter ε. For each value of this parameter, we compute the features of the simplicial complex and determine at what value ε_{birth} a specific feature (such as a connected component, a 1d hole, etc.) appears and at what value ε_{death} it disappears. From the knowledge of ε_{birth}, ε_{death} we construct the corresponding persistent diagram that summarizes the homology of the data across a range of persistence scales ε. Other authors construct the 'barcode' made of horizontal bars (each bar represents a specific feature) of length $\Delta = \varepsilon_{death} - \varepsilon_{birth}$ for each features (this representation is strictly equivalent to the persistent diagram).

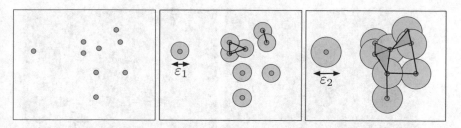

Fig. 8.8 Construction of the simplicial complex from a set of point. Around each point, we construct a disk of diameter ε and connect two nodes if their disks intersect. This allows to construct a simplicial complex at the persistence scale ε (also called the Vietoris-Rips complex in this case). We colored in blue the 2-dimensional simplexes. Figure inspired from [28]

Persistent features are those with a large value of Δ while the others with small Δ are usually considered as 'noise' and represent small scale details.

This description of PH is of course very rough and we refer the interested reader to the excellent paper by Feng and Porter er [30] which contains many references (including mathematical ones) and several empirical examples that we will describe below.

8.4.2   Empirical Results

We discuss here the application of persistent homology and the construction of the persistent diagram for cases taken from [30]. In these examples the network is already known and in order to construct an increasing sequence of simplicial complexes, Feng and Porter used a level-set construction: starting from the network itself, they slowly 'grow' it using a level-set equation. In terms of a geographical map, we can visualize this process as the network being at the emerged top of mountains. When time increases the sea level decreases and we move the entire mountain upward increasing the amount of land above water. The new region above water is the expanded manifold. By definition, we then obtain a sequence of manifolds allowing to construct a filtration. During this evolution, they record the birth and death times of specific features (connected components and holes) and construct the persistent diagram.

8.4.2.1   Street Networks

In this case, as the filtration time increases, more and more city blocks are added to the complex. Very likely, when regular blocks are added homological features won't be affected, and the persistence diagram will in some way record anomalies characterized by blocks and capture information about the size and regularity of blocks. The authors studied morphologies both within a single city and for a variety of different cities.

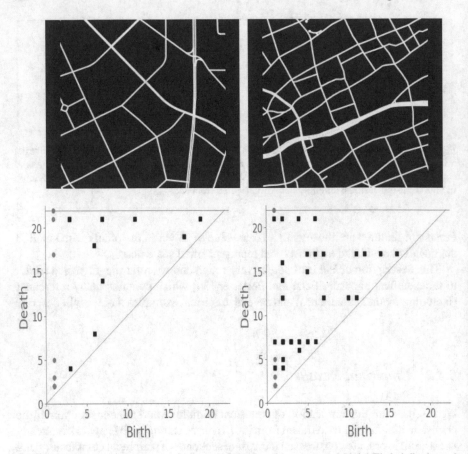

Fig. 8.9 Two sampled street networks from (left) Pudong New Area and (right) Zhabei district and the corresponding persistence diagrams. Figure from [30] and distributed under the terms of the Creative Commons Attribution 4.0 International license

They first sample different areas within Shanghai (China) and computed the corresponding persistence diagrams (two different areas are shown in Fig. 8.9). They then use the standard bottleneck distance for comparing the persistence diagrams obtained for different areas. The bottleneck distance measures the similarity between two persistence diagrams and is basically obtained by computing the minimum cost over all partial matches between the points the length (see for example [30]). Once there is a distance defined between persistence diagrams, clustering methods can be used. Feng and Porter performed a hierarchical clustering in three clusters. Interestingly enough, these three clusters correspond to historical, transition and modern areas.

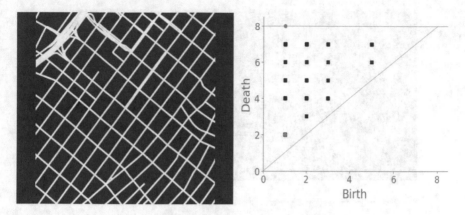

Fig. 8.10 Example of Los Angeles, a city in the first cluster with gridlike street layouts. We show here a sample of its street layout and its persistent diagram. Figure from [30] and distributed under the terms of the Creative Commons Attribution 4.0 International license

Beyond inhomogeneities within a city, it is interesting to compare different cities with each other and to detect various categories (see Chap. 9). Feng and Porter studied 306 cities across the globe and for each of them computed their PD and used hierarchical clustering leading to three clusters of cities. These three clusters comprise (i) gridlike cities, (ii) cities that mix gridlike and non gridlike patches, and (iii) cities with a large number of non-gridlike structures.

The first cluster comprises 99 cities (32% of the dataset) with mostly small gridlike blocks and a typical persistent diagram is shown in Fig. 8.10. In this persistent diagram and in the following the connected components are shown in pink circles while the one-dimensional holes are shown in dark-blue squares. North America is the region of the world with the largest percentage of such cities and Europe the smallest percentage. The distribution of death times of 1d holes tend to be uniform and to occur over a small number of filtration steps. This can be understood as a consequence of the gridlike structure which in general implies a small dispersion of block sizes.

The second major cluster comprises 149 cities (which 49% of the cities in the dataset and is the largest cluster in this study) with a mixture of gridlike and non-gridlike patches. We observe in the persistent diagrams a larger maximum death of features. Gridlike blocks lead to early death times while non gridlike structures lead to later death times. This cluster can be separated into two subclusters and example of cities in each subclusters are shown in Fig. 8.11. The first subcluster - illustrated here with the case of the city of Aleppo (Syria) - consists essentially of cities with a large grid and a small number of large blocks that interrupt the grid. This can be seen on the PD where most of the death times are early and a small number of features with late death times. The second subcluster consists of cities with small patches of gridlike structures mixed with large irregular blocks. This case is illustrated in the Fig. 8.11 with the example of Barcelona (Spain). We observe here the strength of this

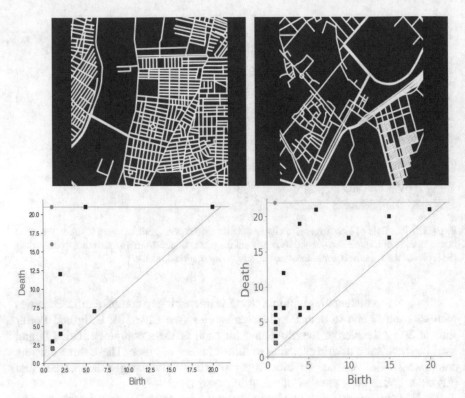

Fig. 8.11 Example of cities (**a** Aleppo and **b** Barcelona) in the second major cluster found in [30]. These cities have patches of gridlike structure that are mixed with large blocks. In the bottom row are shown their associated persistent diagrams. Figure from [30] and distributed under the terms of the Creative Commons Attribution 4.0 International license

method that is able to detect large scale structures which would be otherwise very difficult to find.

Finally, the last cluster contains 58 cities (19% of the dataset) that have a large number of non gridlike structures such as dead ends, rectangular blocks arranged in random order, etc. Here also, by examining the dendrogram from the hierarchical clustering, there are two clear subclusters with one subcluster that contains only two cities (Beirut and Nanyang). We show in Fig. 8.12 the examples of Nanyang and London and their corresponding persistent diagrams. The persistent diagrams are characterized by a large variety of death times which correspond to the large range of block sizes (which are rare in the other clusters). For example, Nanyang has several streets that do not continue through particular blocks which leads to a mixture of block sizes even in areas that tend to be gridlike. The other example of London displays a large diversity of block sizes and a large number of dead ends.

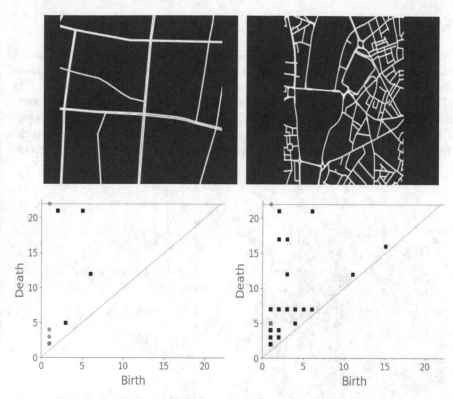

Fig. 8.12 Examples of cities in the third major cluster: **a** Nanyang and **b** London and their associated PDs in the bottom row. Figure from [30] and distributed under the terms of the Creative Commons Attribution 4.0 International license

In summary, North America has the largest proportion of gridlike cities while Europe has the smallest proportion of such cities, in agreement with the common perception that North American cities are more gridlike than European cities (more details and discussions can be found in [30]).

This block based method bears some similarities with the shape and size approach used in [31] (see Chap. 9). In the level-set method used in [30], the irregularity of blocks is encoded in a larger death time in the corresponding persistent diagram. It thus contains the same information as in [31] which used the shape factor to characterize the irregularity of block sizes. However, in [31] correlations between blocks are neglected while in the PH approach information about the spatial relationship between blocks are taken into account.

8.4.2.2   Spider Webs

Another application [30] of the persistent homology method on spatial networks concerns spiderwebs [32, 33]. The main question in this research is how drugs affect the shape of webs built by spiders. Inspired by the first research on this subject by Witt [32], more recent studies [33] suggested that more toxic substances produce more deformed webs. We reproduce in Fig. 8.13 the example of cafferine and refer to [30] for other examples including LSD, peyote, marijuana, etc. Starting with these spider web images, Feng and Porter used the tools of persistent homology and constructed

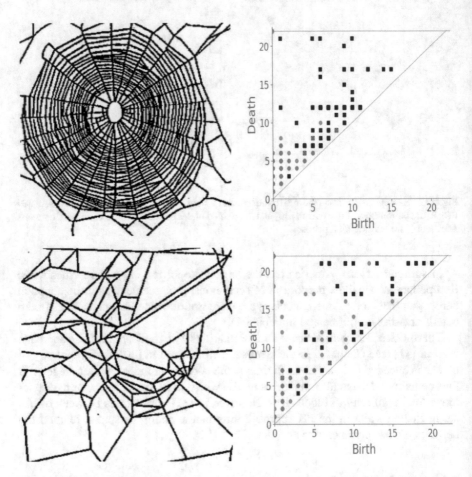

Fig. 8.13 Webs built by drug-free spiders **a** and spiders under the influence of caffeine (other examples are discussed in [30]). This figure is taken from [30] (distributed under the terms of the Creative Commons Attribution 4.0 International license) and the original spider web figures are taken from [33]

Fig. 8.14 Classification of webs built by spiders under the influence of various psychotropic substances. Figure from [30] and distributed under the terms of the Creative Commons Attribution 4.0 International license

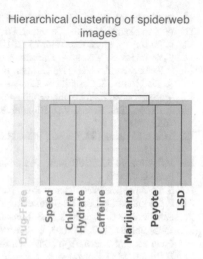

Hierarchical clustering of spiderweb images

Drug-Free | Speed | Chloral Hydrate | Caffeine | Marijuana | Peyote | LSD

the persistent diagrams. They applied a hierarchical clustering algorithm to these PDs and found the result shown in Fig. 8.14. This classification shows that drug-free spider constitute their own cluster characterized by a clear central hole and threads that radiate outward at almost even angles, completed by concentric rings. The effect of marijuana, peyote and LSD are found to be in the same cluster characterized by a clearly identifiable center, branches radiating from it, but with incomplete rings. The last cluster regroups the effects of substances such as speed, chloral hydrate (used in some sleeping pills), and caffeine. These webs are the most disordered with even the absence of a clear center (such as in the caffeine case !), radial threads that are either absent or do not join the center, and the absence of complete rings.

This method clearly demonstrates its ability to capture features at a larger scale. More generally, it would be interesting to apply it to basic spatial networks (such as the ones discussed in Chaps. 15 and 16) in order to reach a more exhaustive interpretation of persistent diagrams.

References

1. A. Okabe, Y. Sadahiro, Environ. Plan. A **28**(9), 1533 (1996)
2. W. Christaller, *Central Places in Southern Germany* (Prentice-Hall, 1966)
3. M. Barthelemy, *The Structure and Dynamics of Cities* (Cambridge University Press, 2016)
4. T. Louail, M. Barthelemy, Submitted (2022)
5. L.P. Kadanoff, *Statistical Physics: Statics, Dynamics and Renormalization* (World Scientific Publishing Company, 2000)
6. C. Cottineau, PloS One **12**(8), e0183919 (2017)
7. V. Verbavatz, M. Barthelemy, Nature **587**(7834), 397 (2020)
8. A. Okabe, B. Boots, K. Sugihara, S.N. Chiu, *Spatial Tessellations: Concepts and Applications of Voronoi Diagrams*, vol. 501 (Wiley, 2009)
9. S. Chiu, Mater. Charact. **34**(2), 149 (1995)

10. D. Weaire, N. Rivier, Contemp. Phys. **25**(1), 59 (1984)
11. A. Bassolas, V. Nicosia, Commun. Phys. **4**(1), 1 (2021)
12. S. Fortunato, Phys. Rep. **486**, 75 (2010)
13. R. Guimerà, S. Mossa, A. Turtschi, L. Amaral, Proc. Natl. Acad. Sci. USA **102**, 7794 (2005)
14. P. Kaluza, A. Koelzsch, M. Gastner, B. Blasius, J.R. Soc, Interface **7**, 1093 (2010)
15. A. De Montis, S. Caschili, A. Chessa, eprint arXiv:1103.2467v1 (2), 19 (2011)
16. M. Sarzynska, E.A. Leicht, G. Chowell, M.A. Porter, J. Complex Netw. **4**(3), 363 (2016)
17. P. Expert, T.S. Evans, V.D. Blondel, R. Lambiotte, Proc. Natl. Acad. Sci. U. S. A. **108**(19), 7663 (2011)
18. M.E.J. Newman, M. Girvan, Phys. Rev. E - Stat. Nonlinear Soft Matter Phys. **69**(2 Pt 2), 16 (2004)
19. V.D. Blondel, J.L. Guillaume, R. Lambiotte, E. Lefebvre, J. Stat. Mech. Theory Exp. **2008**(10), P10008 (2008)
20. S. Fortunato, M. Barthelemy, Proc. Natl Acad. Sci. (USA) **104**, 36 (2007)
21. F. Cerina, V. De Leo, M. Barthelemy, A. Chessa, PloS One **7**(5), e37507 (2012)
22. R. Guimerà, L. Amaral, J. Stat. Mech. **2005**, P02001 (2005)
23. R. Lambiotte, V.D. Blondel, C. de Kerchove, E. Huens, C. Prieur, Z. Smoreda, P. Van Dooren, Phys. A: Stat. Mech. Appl. **387**(21), 5317 (2008)
24. F. Simini, M.C. González, A. Maritan, A.L. Barabási, Nature **484**(7392), 96 (2012)
25. J. Reichardt, S. Bornholdt, Phys. Rev. E Stat. Nonlinear Soft Matter Phys. **74**(1 Pt 2) (2006)
26. S. Erlander, N.F. Stewart, *The Gravity Model in Transportation Analysis: Theory and Extensions*, vol. 3 (Vsp, 1990)
27. P. Shakarian, P. Roos, D. Callahan, C. Kirk, in *Proceedings of the 19th ACM SIGKDD International Conference on Knowledge Discovery and Data Mining* (2013), pp. 1402–1409
28. H. Adams, M.V. Ciocanel, C.M. Topaz, L. Ziegelmeier, SIAM News **53**(1), 1 (2020)
29. M. Feng, M.A. Porter, SIAM News **53**(1), 1 (2020)
30. M. Feng, M.A. Porter, Phys. Rev. Res. **2**(3), 033426 (2020)
31. R. Louf, M. Barthelemy, J. R. Soc. Interface **11**(101), 20140924 (2014)
32. P.N. Witt, Behav. Sci. **16**(1), 98 (1971)
33. D.A. Noever, R.J. Cronise, R.A. Relwani, NASA Tech Briefs **19**, 82 (1995)

Chapter 9
Typology of Planar Graphs

From a theoretical point of view, an important problem amounts to understand the structure of random planar graphs and eventually to propose a classification of these objects. In this chapter, we will discuss different approaches. In the first one, we discuss the statistics of the area and shape of faces, and we apply this to street networks. The possibility of defining a distance between two graphs allows then to propose a first simple typology of street networks. In a second approach, a decimation process is used in order to construct an approximate mapping from a (weighted) planar graph to a tree. In a third part, we will discuss a more mathematical approach which allows to construct an exact bijection between a planar graph and a tree. Finally, in the last part, we will present a recent approach based on machine learning algorithms.

9.1 Area and Shape of Faces

In this first section we discuss a classification method based on the statistics of both the surface area and shape of faces. In the case of planar graph where the geometry counts (this is the case for example for city maps), the visual impression is mostly given by the faces and the pattern they form in the 2d plane (see for example Fig. 9.1).

The method that we will discuss can in principle be applied to any random planar graph but we illustrate here this approach on the example of street networks. These networks can be thought as a simplified schematic view of cities, which captures a large part of their structure and organization [2], and contain a large amount of information about underlying and universal mechanisms at play in their formation and evolution. Extracting common patterns between cities is a way towards the identification of these underlying mechanisms. At stake is the question of the processes behind the so-called 'organic' patterns—which grow in response to local constraints—and whether they are preferable to the planned patterns which are designed under large

© The Author(s), under exclusive license to Springer Nature Switzerland AG 2022
M. Barthelemy, *Spatial Networks*,
https://doi.org/10.1007/978-3-030-94106-2_9

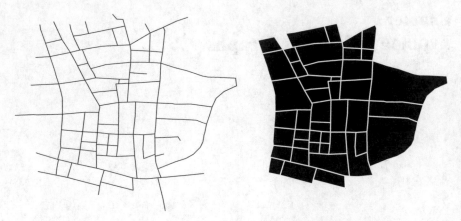

Fig. 9.1 From the street network to blocks. Example of a street pattern taken in the neighbourhood of Shibuya in Tokyo (Japan) and the corresponding set of blocks. Note that the block representation does not take dead-ends into account. Figure taken from [1]

scale constraints. This program is not new [3, 4], but the recent dramatic increase of data availability such as digitized maps, historical or contemporary [5–8] allows now to test ideas and models on large scale cross-sectional and historical data.

9.1.1 Characterizing Blocks

We will thus consider the faces only (also called blocks in the street network context), and we will characterize these objects. Indeed faces are polygons whose properties such as the area or some shape characterization are easily measured.

First, the surface area A of a face is an important indicator and has been measured for various street networks for example [1, 9, 10]. Most of these street networks display a power law behavior for the surface area distribution $P(A)$

$$P(A) \sim \frac{1}{A^\tau} \qquad (9.1)$$

with an exponent always of order $\tau \approx 2$ [1, 6, 7, 9, 11]. This observation is in sharp contrast with the simple picture of an almost regular lattice which would predict a distribution $P(A)$ peaked around some value ℓ_1^2 where ℓ_1 is the typical linear size of a face. It is interesting to note that if we assume that $A \sim 1/\ell_1^2 \sim 1/\rho$ (where ρ is the density of nodes) and that ρ is distributed according to a law $f(\rho)$ with a finite $f(0)$, a simple calculation gives

$$P(A) \sim \frac{1}{A^2} f(1/A) \qquad (9.2)$$

which behaves as $P(A) \sim 1/A^2$ for large A. This simple argument suggests that the observed value ≈ 2.0 of the exponent is 'universal' and reflects the random variation of the density. A consequence of this result is that the distribution $P(A)$ does not allow to distinguish efficiently a graph from another.

A second characterization of a block is through its shape, with the form (or shape) factor Φ, defined in the Geography literature in [12] as the ratio between the area of the block and the area of the circumscribed circle $\mathcal{C}$ (see also Chap. 4)

$$\Phi = \frac{A}{A_{\mathcal{C}}} \tag{9.3}$$

Note that another similar measure is $A/(\pi D^2/4)$ where D is the largest distance found in the polygon. The quantity Φ is in $[0, 1]$, and the smaller its value and the more anisotropic the block is. For example for a rectangle of sides a and b we easily find

$$\Phi_{rectangle} = \frac{4}{\pi} \frac{ab}{a^2 + b^2} \tag{9.4}$$

and indeed varies from 0 for an infinitely elongated rod to $2/\pi \approx 0.636...$ for a square. We note that there is not a unique correspondence between a particular shape and a value of Φ, but this measure gives a good indication about the face's shape (when it is a relatively simple polygon). Empirical observations on street networks for the distribution $P(\Phi)$ display important differences between cities. This would lead to the first naive idea to classify planar graphs according to the distribution of the shape of faces. The shape itself is however not enough to account for visual similarities and dissimilarities between planar graphs. Indeed, in the case of cities, we find for example that for cities such as New-York and Tokyo, the distributions $P(\Phi)$ are similar (see Fig. 9.2), but the visual similarity between both cities's layout is not obvious at all. One reason for this is that blocks can have a similar shape but very different surface areas: if two cities have blocks of the same shape in the same proportion but with totally different areas, they will look different. We thus need to combine the information about both the shape and the area of faces.

In order to construct a simple representation of planar graphs which integrates both area and shape, we rearrange the faces according to their area (on the y-axis) and display their Φ value on the x-axis (Fig. 9.2). We divide the range of areas in (logarithmic) bins and the color of a block represents the area category to which it belongs. We describe quantitatively this pattern by plotting the conditional probability distribution $P(\Phi|A)$ of shapes, given an area bin (Fig. 9.2, right). The colored curves represent the distribution of Φ in each area category, and the curve delimited by the gray area is the sum of all these curves and is the distribution of Φ for all cells, which is simply the translation of the well-known formula for probability conditional distribution

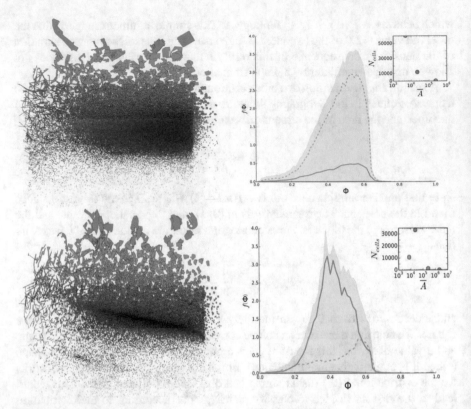

Fig. 9.2 The fingerprints of Tokyo (top) and New-York, NY (bottom). (Left) We rearrange the blocks of a city according to their area (y-axis), and their Φ value (x-axis). The color of each block corresponds to the area category it falls into. (Right) We quantify this pattern by plotting the repartition of shapes per area category, as measured by Φ for each area category—represented by coloured curves. The gray curve is the sum of all the coloured curves and represents the distribution of Φ for all cells. As shown in the inset, we see that intermediate area categories dominate the total number of cells, and are thus enough for the clustering procedure. Figure taken from [1]

$$P(\Phi) = \sum_A P(\Phi|A)P(A) \tag{9.5}$$

These figures give a 'fingerprint' of the planar graph (here the street network of a city) which encodes information about both the shape and the area of faces. We will then base a classification, not on $P(\Phi)$ but rather on the conditional probability $P(\Phi|A)$.

9.1.2 A Typology of Planar Graphs

We first construct bins for the surface areas of faces. Typically for street networks, the block area in the range $A \in [10^3, 10^5]$ (in square meters) dominate the total number of cells, and we can neglect very small blocks (of area $A < 10^3 \text{m}^2$) and very large ones (of area $A > 10^5 \text{m}^2$). We thus sort the blocks according to their area in two distinct bins

$$\alpha_1 = \{\text{cells}|A \in [10^3, 10^4]\}$$
$$\alpha_2 = \{\text{cells}|A \in [10^4, 10^5]\}$$

However, in general, we might have a very different area distribution and the binning should then be done accordingly. We denote by $f_\alpha(\Phi)$ the ratio of the number of faces with a form factor Φ that lie in the bin α over the total number of faces for that graph. We then define a distance d_α between two graphs G_a and G_b characterized by their respective $f_\alpha^{G_a}$ and $f_\alpha^{G_b}$ as

$$d_\alpha(G_a, G_b) = \int_0^1 \left| f_\alpha^{G_a}(\Phi) - f_\alpha^{G_b}(\Phi) \right|^n d\Phi \tag{9.6}$$

with $n = 1$ or $n = 2$ (other choices are also possible but this doesn't seem to be very relevant in the typology construction). We then construct a global distance D between the graphs G_a and G_b by combining all area bins α

$$D(G_a, G_b) = \sum_\alpha d_\alpha(G_a, G_b)^2 \tag{9.7}$$

We note here that from the mathematical point of view the distance between graphs already exists such as the graph edit distance (GED) [13] or the Gromov-Hausdorff distance [14]. These distances are however usually difficult to apply to large graphs and in general do not take into account the geometrical structure of the graph as encoded in the face area distribution.

At this point, we have a distance between two graphs and we can measure the distance matrix between various graphs that will allow us to use a classical hierarchical clustering algorithm [15] leading to a dendrogram displaying a number of groups of graphs with similar properties. This is what was done for 131 cities worldwide [1] and leads to the dendrogram shown in Fig. 9.3. At an intermediate level of this dendrogram we identify 4 distinct categories of street networks, which are easily interpretable in terms of the abundance of blocks with a given shape and with small or large area. On Fig. 9.4 we show the average distribution of Φ for each category and show typical street patterns associated with each of these groups. The main features of each group are the following.

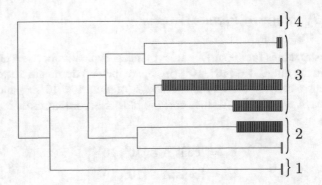

Fig. 9.3 Dendrogram for street networks. We represent the structure of the hierarchical clustering at a given level. Interestingly, 68% of american cities are present in the second largest sub-group of group 3 (fourth from the top of al groups). Also, all european cities but Athens are in the largest subgroup of the group 3 (third from top). This result gives a quantitative basis to the feeling that European and most American cities are laid out differently. Figure taken from [1]

- In the group 1 (comprising Buenos Aires, Argentina only) we essentially have blocks of medium size (in the bin α_2) with shapes that are dominated by the square shape and regular rectangles. Small areas (in bin α_1) are almost exclusively squares.
- Athens, Greece is a representative element of group 2, which comprises cities with a dominant fraction of small blocks with shapes broadly distributed.
- The group 3 (illustrated here by New Orleans, USA) is similar to the group 2 in terms of the diversity of shapes but is more balanced in terms of areas, with a slight predominance of medium size blocks.
- The group 4 which contains for this dataset the interesting example of Mogadishu, Somalia displays essentially small, square-shaped blocks, together with a small fraction of small rectangles.

The advantage of this method is that it is relatively simple and allows for an interpretation in terms of shape and area of faces. A drawback however is that correlations between faces are completely neglected. This is particularly important in the case of city street networks where there is usually a certain level of homogeneity at a certain scale (what we usually call a 'neighborhood'). In the case of New York City we can isolate the different Boroughs (the Bronx, Brooklyn, Manhattan, Queens and Staten Island), and extract their fingerprint shown on Fig. 9.5. The fingerprint of New York, NY (bottom Fig. 9.2) results therefore from the combination of different fingerprints for each of the boroughs. While Staten Island and the Bronx have very similar fingerprints, the others are different. Manhattan exhibits two sharp peaks at $\Phi \approx 0.3$ and $\Phi \approx 0.5$ which are the signature of a grid-like pattern with the predominance of two types of rectangles. Brooklyn and the Queens exhibit a sharp peak at different values of Φ, also the signature of grid-like patterns with different rectangles for basic shapes.

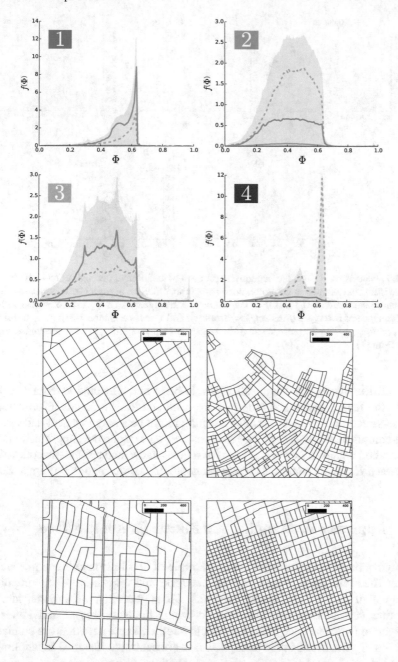

Fig. 9.4 Four groups of street networks. (Top) Average distribution of the shape factor Φ for each group found by the clustering algorithm (the color of each curve corresponds to the area bin from small to large: dashed green, orange, blue). (Bottom) Typical street pattern for each group (plotted at the same scale in order to observe differences both in shape and areas). Group 1 (top left): Buenos Aires | Group 2: Athens | Group 3: New Orleans | Group 4: Mogadishu. Figure taken from [1]

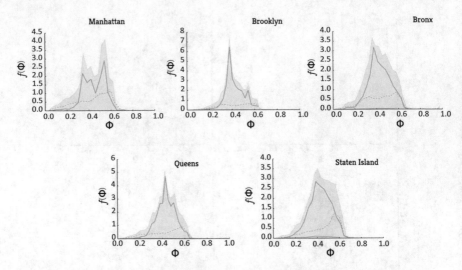

Fig. 9.5 New-York, NY and the fingerprints for each of its different boroughs. Only Staten Island and the Bronx have similar fingerprints and the others are different. In particular, Manhattan exhibits two sharp peaks at $\Phi \approx 0.3$ and $\Phi \approx 0.5$ which are the signature of a grid-like pattern with the predominance of two types of rectangles. Brooklyn and Queens exhibit a sharp peak at different values of Φ, signalling the presence of grid-like patterns made of different basic rectangles. Figure taken from [1]

A further step in this classification would be to find a method to extract regions where the faces features are strongly correlated and to include this information in the construction of classes. Machine learning might be a way to detect and classify these complicated correlations (see sections below). Despite the simplifications that this method entails, it however allows for a quantitative and systematic comparison between different planar graphs where the geometrical information is of importance.

9.2   Approximate Mapping of a Planar Graph to a Tree

Biological networks constitute a prime example of networks naturally embedded in space. The brain, insect wings, vascular networks and veins in leaves are important examples of such natural spatial networks. More generally, we can consider the important case of transportation networks where fluids, energy or individuals are transported from one node of the network to another. These networks are crucial for societies in general and in the biological case are central to the good functioning of living beings.

Interesting examples of such biological networks are founds in leaves where veins constitute planar graphs that often contain loops [16, 17]. However the organization of loops in leaves display variations (see the examples shown in Fig. 9.6) and an important question in botanics is how to compare the leaves of different plants.

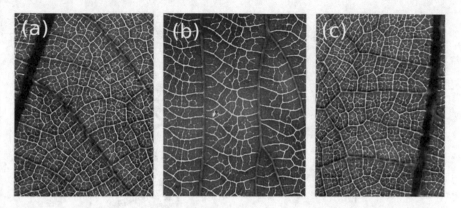

Fig. 9.6 Various patterns in leaves. **a, b** Leaf vasculature of two dicotyledonous species. **c** Detail of leaf collected from the same plant as leaf (**a**). The scale is 1 cm. Figure taken from [16]

Also comparing networks between them or with results of simulations implies the capacity to test the null hypothesis that two networks have been drawn from the same distribution. The veination pattern is a planar graph and the question thus amounts to the more general one of the comparison and the classification of planar graphs. In order to address this question, two groups [16, 17] proposed simultaneously an approximate method for mapping the architectural organization of a planar graph to a binary tree. Both these methods rely on the idea of deleting edges and to observe how the faces merge together. The method essentially consists in deleting successively edges and to represent the hierarchical structure of faces by a tree. It is then possible to use standard metrics developed for trees in order to characterize the original planar maps.

We will essentially follow the paper [16] and present their construction for this approximate mapping. We won't however discuss the structure of binary trees and the consequences for botanics and refer the interested reader to the original articles [16, 17]. The process is described in Fig. 9.7. The first step is to rank edges according to their weight (in the leaf case, the natural weight is the width of veins). In this planar map, each edge separates two faces and its deletion creates a new face. The idea is then to describe this process by a tree where the nodes represent the faces at a certain level of the decimation. The merging process of faces F_1 and F_2 at level ℓ is then represented by a new face $F_1 \cup F_2$ whose node is at level $\ell + 1$ (see Fig. 9.7).

This decimation will decompose in a hierarchical way the original graph and encode this organization in a binary tree—called 'nesting tree' by the authors of [16]. This tree contains some topological information about the original graph and allows the use of standard tools for characterizing trees. The nodes of the nesting tree at different levels that correspond to the neighborhoods of the original graph at different coarse-graining level. This simple decimation also allows us to distinguish different 'building blocks' that can help characterizing the loopy architecture of planar maps. For example, the authors of [16] identified a few types of such blocks. In

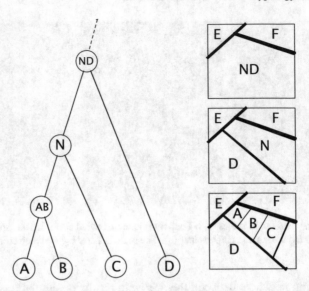

Fig. 9.7 Decimation process and construction of the 'nesting tree' (shown on a small part of a graph). The edges are ranked according to their width (or weight in general) and the faces are labeled. The smallest edge here separates faces A and B and its deletion implies the merging of these faces and the creation of a new face denoted by AB. This decimation is represented on the nesting tree by A and B on the same level and AB on the upper level. The next edge to delete is the one separating AB and C and its deletion gives rise to a new face $N = ABC$. The process is continued until we reach a tree and there is no face to merge anymore. Figure redrawn from [16]

particular, they distinguish blocks that correspond to an 'additive' nestedness where loops join consecutively, and 'multiplicative' nestedness where nested loop merge hierarchically. We also see here that this mapping to a nesting tree is not a bijection in general. Indeed as shown in Fig. 9.8 the two different graphs (top and bottom) leads to the same nested tree. Even if this is in principle a problem, it seems that the decimation however keeps important elements of the architectural organization of loops.

Different synthetic examples are shown in Fig. 9.9. Although it is very exploratory at this stage, we can guess from the different examples proposed in this figure, some specific features. In particular, the hierarchical structure of the nesting tree reflects some order in the relative organization of faces, while randomness creates links between nodes at a low level of the tree and higher order nodes. As another example but coming from the real-world, Katifori and Magnasco [16] could extract the arterial vasculature of the neocortex of the rat (see Fig. 9.10), which forms a planar graph with multiple loops and with veins of various diameters (that could also be measured), allowing them to obtain a weighted map of the arterial vasculature of the rat brain. Using this information, they could identify major vascular sections and determine from this dataset that the architecture of this network is essentially additive rather than

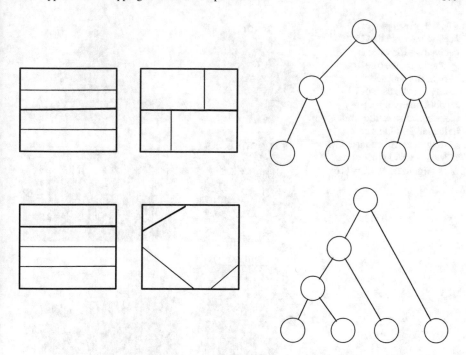

Fig. 9.8 Four building blocks for loops identified in [16] and their corresponding trees. On top, we have a 'multiplicative' nestedness while the bottom graph leads to 'additive' nestedness. Figure inspired from [16]

multiplicative (we refer the interested reader to [16] for more details and discussions about this case).

This approximate mapping to the nesting tree contains however no information about the geometry of the original planar map—it just tells us how the different cells are organized with respect to the width of edges. Katifori and Magnasco proposed more systematic investigations of the structure of the nesting tree [16]. In particular they proposed to characterize the asymmetry of the tree (and the Strahler index) and also the size distribution of the cells obtained by successive merging. More generally, this method proposed simultaneously by two groups [16, 17] has the advantage of being simple and to provide a quantitative characterization of the loop organization in planar maps. It is certainly a interesting direction for future research in this field. An important step would be achieved by generalizing this method to non-weighted networks, and to integrate naturally more geometrical information in the nesting tree.

Fig. 9.9 Examples of
nesting trees obtained by
decimation for various
synthetic graphs (the trees
have been truncated for
clarity). In the case of
random lines we keep a
hierarchical structure while
in the random links case,
nesting trees can connect low
order nodes to higher order
ones. Figure taken from [16]

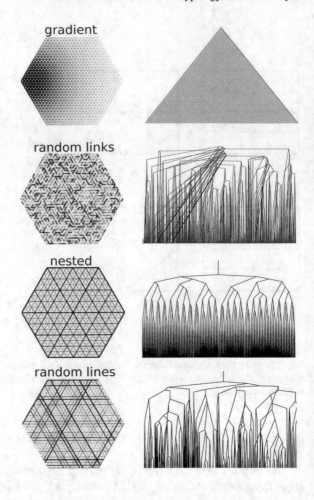

9.3   An Exact Bijection Between a Planar Graph and a Tree

In 2004, Bouttier, Di Francesco and Guitter [18] extented a bijection between a
planar graph and a tree originally proposed by Schaeffer in his Ph.D. thesis for
quadrangulations (i.e. planar maps with faces made of four edges) [19, 20]. This
bijection is in fact a process to encode a planar map in a simple way and even if
this method was originally developed mostly for combinatorial reasons, it has the
avantadge to be simple and could be useful for other, more practical purposes.

We will follow the construction[1] proposed in [18] and we won't provide all the
details and proofs (all details can be found in [18]). In particular, we will provide
some level of detail about the case of a planar map G whose faces have even degrees
(here the degree of a face is meant as the number of edges belonging to this face). The

---

[1] I thank Emmanuel Guitter for numerous discussions about this work.

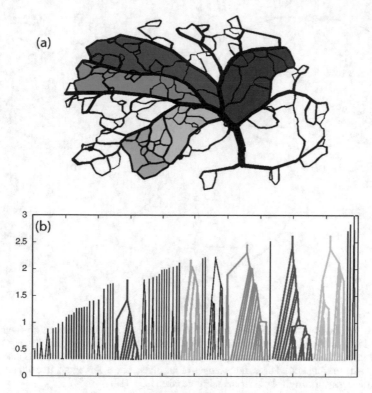

Fig. 9.10 a Digitalized arterial vasculature of the rat neocortex. The arterial network forms a planar graph and different segments of the network, as identified by hierarchical decomposition are represented by different colors. **b** Nesting tree of the corresponding graph (the highlighted segments of the network are color-coded). Figure taken from [16]

main ingredient for constructing the tree starting from a planar map is to choose an origin. The obtained tree will depend on the choice of this root and this is somehow a limitation of this method but further work is actually needed in order to understand the impact of this origin choice on the statistical structure of the tree (we note that many quantities won't depend on this origin choice which makes the method anyway very useful for computing various quantities).

Once an origin has been chosen, we can label all the nodes by a number that represents their shortest path distance to this origin (see Fig. 9.11 for an example). The next step is then to assign to each face a 'centroid' node that is represented by a black circle (see Fig. 9.11). We then choose a rotation order (for example clockwise) and for each face we connect a node to the centroid if its consecutive node (along the face and following the clockwise rotation) has a smaller label. We reproduce from [18] a figure (Fig. 9.12) which illustrates this process for a given face. Once we have explored all faces, we remove all edges of the original graph and we are left (see Fig. 9.11) with the original nodes (except the origin) together with the centroid nodes

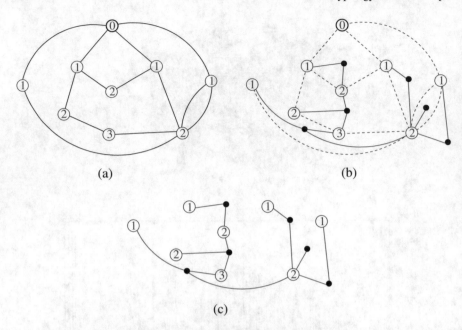

(a)   (b)

(c)

Fig. 9.11 Typical illustration of the construction of the tree. **a** A typical planar map with a node selected as the origin (0). All the vertices are then labelled by their shortest path distance to (0). **b** For each face, a node is added (in black) and the construction of link is performed as discussed in the tex. **c** After having erased the original edges of the graph, one is left with a tree which encodes in a univoque way the original planar map. Figure taken from [18]

which all form a graph denoted by T (with links connecting centroids to original nodes).

Bouttier, Di Francesco and Guitter demonstrated that this graph is a single connected tree (an example is shown in Fig. 9.11) and we outline here their interesting proof. The first thing they showed is that T does not contain any cycle. Indeed if we assume that it does contain a cycle, it separates the plane into two regions (see Fig. 9.12b), one which contains the origin and the other not—we call the latter the 'interior'. We then choose a labeled node (i.e. a node that is not a centroid) on this cycle and we denote its label by n. This node has two unlabeled neighbors (i.e. centroids) and we can examine the nodes that belong to the corresponding original faces of the planar map. In particular, by construction we must have two nodes labeled by $n - 1$ (see Fig. 9.12b) leading to the contradiction that there is a node labeled $n - 1$ in the interior. This argument shows that T has no cycle but it doesn't mean that it is a single connected tree and it could be a 'forest' (i.e. a collection of disconnected trees). Following [18], we denote by N, F, and E the numbers of vertices, faces and edges of the original graph G. The graph is planar and the Euler relation is satisfied (see Chap. 2)

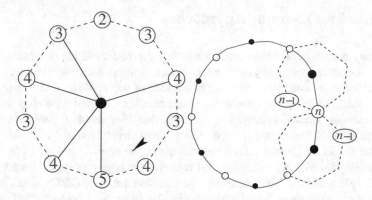

Fig. 9.12 a A face with nodes and their label that represent the distance to the origin. The black node represents the 'centroid' of the face and we explore this face by following the nodes clockwise here. If a node is followed by a node with label that is smaller, the rule is then to connect this first node to the centroid. **b** Sketch of the proof that T has no cycle (see text). Figures taken from [18]

$$N - E + F = 2 \tag{9.8}$$

Denoting by $N(T)$, $E(T)$, $F(T)$ the numbers of vertices, edges, and faces of T, these number satisfy the Euler relation for a planar graph with C disconnected components

$$N(T) - E(T) + F(T) = C + 1 \tag{9.9}$$

The total number of vertices of the tree T is $N(T) = N + F - 1$ (the original nodes minus the origin plus one centroid per face), the number of edges is $E(T) = N(T) - 1$, and the number of faces $F(T) = 1$. Plugging all this in the Euler formula then gives $C = 1$. This proves that the construction leads to a single tree, but it is not enough to prove that it is a bijection. We refer the interested reader to [18] for the demonstration that it is indeed a bijection and that there is a converse construction allowing to go from T to G in a unique way.

Compared with the approximate mappings discussed above, this construction has a certain number of avantages. First, it can be applied to any planar map and does not need weighted edges. Second, it is not an approximate mapping but an exact encoding of the planar map. It then gives access to the statistics of the local environment of a vertex (chosen then as the origin in this construction). Using generating functions, it is then possible to compute various things such as the average number of edges or nodes at a given distance [18]. However, the constructed tree depends on the particular origin that is chosen and further studies are needed to investigate this point and the possibility to apply this method to more applied questions and problems.

9.4 Machine Learning Approaches

We briefly discuss here some machine learning approaches to the problem of street
network—and more generally to the cities—classification. As it is oftern the case with
machine learning, one usual criticism is the lack of any control and understanding
of the output. However, we believe that these results combined with more standard
approaches developped in network studies could lead to a wealth of important results.

In the work of Xue et al. [21], the authors use a Graph Neural Networks (GNN)
to model the 11,790 urban road networks across 30 cities worldwide and use its
predictability in order to understand the inner-city and inter-city spatial homogeneity
of urban road networks. More precisely, the authors train these GNN models in order
to predict connections on street network samples from 30 cities in the US, Europe
and Asia. They address a 'link-prediction' problem which consists in predicting the
presence of an edge based on the structural roles of its endpoints. It is then expected
that the spatial homogeneity of the graph will be correlated to a good prediction
of edges. They use then a F1 score (a measure of accuracy in [0, 1] calculated as
the harmonic mean of the precision and recall of the test) to construct a measure of
the spatial homogeneity of the street network. These authors then assume that this
predictability capture many inherent properties of the road network. In order to check
this they first use a k-clustering algorithm based on 11 different measurements (such
as the proportion of nodes with various degrees, circuity, etc.) and found 4 different
families of street networks (Fig. 9.13a). Here, circuity is another word for the detour
index and dendricity represents the fraction of dead end edges. Typically, the first
type of street network has the largest average degree and the largest fraction of nodes
with degree $k \geq 4$. The type 2 is characterized by a large fraction of $k = 3$ nodes,
and the type 3 has many dead ends ($k = 1$ nodes). Finally, type-4 networks have the
largest value of circuity and fraction of bridges.

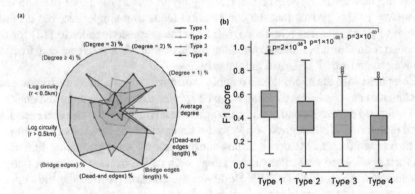

Fig. 9.13 **a** Coordinates of centroids for the four types of street networks found in [21]. **b** Boxplots
of F1 scores for the four types we obtained in **a**. Figures taken from [21]

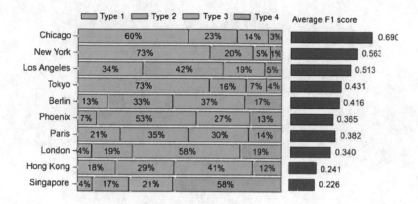

Fig. 9.14 Type compositions and the average F1 scores of street networks in ten representative cities. Figure taken from [21]

In order to test their assumption, they computed the F1 measure for each type of networks and the results are shown in Fig. 9.13b. We observe on this figure that the F1 score associated with type 1 networks is significantly larger than the other types, suggesting that grid like areas correspond indeed to more predictable patterns. This can be confirmed by calculating the composition of each city shown in Fig. 9.14.

As expected, the average F1 score of the street network of a city is positively correlated to the proportion of gridlike areas. Also, we observe here that is a negative correlation between the proportion of gridlike areas and the large detour index revealing their opposite effects on the F1 score.

This type of approach certainly represents an interesting direction for the future and more work is needed in order to reach a more general picture integrating classical network measures and classification and metrics obtained with machine learning techniques.

Another related work concerns the classification of cities [22] where the authors used a convolutional neural network combined with a graph-based approach on a dataset containing 1692 worldwide cities. They used sample maps for these cities and thus include all elements, not only streets but also the presence of green spaces, rivers, etc. In this respect, this work doesn't concern spatial networks *stricto sensu*, but their method could certainly be applied to this (simpler) case. The algorithm is trained over some samples and tested on the rest. When two samples of two different cities cannot be distinguished by the algorithm, the cities are connected by a link in the so-called confusion matrix. A link in this matrix thus indicates that shared features exist between maps from each location. This confusion matrix is shown in Fig. 9.15.

A modularity analysis is then applied on this network and leads to 9 different groups of cities (in this method the number of modules is not predetermined, see for example the review [23]). The difficulty at this point is how to understand these

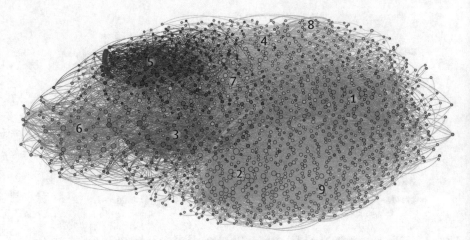

Fig. 9.15 Network graph of the model's confusion matrix with all 1632 cities. The different colors represent the city types identified (1 = informal, 2 = irregular, 3 = large block, 4 = cul de sac, 5 = high transit, 6 = motor city, 7 = chequerboard, 8 = intense, 9 = sparse). Figure from [22]

categories and what are the features that characterize them (in the study [22], the authors were mainly interested in correlating injuries with the street design). This is however a very interesting study that could serve as a meeting point halfway between machine learning results and more traditional network tools.

References

1. R. Louf, M. Barthelemy, J. R. Soc. Interface **11**(101), 20140924 (2014)
2. M. Southworth, E. Ben-Joseph, *Streets and the Shaping of Towns and Cities* (Island Press, 2003)
3. P. Haggett, R.J. Chorley, *Network Analysis in Geography*, vol. 67 (Edward Arnold London, 1969)
4. F. Xie, D.M. Levinson, *Evolving Transportation Networks*, vol. 1 (Springer, 2011)
5. Open street map project. http://www.openstreetmap.org
6. E. Strano, V. Nicosia, V. Latora, S. Porta, M. Barthelemy, Sci. Rep. **2** (2012)
7. M. Barthelemy, P. Bordin, H. Berestycki, M. Gribaudi, Sci. Rep. **3** (2013)
8. S. Porta, O. Romice, J.A. Maxwell, P. Russell, D. Baird, Urban Studies p. 0042098013519833 (2014)
9. S. Lämmer, B. Gehlsen, D. Helbing, Phys. A **363**(1), 89 (2006)
10. M. Fialkowski, A. Bitner, Landscape Ecol. **23**(9), 1013 (2008)
11. M. Barthelemy, Phys. Rep. **499**(1), 1 (2011)
12. P. Haggett, A.D. Cliff, A. Frey, Tijdschrift Voor Economische En Sociale Geografie **68**(6) (1977)
13. X. Gao, B. Xiao, D. Tao, X. Li, Pattern Anal. Appl. **13**(1), 113 (2010)
14. F. Mémoli, (2007)
15. L. Kaufman, P.J. Rousseeuw, *Finding Groups in Data: An Introduction to Cluster Analysis*, (The University of Chicago Press, New York, 1990)

16. E. Katifori, M.O. Magnasco, PloS one **7**(6), e37994 (2012)
17. Y. Mileyko, H. Edelsbrunner, C.A. Price, J.S. Weitz, PloS one **7**(6), e36715 (2012)
18. J. Bouttier, P. Di Francesco, E. Guitter, Electron. J. Combin. **11**(1), R69 (2004)
19. G. Schaeffer, Ph.D. Thesis (1999)
20. G. Chapuy, M. Marcus, G. Schaeffer, SIAM J. Discret. Math. **23**(3), 1587 (2009)
21. J. Xue, N. Jiang, S. Liang, Q. Pang, S.V. Ukkusuri, J. Ma (2021), arXiv:2101.00307
22. J. Thompson, M. Stevenson, J.S. Wijnands, K.A. Nice, G.D. Aschwanden, J. Silver, M. Nieuwenhuijsen, P. Rayner, R. Schofield, R. Hariharan et al., Lancet Planet Health **4**(1), e32 (2020)
23. S. Fortunato, Phys. Rep. **486**(3–5), 75 (2010)

Chapter 10
Measuring the Time Evolution of Spatial Networks

We discussed in the previous chapters how to characterize the structure of static spatial networks. In many instances however, these networks are evolving in time, growing and expanding in space. This is typically the case of transportation networks such as roads, subways and railways, but also for biological networks. It is therefore important to be able to characterize the evolution of these networks, and to detect crucial changes and distinguish them from ordinary growth. In this chapter, we will address such problems for the road networks and we will try to highlight the major differences between an 'organic' growth from systems that experiences major changes due to planning decisions. We will illustrate these two types of evolution on the example of the region of Groane (Italy) and the example of central Paris which experienced major large-scale planning operations during the 19th century (the 'Haussmann' period). In this latter case, usual network measures display a smooth behavior and the most important quantitative signatures of central planning is the spatial reorganization of centrality and the modification of the block shape distribution. Such effects can only be obtained by structural modifications at a large-scale level, with the creation of new roads not constrained by the existing geometry. The evolution of the street network of a city thus appears here as resulting from the superimposition of continuous, local growth processes and punctual changes operating at large spatial scales.

In a second part of this chapter, we will consider the evolution of the subway network and illustrate the difficulty of characterizing in a relevant way the time evolution of a spatial network. We will show that the use of a template simplifies this problem and reveals itself as a precious guide that allows us to avoid to get lost in the vast number of possible measures. In particular, it allows us to show that all subway networks in the world seem to converge when their size grows to a common shape described by the same features, despite their geographic and economic differences. This limiting shape is made of a core with branches radiating from it, and with similar numerical characteristics (average degree of a node within the core, fraction of nodes with degree equal to 2, number of branches, etc.).

© The Author(s), under exclusive license to Springer Nature Switzerland AG 2022   187
M. Barthelemy, *Spatial Networks*,
https://doi.org/10.1007/978-3-030-94106-2_10

10.1   Road Networks

Road networks play a crucial role in the development of urban areas and understanding the main mechanisms that govern the evolution of this network could shed some light on the urbanization process. Urbanization is one of the fundamental process in human history, and is increasingly affecting our environment and societies. Understanding it and how land use change under different circumstances, and what are the dominant mechanisms and, if any, the 'universal' features of such large-scale self-organized processes, is more important than ever as policy makers, professionals and the scientific community are actively looking for new paradigms in urban planning and land management (see [1] and references therein). Generally speaking, urbanization includes a complex set of physical modifications of the environment. In particular, the spatial organization of the transportation network plays a central role in the evolution of urban areas [2, 3], and is an important ingredient for the dynamical processes occurring on them [4–6]. Admittedly, the evolution of road and street networks constitutes only one facet of urbanization dynamics, but it is a relevant aspect of the temporal evolution of an urban area, and certainly provides a solid framework for quantitative studies of urbanization phenomena and land fragmentation [7, 8].

The evolution of street networks (and other transportation systems) was discussed some time ago, in particular by quantitative geographers [7], but new studies appeared with the availability of new datasets. In particular, the evolution of the street network in a region of Italy was studied in [9] with 7 points in time between 1833 and 2007. In another study, the authors of [10] considered the temporal evolution of the street network of central Paris between 1789 and 2010, notoriously reorganised on a large scale by Haussmann in the 19th century. Finally, the authors of [11] studied the evolution of the Greater London Area (GLA) street networks between 1786 and 2010, distinguishing the GLA network from the network of central London.

We won't discuss all details of these studies here, but we will highlight some results which are the most useful and illuminating about the processes that govern the growth of these spatial networks.

10.1.1   Organic Growth

The study [9] focused on the evolution, over two centuries, of the road network in a large area of $125\,km^2$, located north of Milan, Italy. In Fig. 10.1 we show the street networks obtained at seven different times from 1833 to 2007. These networks were extracted from historical topographic and photogrammetric maps imported into a Geographical Information System (GIS) environment. This area includes 29 urban centers within 14 municipalities that have developed along two main radial paths, connecting Milan to Como and Varese. The first path was constructed by the Romans during the II century B.C. while the latter was created during the 16th century. This area faced a process of urban transition from a polycentric region into a

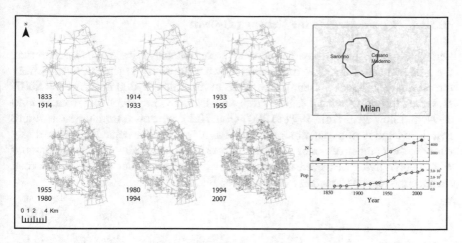

Fig. 10.1 Evolution of the road network from 1833 to 2007. On each map we report in grey all the nodes and links already existing at the previous time, while we indicate in color the new streets added in the time window under consideration. The bottom right panel reports, as a function of time, the total number of nodes N of the graph and the total population in the area obtained from census data. The map on the top-right panel shows the location of the area under study in the metropolitan region of Milan. Figure taken from [9]

so-called urban sprawled structure, a process that is common to many large European metropolitan regions.

This urban system is sampled at seven different points in time, namely at $t = 1833, 1914, 1933, 1955, 1980, 1994, 2007$, obtaining one snapshot of the street network for each of these seven years. For each snapshot, the corresponding primal graph is constructed, where the junctions are represented as nodes and the roads (or streets) are the links of the networks. We denote by $G_t \equiv G(V_t, E_t)$ the obtained graph at time t, where V_t and E_t are respectively the set of nodes and links at time t. The number of nodes at time t is then $N(t) = |V_t|$ while the number of links is $K(t) = |E_t|$. By definition, we have $V_t = V_{t-1} \cup \Delta V_t$ and $E_t = E_{t-1} \cup \Delta E_t$, where ΔV_t and ΔE_t are respectively the set of new nodes and new links added in the time window $]t - 1, t]$ to the network existing at time $t - 1$. We note that there are many technical difficulties here. In particular, matching two maps at different times can be a complex operation as maps are not always very faithful. Some road segments can be represented or not, can disappear or appear in reality and it is not always easy to understand the reality of the field from maps only. We won't however discuss all these details and refer the interested reader to specialized papers on the subject (see for example [12, 13] and references therein).

10.1.1.1 Characterizing the Network Growth

During the period under study. The network is continuously growing: in two centuries, the total number of nodes N in the road network grows by a factor of twenty, from the original 255 nodes present at $t = 1833$ to more than 5,000 nodes at $t = 2007$. However, this growth is not regular: it is slow from 1833 to 1933, fast from 1933 to 1994, and slow again from 1994 to 2007 (Fig. 10.1). We notice from the same figure that the evolution of population over time has a very similar behavior. Interestingly enough, the number of nodes N in the network is a linear function of the number P of people living in the area or, in other words, the number of people per road intersection remains constant over time (see Fig. 10.2a)

$$N \simeq aP \tag{10.1}$$

with $a \approx 0.02$.

In order to exclude complex 'exogenous' socio-economical factors, it seems natural to use the number of nodes N as the time clock. The number of links (here denoted by K) grows almost linearly with N (Fig. 10.2b, top), showing that the average degree is roughly constant despite massive historical changes, with a slight increase from $\langle k \rangle \simeq 2.57$ to 2.8 when going from 1914 to 1980 (Fig. 10.2b, bottom). Moreover, in Fig. 10.2c we observe that the total network length L_{tot} increases as N^γ where $\gamma \simeq 0.54$ and, accordingly, the average length of links decreases as $N^{\gamma-1}$. This result is consistent with the evolution of two-dimensional lattices with a peaked link length distribution [8, 14] which are described by a value $\gamma = 1/2$. Indeed if the surface area is A, the density of nodes is $\rho = N/A$ and the typical link length given by

$$\ell_1 \sim \frac{1}{\sqrt{\rho}} \tag{10.2}$$

which implies that the total length

$$L_{tot} \sim K\ell_1 \sim \sqrt{N} \tag{10.3}$$

Another interesting quantity is given by

$$r_N = \frac{N(1) + N(3)}{N} \tag{10.4}$$

where $N(k)$ is the number of nodes of degree k, and provides additional information on the structure of the network. It indicates the fraction of 'unfinished' intersections and measures the degree of evolution of the system. The plot of r_N versus N (Fig. 10.2d) shows that it steadily decreases from $r_N \simeq 0.87$ at $t = 1833$ to $r_N \simeq 0.835$ at $t = 2007$. In the inset, the relative abundance of four-ways crossings, i.e. N_4/N is shown. We notice a substantial increase from $N_4/N \simeq 11\%$ at $t = 1833$

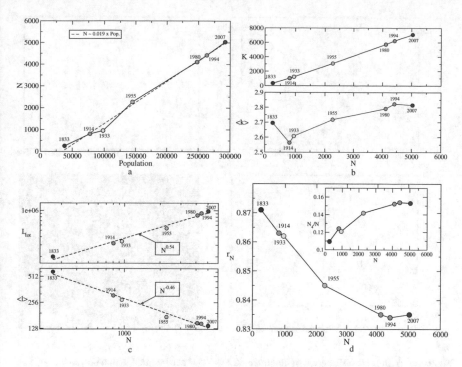

Fig. 10.2   a Number of nodes N versus total population (continuous line with circles) and its linear fit (dashed line). **b** The total number of edges K, and the average node degree $\langle k \rangle$ as a function of N. The number of edges increases slightly faster than linearly with the number of nodes, as shown by the slight increase in the average node degree. **c** Both the total network length L_{tot} (upper panel) and the average link length $\langle l \rangle$ (bottom panel) scale as power-law functions of N (the corresponding fits are reported as dashed lines). **d** As the network grows, the value of the ratio r_N between the number of nodes with degree $k = 1$ and $k = 3$, and the total number of nodes, decreases, indicating the presence of a higher number of 4-ways crossings. In the inset we report the percentage of nodes having degree $k = 4$ as a function of N. Notice that the relative abundance of four-ways crossings increases by 5% in two centuries. Figure taken from [9]

to $N_4/N \simeq 15.5\%$ at $t = 2007$. This trend is the signature of a historical transition from a pre-urban to an urban phase.

We now consider the blocks (or cells or faces) of the planar street network (see Chap. 4). The statistics on area and shapes of cells can be used to distinguish regular lattices from very heterogeneous patterns. As we saw in Chap. 4, the distribution $P(A)$ of the area of blocks is in general a power law of the form

$$P(A) \sim A^{-\tau} \tag{10.5}$$

where the exponent is $\tau \simeq 2.0$ [8, 15]. In Fig. 10.3a we show that the distribution of the cell areas at $t = 2007$ is a power law with the same exponent $\tau = 1.9 \pm 0.1$. As reported in the inset, the exponent however changes in time: it takes a value $\tau \simeq 1.2$

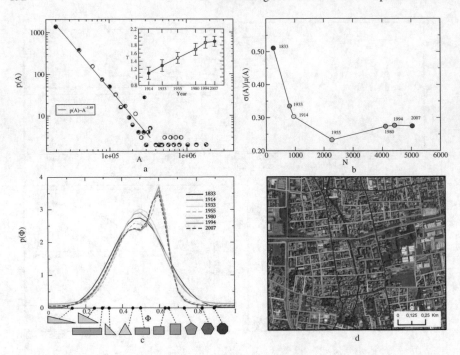

Fig. 10.3 **a** The size distribution of cell areas at $t = 2007$ can be fitted with a power-law $p(A) \sim A^{-\tau}$, with an exponent $\tau \simeq 1.9$. The values of τ increase over the years as shown in the inset. **b** Relative dispersion in the distribution of areas as a function of the network size N. **c** Distribution of cell shapes at each time, as quantified by the shape factor Φ. We also report, as a reference, the values of shape factors corresponding to various convex regular polygons. **d** The map shows some typical cell shapes at different times with the same color-code as in the previous panels. Figure taken from [9]

at $t = 1833$ and converges towards $\tau \simeq 1.9$ as the network grows. Because a larger exponent indicates a higher homogeneity of cell areas, we are thus witnessing here a process of homogenization of the size of cells. This appears to be a clear effect of increasing urbanization in time, with the fragmentation of larger cells of natural land into smaller urbanized ones. Accordingly, the relative dispersion of cell areas, shown in Fig. 10.3b, decreases from 0.5 at $t = 1833$ to 0.26 at $t = 2007$, indicating that the variance of the distribution becomes smaller as N increases.

As also discussed in Chap. 4, the diversity of the cells shape can be quantitatively characterized by the shape factor Φ, defined as the ratio between the area of the cell and the area of the circle circumscribed to the cell [7]. The value of the shape factor is in general higher for regular convex polygons, and tends to 1 when the number of sides in the polygons increases. The distributions $P(\Phi)$ reported in Fig. 10.3c clearly reveals the existence of two different regimes: for $t \le 1933$ the distribution is well approximated by a single Gaussian function with an average of about 0.5 and a standard deviation of 0.25. Conversely, for $t \ge 1955$ the distribution of shape factors displays two peaks and can be fitted by the sum of two Gaussian functions. The first peak coincides roughly with the one obtained for $t \leq 1933$, while the second peak, centered at 0.62, signals the appearance for $t \geq 1955$ of an important fraction of regular shapes such as rectangles with sides of similar lengths. In Fig. 10.3d we show some examples of the cell shapes at different times.

10.1.1.2 Betweenness Impact and Elementary Processes of Urbanization

A road network grows by the addition of new streets (edges) and new junctions (nodes). We focus here on the properties of these new links by looking at their length and centrality. In Fig. 10.4 we show the cumulative distribution of the length of new links according to the time-section in which they appeared first. The inset shows that the average length of new links steadily decreases over time. More precisely, let us consider at each time t the length value $\ell_{90\%}(t)$ such that 90% of new links at time t are shorter than $\ell_{90\%}(t)$ (i.e. such that $P(\ell \leq \ell_{90\%}) = 0.9$). We notice that $\ell_{90\%}$ decreases in the period 1833–1933 from 625 m down to 325 m, while no sensible variation is observed from 1933 to 1994, even if the network keeps growing. In the last period, i.e. from 1994 to 2007, we observe another decrease of $\ell_{90\%}(t)$ from 325 to 225 m. The relative dispersion of the length of the new links is almost constant and of order one, and the distribution does not vary too much after 1955.

In order to evaluate the impact of a new link on the overall distribution of BC in the graph at time t, we first compute the average betweenness centrality (see Chap. 5) of all the links of G_t as

$$\bar{b}(G_t) = \frac{1}{(N(t) - 1)(N(t) - 2)} \sum_{e \in E_t} b(e) \qquad (10.6)$$

Fig. 10.4 Region of Groane case. The cumulative distributions of the length of links added at different times indicate that the typical length of the links decreases with time. In the inset we report the average length of new links. Figure taken from [9]

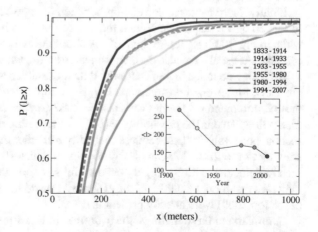

where $b(e)$ is the betweenness centrality of the edge e in the graph G_t. Then, for each link $e^* \in \Delta E_t$, i.e. for each newly added link in the time window $]t - 1, t]$ we consider the new graph obtained by removing the link e^* from G_t and we denote this graph by $G_t \setminus \{e^*\}$. We compute again the average edge betweenness centrality, this time for the graph $G_t \setminus \{e^*\}$. Finally, the betweenness centrality impact $\delta_b(e^*)$ of edge e^* at time t is defined as

$$\delta_b(e^*) = \frac{\overline{b}(G_t) - \overline{b}(G_t \setminus \{e^*\})}{\overline{b}(G_t)} \tag{10.7}$$

i.e. as the relative variation of the graph average betweenness due to the removal of the link e^*. We will thus quantity the relative impact $\delta_b(e)$ of each new link e on the overall BC of the network at time t. Remarkably, in the empirical case studied here, the distribution of this quantity $\delta_b(e)$ displays two well-separated peaks (Fig. 10.5), with the first peak tending to increase in time while the second peak decreases, until they merge into only one peak in the last time-section (1994–2007). In order to understand this remarkable dynamics and the nature of these two peaks, we focus on the geographical location of new links according to their impact on BC. Basically, we observe that links corresponding to the first peak (small δ_b) tend to bridge already existing streets while the links corresponding to the second peak (large δ_b), usually connect existing edges to new nodes. The distribution of BC impact thus suggests that the evolution of the road network is essentially characterized by two distinct, concurrent processes: one of 'densification' (first peak, lower impact on centrality) which corresponds to an increase of local density of the urban texture, and one of 'exploration' (second peak, higher impact on centrality) which corresponds to the expansion of the network towards previously non-urbanized areas. Obviously, since the amount of available land decreases over time, at earlier time–sections (such as in 1833) the fraction of exploration is higher, while in the 80s it becomes smaller until it completely disappears in 2007.

Finally, there is strong correlation between the age of a street and its centrality. In Fig. 10.6a we display, in colors, the age of the links in the network at $t = 2007$, while in Fig. 10.6b we report their BC. A simple visual inspection shows that highly central links usually are also the oldest ones. In particular, the links constructed before $t = 1833$ have a much higher centrality than those added at later time–sections. More precisely, the seven curves in panel Fig. 10.6c report the cumulative distribution of the BC computed on the network at $t = 2007$ for the links added at the different time points. That is the probability $P(b \leq x)$ that a link, appearing at a certain time-section, has a value of betweenness centrality b smaller than or equal to x in the final network at $t = 2007$. We can clearly see that the historical structure of oldest links mostly coincides with the highly central links at $t = 2007$. The figure in the inset indicates that more than 90% of the 100 most central links in 2007 (and almost 60% of the top 1000) were already present in 1833.

Summarizing these findings, the evolution of this street network is characterized by

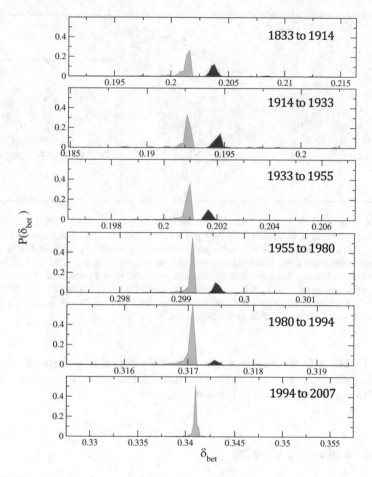

Fig. 10.5 Probability distribution of the BC impact $\delta_b(e)$ for the different time snapshots. The red peak corresponds to exploration, and the green peak to densification. Notice that the red peak becomes smaller and smaller with time, and completely disappears in the last snapshot. Figure taken from [9]

- An increase of the fraction of intersections of degree four.
- Important structural changes can be identified with the shape factor.
- There are two elementary processes through which urbanization fills out space. The first process of 'densification' corresponds to an increase in the local density of roads around the main existing central points and directions. The second process of 'exploration' consists in new roads triggering the spatial evolution of the urbanization front. These two processes are clearly identified by the betweenness centrality impact of new links. The quantitative identification of such simple elementary mechanisms suggests the existence of universal properties of urbanization, and opens up new possibilities to conceive its modelling.

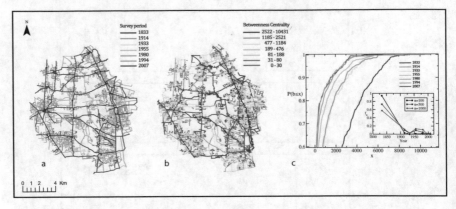

Fig. 10.6 The maps show **a** the time of creation of each link and **b** the spatial distribution of betweenness centrality (BC) at 2007. On the right panel we show the cumulative distribution of BC of links added at different times. We notice a correlation between the age of links and their BC. **c** In particular, the edges constructed before $t = 1833$ have a much higher BC than the edges added at later times, indicating that the oldest streets tend to remain central over time. Accordingly, the value of BC seems to be a good predictor for the age of an edge. The inset reports the percentage of edges added at a certain time step which are ranked in the top n positions according to BC. Different curves correspond to $n = 100, 500, 1000$. Figure taken from [9]

- The evolution of the road network relies on a set of central points which are stable throughout time, and which constitute the backbone of the urban structure, confirming the importance of historical paths.

10.1.2   Effect of Planning

The existence of central planning is often invoked as a counter-argument to the possibility of understanding the growth of cities as the result of self-organized processes. From a very general perspective, the large number and the diversity of agents operating simultaneously in a city suggest the intriguing possibility that cities are an emergent phenomenon ruled by self-organization [1, 3]. On the other hand, the existence of central planning interventions might minimize the importance of self-organization in the course of evolution of cities. Central planning—here understood as a top-down process controlled by a central authority—plays an important role in the city, leaving long standing traces, even if the time horizon of planners is limited and much smaller than the age of the city. One is thus confronted with the question of the possiblity of modelling a city and its expansion as a self-organized phenomenon. Indeed central planning could be thought of as an external perturbation, as if it were foreign to the self-organized development of a city.

    In order to bring some elements to this discussion, we consider here the evolution of the street network of Paris over more than 200 years with a particular focus on

Fig. 10.7 Illustration on a small area of the impact of Haussmann's transformations. On the yellow background, we show the parcel distribution before Haussmann (extracted from the Vasserot cadastre, 1808–1836), and in brown we show the new buildings delineating the new streets as designed by Haussmann and as they appeared in 1888. We can see on this example that the Haussmann plan implied a large number of destruction and rebuilding: approximately 28,000 houses were destroyed and 100,000 were built [16]. Figure taken from [10]

the 19th century, period when Paris experienced large transformations under the guidance of Baron Haussmann [10]. It would be difficult to describe in a few lines the social, political, and urbanistic importance and impact of Haussmann works and we refer the interested reader to the existing abundant literature on the subject (see for example [16] or the references cited in [10]). Essentially, until the middle of the 19th century, central Paris has a medieval structure composed of many small and crowded streets, creating congestion and, according to some contemporaries, probably health problems. In 1852, Napoleon III commissioned Haussmann to modernize Paris by building safer streets, large avenues connected to the new train stations, central or symbolic squares (such as the famous place de l'Etoile, place de la Nation and place du Panthéon), improving the traffic flow and, last but not least, the circulation of army troops. Haussmann also built modern housing with uniform building heights, new water supply and sewer systems, new bridges, etc. (see Fig. 10.7 where we show on a small area how dramatic the impact of Haussmann transformations are).

   The case of Paris under Haussmann provides thus an interesting example where changes due to central planning are very important and where a naive modelling is a priori bound to fail. By digitilizing historical maps into a Geographical Information

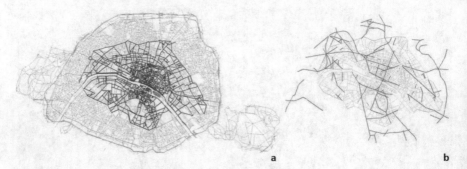

Fig. 10.8 **a** Map of central Paris in 1789 superimposed on the map of current 2010 Paris. In the whole study, we focus on the Haussmann modifications and limited ourselves to the 1789 portion of the street network. **b** Map of Haussmann modifications. The grey lines represent the road network in 1836, the green lines represent the Haussmann modifications which are basically all contained in the 1789 area. Figure taken from [10]

System (GIS) environment, the detailed road system (including minor streets) and the corresponding primal graph can be constructed at six different moments in time, $t = 1, 2, \ldots, 6$, respectively corresponding to years: 1789, 1826, 1836, 1888, 1999, 2010. It is important to note here that there are snapshots before Haussmann works (1789–1836) and after (1888–2010) which allows to study quantitatively the effect of such central planning.

As above, we denote by $G_t \equiv G(V_t, E_t)$ the obtained primal graph at time t, where V_t and E_t are respectively the set of nodes and links at time t. The number of nodes at time t is then $N(t) = |V_t|$ and the number of links is $E(t) = |E_t|$. In Fig. 10.8a, we display the map of Paris as it was in 1789 on top of the current map (2010). In order to use a single basis for comparison, this study is limited to the central portion of Paris. We then have 6 maps for different times and for the same area (of order 34 km^2). We also represent on Fig. 10.8b, the new streets created during the Haussmann period which covers roughly the second half of the 19th century. Even if we observe some evolution outside of this portion, most of the Haussmann works are comprised within this portion.

10.1.2.1   Simple Measures

In Fig. 10.9, we show the evolution of the number of nodes and of the population of Paris (for the 12 districts delimited by the 'fermiers generaux' for the period 1789–1851 and after for the 20th districts of Paris). The area under consideration for the calculation of the population is not exactly the same, and only the order of magnitude can be trusted here. We can compute the number of nodes N versus the population P and as in the Groane case discussed above, we observe a linear dependence

$$N \simeq aP \tag{10.8}$$

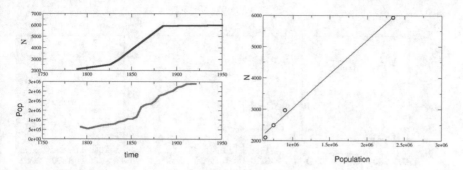

Fig. 10.9 Left (top panel): Evolution of the number of nodes versus time for Paris. Left (bottom panel) Evolution of the Paris population. Right panel: Number of nodes versus population. The line is a linear fit ($r^2 > 0.99$). Figure taken from [10]

with coefficient $a = dN/dP = 0.0021$ (we note here that this prefactor is 10 times smaller than the one observed for Groane, cf. Eq. (10.1)). It is thus clear that the number of nodes follows the demographic population and that the large increase observed during the Haussmann period is largely due to the demographic pressure.

Basic measures include the evolution of the number of nodes N, edges E, and total length L_{tot} of the networks (restricted to the area corresponding to 1789). In Fig. 10.10a–c we show the results for these indicators which display a clear acceleration during the Haussmann period (1836–1888). As discussed above, in order to exclude exogeneous effects and focus on the structure of networks, we plot the various indicators such as the number of edges and the total length versus the number of nodes taken as a time clock. The results shown Fig. 10.10d–f display a smoother behavior. In particular, E is a linear function of N, demonstrating that the average degree is essentially constant $\langle k \rangle \approx 3.0$ since 1789. The total length versus N also displays a smooth behavior consistent with a perturbed lattice [8] and scaling as $L_{tot} \sim \sqrt{N}$. A square root fit is shown in Fig. 10.10e and the value of the prefactor leads to an estimate of the area of the order $A \simeq 29.7\,\text{km}^2$, in agreement with the actual value $A = 33.6\,\text{km}^2$ (for the 1789 portion). This agreement demonstrates that all the networks at different times are not far from a planar graph with a peaked distribution of link length.

We also show the average route distance d_R defined as the average over all pairs of nodes of the shortest route between them. For a two dimensional spatial network, we expect this quantity to scale as $d_R \sim \sqrt{N}$ (see Chap. 4) and thus increases with N. The ratio $d_R/\sqrt{N}$ is thus better suited to measure the efficiency of the network and we observe (Fig. 10.10c, f) that it decreases with time and N. This result simply demonstrate that if we neglect delays at junctions, it becomes easier to navigate in the network as it gets denser.

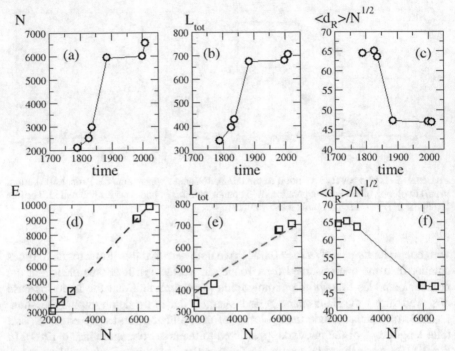

Fig. 10.10 Top panels: Number of **a** nodes, **b** total length (kms), and **c** rescaled average route distance versus time. Bottom panels: Number of **d** edges, **e** total length (kms), and **f** the rescaled average route distance versus the number of nodes N. In **d** the dashed (blue) line is a linear fit with slope 1.55 ($r^2 = 0.99$) consistent with constant average degree of order $\langle k \rangle \approx 3$, and in **e** the dashed (green) line a square root fit of the form $a\sqrt{N}$ with $a = 8.44$ kms ($r^2 = 0.99$). Based on a perturbed lattice picture this gives an area equal to $A \simeq 29.7$ km^2 consistent with the measured value ($A = 33.6$ km^2). In **f**, we show the rescaled average shortest route versus N which decreases showing that the denser the network and the easier it is to navigate from one node to the other (if delays at junctions are neglected). Figure taken from [10]

10.1.2.2   Typology of New Links

We can have three different types of new links depending on the number of new nodes they connect and we denote by E_i ($i = 0, 1, 2, 3$) the number of new links appearing at time $t + 1$ and connecting i new nodes. For example E_0 counts the new links appearing at time $t + 1$ connecting two nodes existing at time t. In the Paris case, we observe that in the first period, the majority of new links are of the E_2 type and correspond to construction of new streets with new nodes. We see that the Haussmann transition period (1836–1888) is not particularly different from the other previous periods. In the modern period (after 1999), E_0 becomes dominant and consistent with the idea of a mature street network where densification dominates the evolution of the urban tissue. Obviously, this is also an effect of limiting ourselves to the central Paris portion: in a wider area, many new roads were created and both

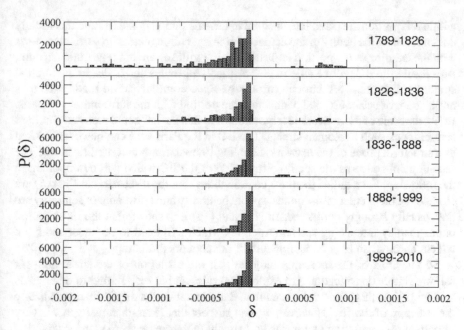

Fig. 10.11 Betweenness centrality impact distribution for the periods $1789 \to 1826$, $1826 \to$ 1836, $1836 \to 1888$, $1888 \to 1999$, $1999 \to 2010$. This figure shows that densification is the main process for this portion of Paris and that from this point of view, the Haussmann period seems to be rather smooth and comparable to other periods. Figure taken from [10]

densification and exploration coexist. We note here that the structure of the street network of central Paris remained remarkably stable from 1888 until now (and in this period also, densification was the main process in this area).

In order to categorize more precisely these new links, we use the betweenness centrality impact δ defined above Eq. (10.7). The distribution of the BC impact is shown on Fig. 10.11. These figures show that for all periods most new links belong to the densification process with a small peak of exploration in the period 1836–1888. In well-developed, mature systems, it is expected that densification is the dominant growth mechanism. Here also, we see that the Haussmann period is not significantly different from previous periods (from this point of view).

10.1.2.3 Evolution of the Spatial Distribution of Centrality

The betweenness centrality (BC) $g(i)$ of a node i is defined in Chap. 5, and essentially measures the fraction of times a given node is used in the shortest paths connecting any pair of nodes in the network, and is thus a measure of the contribution of a link in the organisation of flows in the network [17].

In the case where we consider a limited portion of a street network, two important effects need to be taken into consideration. First, as we consider a portion, only paths within this portion are taken into account in the calculation of the BC and this usually does not reflect the reality of the actual origin-destination matrix. In particular, flows with the exterior of the portion and surrounding villages are not taken into account. As a result, the BC will be able to detect important routes and nodes in the internal structure of the network but can miss large-scale communication roads such as a north-south or east-west road connecting the portion with the surroundings of Paris. In [9], the scale of the network is large enough so that the BC could recover important central roads such as Roman streets. The BC in the Paris case can however be used as a structural probe of the network, enabling to track important modifications. The second point concerns the spatial distribution of the BC which will be important in the following. For a lattice the most central nodes (see the discussion in Chap. 5) are close to the barycenter of the nodes: spatial centrality and betweenness centrality are then usually strongly correlated. In [15] and [18] it is shown that the most central points display interesting spatial structures which still need to be understood, but which represent an important signature of the networks' topology.

So far, most of the measures indicate that the evolution of the street network follows simple densification and exploration rules and is very similar to other areas studied [9]. At this point, it appears that Haussmann works didn't change radically the structure of the city. However, we can suspect that Haussmann's impact is very important on congestion and traffic and should therefore be seen on the spatial distribution of centrality. In the Fig. 10.12, we show the maps of Paris at different times and we indicate the most central nodes such that their centrality $g(i)$ is larger than max g/α with $\alpha = 10$ (similar results are obtained for other values of α). We can clearly see here that the spatial distribution of the BC is not stable, displays large variations, and is not uniformly distributed over the Paris area (we represented here the node centrality, and similar results are obtained for the edge centrality). In particular, we see that between 1836 and 1888, the Haussmann works had a dramatical impact on the spatial structure of the centrality, especially near the heart of Paris. Central roads usually persist in time [9], but in this case, the Haussmann reorganization was acting precisely at this level by redistributing the shortest paths which had certainly an impact on congestion inside the city. After Haussmann we observe a large stability of the network until nowadays.

In order to analyse the spatial redistribution effect more quantitatively, various quantities inside a disk of radius r centered on the barycenter of all nodes (which stays approximately at the same location in time) can be computed. We first study the number of nodes $N(r)$ (Fig. 10.13), its variation $\delta N(r)$ between t and $t + 1$, and the number of central nodes (such that $g(i) > $ max $g/10$). We see that the largest variation of the number of nodes (see 10.13b) is indeed in the Haussmann period 1836–1888, especially for distance $r > 1, 500$ m. Maybe more interesting, is the variation of the most central nodes (Fig. 10.13d). In particular, we observe that during the pre-Haussmann period, even if in the period 1789–1826 there was an improvement of centrality concentration, there is an accumulation of central nodes both at short distances ($r < 2, 500$ m) and at long distances ($r > 2, 500$ m) in the following period (1826–1836). As a result, visually clear in Fig. 10.12, there is a large concentration of centrality in the center of Paris until 1836 at least. The natural consequence of this

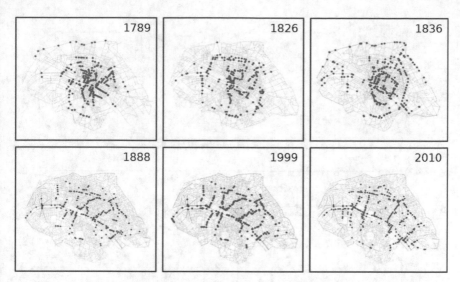

Fig. 10.12 Spatial distribution of the most central nodes (with centrality $g(i)$ such that $g(i) > \max g/10$). We observe for the different periods important reorganizations of the spatial distribution of centrality, corresponding to different specific interventions. In particular, we observe a very important redistribution of centrality during the Haussmann period with the appearance of a reticulated structure on the 1888 map. Figure taken from [10]

concentration is that the center of Paris was probably very congested at that time. In this respect, what happens under the Haussmann supervision is natural as he acted directly on the spatial organization of centrality. We see indeed that in 1888, the most central nodes form a more reticulated structure excluding concentration of centrality. A structure which remained stable until now. Interestingly, we note that Haussmann's new roads and avenues represent in this area approximately 6% of the total length only (compared to nowadays network), which is a small fraction, considered that it has a very important impact on the centrality spatial organization. Finally, we note that these results were confirmed in [19].

10.1.2.4 Evolution of the Shape Factor

This reorganization of centrality was undertaken with the creation of new roads and avenues destroying parts of the original pattern (see Figs. 10.7 and 10.8b) resulting in the modification of the geometrical structure of blocks. The effect of Haussmann modifications on the geometrical structure of blocks can be quantitatively measured by the distribution of the shape factor Φ (see Eq. (9.3) in Chap. 9) shown in Fig. 10.14. We see that before the Haussmann modifications, the distribution of Φ is stable and is essentially centered around $\Phi = 0.5$ which corresponds to rectangles. From 1888, the distribution is however much flatter showing a larger diversity of shapes. In

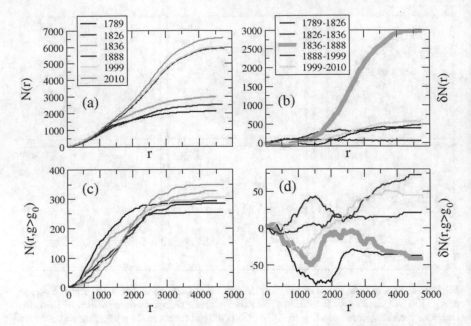

Fig. 10.13 Top panels: **a** number of nodes in a disk of radius r measured from the barycenter of Paris and **b** its variation versus r. As expected the largest variation occurred during the Haussmann period. **c** Number of nodes at distance r and with centrality larger than g_0 ($g_0 = \max g/10$) and **d** its variation. The thick green line in the right panels indicate the Haussmann transition 1836–1888. We see here that during the Haussmann period (and also in the 1789–1826 period), there is a large decrease of the number of central nodes in the central region of Paris ($r < 2,000\,$m). Figure taken from [10]

particular, we see that for small values of $\Phi < 0.25$ there is an important increase of $P(\Phi)$ demonstrating an abundance of elongated shapes (triangles and rectangles mostly) created by Haussmann's works. These effects can be confirmed by observing the angle distribution of roads shown on Fig. 10.15 where we represent on a polar plot $r(\theta) = P(\theta)$ with $P(\theta)$ the probability that a road segment makes an angle θ with the horizontal line. Before Haussmann's modifications, the distribution has two clear peaks corresponding to perpendicular streets and in 1888 we indeed observe a more uniform distribution with a large proportion of various angles such as diagonals.

10.1.3   Simplicity Measures

In this section, we use the simplicity measure discussed in Chap. 7 to the time evolution of road networks. More precisely, we will consider the time evolution the road networks discussed so far (the Groane network and Paris), and for comparison

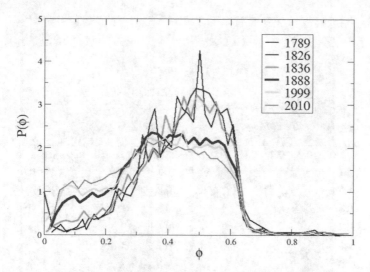

Fig. 10.14 Probability distribution of the shape factor Φ for the blocks at different years. Until 1836, this distribution is stable and we observe a dramatical change during the Haussmann period with a larger abundance of blocks with small values of Φ. These small values correspond to elongated rectangle or triangles created by streets crossing the existing geometry at various angles. Figure taken from [10]

also for an example of a simple biological network, the slime mould network (see Fig. 10.16).

We first consider the evolution of the Groane region network, which is a good example of an 'organic' evolution of urban systems. The simplicity profile shown in Fig. 10.16a allows us to distinguish two different periods. The first period from 1833 to 1955 displays a relatively small simplicity at all scales, while a distinct second regime appears from 1980 until now. In this latter regime, the simplicity profile is substantially larger for all scales. This is an effect of the massive urban densification, leading to a polycentric structure where the readability and the ease to navigate are drastically lowered.

At a smaller scale, we have the evolution of central Paris between 1789 and 1999, which, as discussed above, provides an opportunity to observe quantitatively the effect of top-down planning: until 1836, we are in the pre-Haussmann Paris, while from 1888 until now we are in the post-Haussmann period. The effect of Haussmann's central planning is clearly visible on the network shown in Fig. 10.16b. From 1789 to 1836, we have a relatively large simplicity at all scales and we observe a decrease in that period at small scales ($d/d_{max} < 0.4$) which corresponds well to the fact that many religious and aristocratic domains and properties were sold and divided in order to create new houses and new roads, improving congestion inside Paris. The 1826–1836 transition displays a decrease of the simplicity for distance larger than roughly 5 kms (corresponding to $d/d_{max} \approx 0.6$) indicating that long distance routes were simplified. It is interesting to note that during this period the eastern part

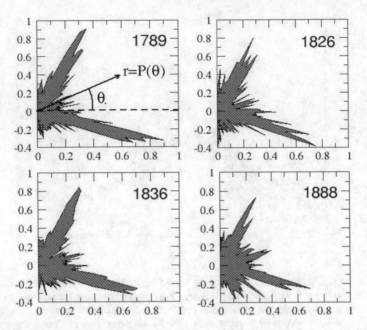

Fig. 10.15 Radial representation of the angle distribution of road segments for 1789, 1826, 1836, 1888. The radial distance r in this plot represents the probability to observe a street with angle θ: $r = P(\theta)$ with $\theta \in [-\pi/2, \pi/2]$ and $P(\theta)$ is the probability to observe an oriented road with angle θ with the horizontal line (see first panel, top left). Until 1836, the distribution is peaked around two values separated by approximately 90° and in 1888, we observe an important fraction of diagonals and other lines at intermediate angles. Figure taken from [10]

of Paris experienced large transformations with the construction of the channel St. Martin. Finally in the period 1836 to 1888, when Paris experienced Haussmann's transformation, the simplicity profile is strongly affected: compared to 1836, the simplicity is improved in the range $d/d_{max} \in [0.3, 0.8]$, which can be attributed to the construction of large avenues connecting important nodes of the city. In addition, we observe the surprising effect that at large scales $d/d_{max} > 0.8$, the simplicity is degraded by Haussmann's work: this however could be an artifact of the method and the fact that we considered a portion of Paris only and neglected the effect of surroundings. Finally, we note that differences between Groane and Paris might be explained in terms of a sparse, polycentric urban settlement (Groane) versus a dense one (Paris). In particular, in the 'urban' phase for Groane (after 1955), the simplicity profile becomes similar to the one of a dense urban area such as Paris.

Finally, we also show the results in Fig. 10.16c for the *Physarum Policephalum*, a biological system evolving at the centimeter scale. *Physarum* is a unicellular multinucleated amoeboid that during its vegetative state takes a complex shape. Its plasmodium viscous body whose goal is to find and connect to food sources, crystallizes in a planar network-like structure of micro-tubes (see for example [21] and references therein). In simple terms, Physarum's foraging strategy can be summarized

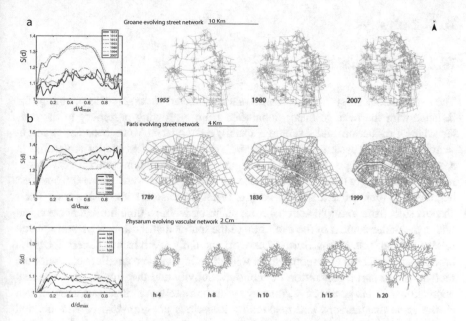

Fig. 10.16 Simplicity profiles for time-varying networks. We represent here the profiles for **a** the road network of the Groane region (Italy), **b** the street network of Paris (France) in the pre-Haussmannian (1789, 1836) and post-Haussmannian (1999) periods, and in **c** the *Physarum* network growing on a period of one day approximately. We observe on **a** and **b** that the evolution of the profile is able to reveal important structural changes. In **c** the evolution follows closely the one obtained with the null model. Figure taken from [20]

in two phases: (i) the exploration phase in which it grows and reacts to the environment and (ii) the crystallization phase in which it connects to food sources with micro-tubes. Active plasmodium was inoculated over a single food source and the micro-tube network was observed at six phases of its growth. Under these conditions, we observe that the network is statistically isotropic around the food source as shown in Fig. 10.16c and develops essentially radially. The simplicity profile for the Physarum is relatively low (less than ~ 1.2), suggesting that simplicity could be an important factor in the evolution of this organism. A closer observation shows that during its evolution, the Physarum adds new links to the previous network and also modifies the network on a larger scale, as revealed by the changes of the simplicity profile. The evolution of the profile is similar to the one obtained for the null model when the density is increased (this null model is obtained by adding random lines with length distributed according to $P(\ell) \sim \ell^{-\alpha}$, to a Voronoi tessellation, see [20] for details), suggesting that the statistics of straight lines in this case could be described as essentially resulting from the random addition of straight lines of random lengths with the value $\alpha = 2$.

10.2   Subways

10.2.1   Generalities

The early history of subways is sometimes connected to large-scale planning, for instance with the need to bring population from a growing periphery to the center where production and exchange usually take place. More broadly, it might seem that subway systems are engineered systems and intentionally structured in a core/periphery shape with their self-organization thus playing only a very minor role. This actually would be true if these subway systems were planned from their beginning to their current shape, but this is not the case for most networks. Their shape results from multiple actions, from planning within a time limited horizon, set within the wider context of the evolution of the spatial distribution of population and related economic activities. Subway networks actually result from a superimposition of many actions, both at a central level with planning and at a smaller scale with the reorganization and regeneration of economic activity and the growth of residential populations. In this perspective, subway systems are self-organizing systems, driven by the same mechanisms and responding to various geographical constraints and historical paths. This self-organized view leads to the idea that—beside local peculiarities due to the history and topography of the particular system—the topology of world subway networks should display general, universal features, within the limits of the physical geometry and cultural context in which their growth takes place.

For some cities, subway systems have existed for more than a century. Fascination with the apparent diversity of their structure has led to many studies and to particular abstractions of their representation (see Fig. 10.17 for some examples) in the design of idealized transit maps [22], and although these might appear to be planned in some centralized manner, we will see that subway systems like many other features of city systems evolve and self-organize themselves as the product of a stream of rational but usually uncoordinated decisions taking place through time.

Static properties of transportation networks have been studied for many years [7] and in particular simple connectivity properties were studied in [23] while fractal aspects were considered in [24]. With the recent availability of new data, studies of transportation systems have accelerated [8] and this is particularly so for subway systems [25–34]. These studies have revealed some significant similarities between different networks, despite differences in their historical development and in the cultures and economies in which they have been developed. In particular, their average shortest path seems to scale with the square root of the number of stations and the average clustering coefficient is large, consistent with general results associated with two-dimensional spatial networks (see Chap. 4). In [27], a strong correlation between the number of stations (for bus and tramway systems) and population size has been observed for 22 Polish cities, but such correlation are not observed at the world level (for all public transportation modes [31]).

Generally speaking, subway systems have been developed to improve movement in urban areas and to reduce congestion. Individual transportation increases in cost

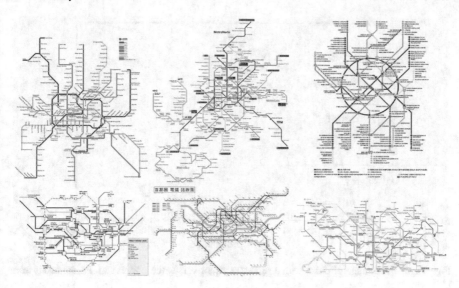

Fig. 10.17 A sample of large subway networks in large urban areas. From left to right and top to bottom: Shanghai, Madrid, Moscow, Tokyo, Seoul, Barcelona. Figures from Wikimedia Commons

as cities grow larger, and mass transit such as subway networks, become central to the evolution of cities, their spatial organization [2, 3, 35] and dynamical processes occurring on them [4, 6]. The percentage $s(P)$ of cities with population less or equal than P and with subway system versus P is shown in Fig. 10.18 (the data were obtained for cities with population larger than 10^5 [36]) which confirms that the larger a city, the more likely it is to have some form of mass transit system (see also [37]). Approximately 25% of the cities of more than one million individuals have a subway system, 50% of those of more than two million, and all those above 10 million have a subway system (as an indication, an exponential fit of the plot in Fig. 10.18 gives $s(P) = 1 - \exp(-P/P_0)$ where the typical population P_0 is of order 3 million).

Here we will focus on the largest networks in major world cities and thus ignore currently developing, smaller networks in many medium-sized cities (in the Chap. 19 we will discuss how the shape of the subway networks evolve when their total length increases). We will consider the largest metro networks, with at least one hundred stations and which are in: Barcelona, Beijing, Berlin, Chicago, London, Madrid, Mexico, Moscow, New York City (NYC), Osaka, Paris, Seoul, Shanghai, and Tokyo, for which we show a sample in Fig. 10.17. For details about the data and how it was gathered and organized we refer the interested reader to [38].

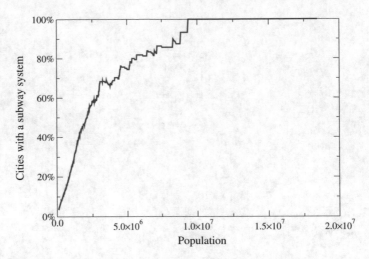

Fig. 10.18 Percentage of cities with a subway system versus the population. Figure taken from [38]

10.2.2   Typical Numbers

The main characteristics of these networks are shown in Table 10.1 where we first observe that the number of different lines appears to increase incrementally with the number of stations and that on average for these world networks, there are approximately 18 stations per line. Also, the mean interstation distance is on average $\overline{\ell_1} \approx 1$ km with Beijing and Moscow showing the longest ones (1.79 kms and 1.67 kms, respectively) and Paris displaying the shortest one (570 m), a diversity which probably finds its origin in the different historical paths of these networks.

Other quantities such as the catchment area (the average number of individuals served by one station) could be computed but should be used with care: residential and economic activity density vary strongly across space and back-of-the-envelop arguments should only serve as a guide. Generally speaking, many parameters such as the population density, land use activity distribution, and traffic are important drivers in the evolution of those networks, but we will focus here on the characterization of these networks in terms of space and topology, independently from other socio-economical considerations (in Chap. 20, we will examine the relations between the network properties and socio-economical features of the corresponding cities).

In order to get some initial insight into the topology of these networks, one can first compare the total length ℓ_T of these networks to the corresponding quantity computed for an almost regular graph ℓ_T^{reg} with the same number of stations, area, and average degree. For a random planar graph with small degree fluctuations ($k \approx \langle k \rangle$) and small fluctuations of the spatial distribution of nodes, we can consider that the internode spacing is roughly constant and given by $\ell_0 \sim 1/\sqrt{\rho}$ where $\rho = N/A$ is the density of nodes defined as the number of nodes over the total area comprising all the nodes.

Table 10.1 List of various indicators (for the year 2009) for the major subway networks considered in this study (and sorted according to their metro population). P is the metropolitan area population (for 2009). N_L is the number of lines, N the number of physical stations, $\overline{\ell_1}$ is the average inter-station distance, ℓ_T total route length, ℓ_T^{reg} the total route length for a regular graph with same average degree, area, and number of stations, and β the final ratio between branch and core stations. Table taken from [38]

City	P (millions)	N_L	N	$\overline{\ell_1}$ (kms)	ℓ_T (kms)	ℓ_T/ℓ_T^{reg}	β (%)
Beijing	19.6	9	104	1.79	204	0.14	39
Tokyo	12.6	13	217	1.06	279	0.13	43
Seoul	10.5	9	392	1.39	609	0.39	38
Paris	9.6	16	299	0.57	205	0.18	38
Mexico City	8.8	11	147	1.04	170	0.15	39
New York City	8.4	24	433	0.78	373	0.12	36
Chicago	8.3	11	141	1.18	176	0.08	71
London	8.2	11	266	1.29	397	0.20	47
Shanghai	6.9	11	148	1.47	233	0.21	61
Moscow	5.5	12	134	1.67	260	0.16	71
Berlin	3.4	10	170	0.77	141	0.30	60
Madrid	3.2	13	209	0.90	215	0.42	46
Osaka	2.6	9	108	1.12	137	0.88	43
Barcelona	1.6	11	128	0.72	103	0.32	38

The total length is then the number of edges $E = N\langle k \rangle/2$ times ℓ_0 which leads to [8]

$$\ell_T^{reg} \sim \frac{\langle k \rangle}{2}\sqrt{AN} \qquad (10.9)$$

In real-world applications, the determination of the quantity A is a difficult problem, but here we choose to use the metropolitan area as given by the various data sources. As shown in the Table 10.1, the ratio ℓ_T/ℓ_T^{reg} varies from 0.08 to 0.88, has an average of order 0.29 and displays essentially three outliers. First, Osaka (and also Madrid and Seoul) has a very large value indicating a highly reticulated structure. In contrast, Chicago and NYC have a much smaller value (≈ 0.1) signaling a more heterogeneous structure which in both these cases is probably due to their strong geographical constraints.

The total length and the comparison with a regular structure gives a first hint about the structure of these networks but other indicators are needed to get a more focused view. There exist many different indicators and variables that describe these networks and their evolution. An important difficulty thus lies in the choice among many possible indicators and how to extract useful information from them. In addition, the largest networks have a relatively small number of stations (always smaller than 500) which implies that we cannot expect to extract useful information from the probability distributions of various quantities as the results are too noisy. We thus have to compute more globally structured indicators which are, however, sensitive to the usually small temporal variations associated with these networks. In the following, we will focus on a certain number of these indicators, which we consider to be the most informative at this point.

10.2.3   Network Evolution

We first show various maps showing the evolution of these subways (the data is from Wikipedia and all these networks are available at www.quanturb.com). We start with Paris (the scale is not the same for the different figures) shown in Fig. 10.19. In this case we see an important densification in the center and the appearance of branches and their subsequent growth. The same phenomenon can be seen on other networks such as Moscow, London, or Tokyo (see Figs. 10.20, 10.21, and 10.22). At this point we can draw some preliminary conclusions from the observation of these maps:

- There are essentially two growth modes: the center grows and densifies and the branches multiply and grow.
- There are no clear similarities between the two processes, although the resulting networks are similar: a dense core, a circular line encircling the core and 'dendrites' reaching out to suburbs.
- It seems that for Moscow and also for London the 'branches' to the suburbs grow before the densification of the center. In Paris, it is essentially the opposite.

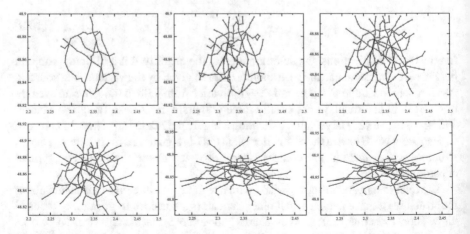

Fig. 10.19   From **a** to **f**: Paris subway for the years 1910, 1930, 1950, 1970, 1990, 2009

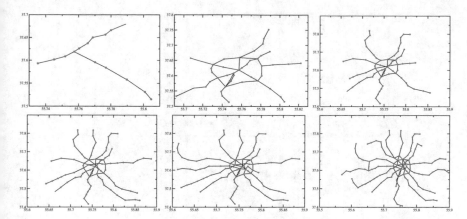

Fig. 10.20 From **a** to **f**: Moscow subway for the years 1940, 1960, 1970, 1980, 1990, 2009

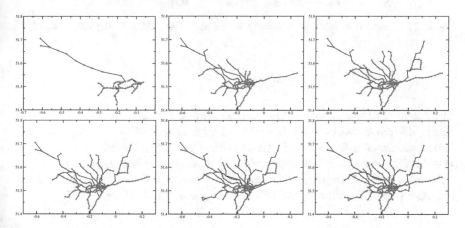

Fig. 10.21 From **a** to **f**: London subway for the years 1900, 1930, 1950, 1970, 1990, 2009

In the following we will give a more precise definition of the core and branches and we will show that indeed these various networks seem to converge to the same shape.

10.2.4 Standard Measures

We first have a look on some 'standard' quantities such as the number of stations $N(t)$, the average degree $\langle k \rangle(t)$, or the meshedness versus time t. We recall here that the meshedness is defined as the ratio of the number of loops divided by the maximum number of loops for planar graphs (see also Chap. 4)

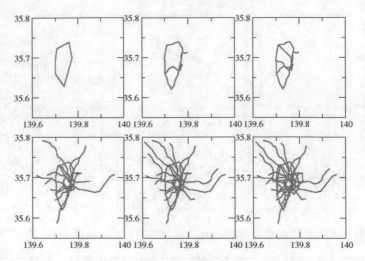

Fig. 10.22   From left to right: Tokyo Subway for the years 1900, 1920, 1960, 1980, 2000, 2009

$$M = \frac{E - N + 1}{2N - 5} \tag{10.10}$$

The results for these quantities (Fig. 10.23) suggest that there is a convergence of these different networks to a unique structure characterized by the same values of indicators. Indeed, despite the fact that time scales are different we can see that $\langle k \rangle$, the dispersion of k (not shown), the cyclomatic number per node C/N, and the meshedness M converge to the same value for these three cities. The core and branches picture described below will help us to confirm this idea.

10.2.5   Efficiency

An important point aspect these transportation systems is the efficiency. It can be measured by different indicators (see Chap. 4) such as the average detour index

$$\langle \eta \rangle = \frac{1}{N_p} \sum_{i,j} \frac{d_{tot}(i, j)}{d_e(i, j)} \tag{10.11}$$

where N_p is the number of pairs, $d_e(i, j)$ ($d_{tot}(i, j)$) the euclidean (network) distance between i and j. The closer $\langle \eta \rangle$ to one and the more efficient the network (from a user's perspective). Its variation with time is shown for different subways in Fig. 10.24

Another important indicator is the performance P and defined as the ratio of the average shortest path on the network, divided by the average shortest path calculated on the minimum spanning tree.

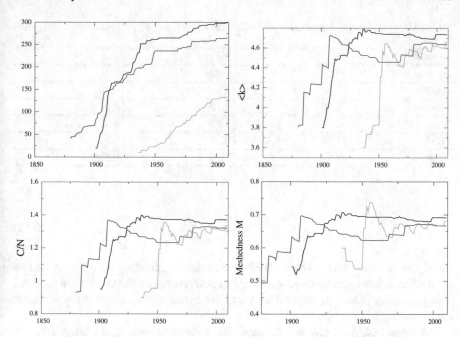

Fig. 10.23 Black: Paris, Red: London, Green: Moscow. **a** Number of stations versus time. **b** Average degree versus time (the degree dispersion also converges to the same value ≈ 1 for the three cities). **c** Cyclomatic number per station. **d** Meshedness for the three cities. We note a convergence to the same value

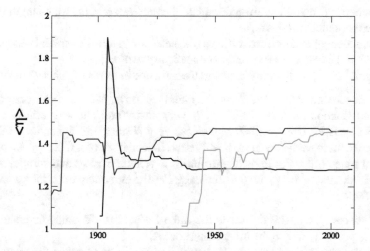

Fig. 10.24 Average detour index versus time for Paris (black), London (red), and Moscow (green)

Table 10.2 Cost, average detour index, and efficiency (C, $\langle \eta \rangle$, and P) for six networks (computed for the year 2009)

Network	C	$\langle \eta \rangle$	P
Paris	1.35	1.88	0.47
London	1.52	1.85	0.49
Moscow	1.40	2.10	0.56
Tokyo	1.67	1.90	0.395
NYC	1.74	2.51	0.33
Seoul	1.64	1.98	0.555

$$P = \frac{\overline{\ell}}{\ell_{MST}} \tag{10.12}$$

Finally, the relative cost C is computed as the total length of the network divided by the total length of the minimum spanning tree constructed on the same set of stations. This parameter is always larger than 1 and the closer to one, the more economical the network.

All these quantities can be measured for various networks and the results are summarized in the Table 10.2 where we give the final values (for the year 2009) for 6 different networks.

In order to understand this table we have to keep in mind that:

- The smaller C and the more economical the network (i.e. the total length is close to the theoretical minimum).
- The smaller $\langle \eta \rangle$ and the most efficient the network-in terms of spatial distance (i.e. for $\langle \eta \rangle = 1$ no detour is needed to connect any two points).
- The smaller P and the most efficient the network in terms of shortest paths.

The most economical networks are therefore Paris, Moscow and London, but Paris and London are more efficient. A very costly and not very efficient seems to be NYC. Even if these 2009 values are very similar, the dynamics in time to converge to these values seems to be different. Indeed if we plot $C(t)$, $\langle \eta \rangle$ we observe different trajectories. For Paris, the evolution is relatively clear and simple, the cost increases simultaneously with the efficiency. For the other networks, things are more complicated:

- For Moscow, the cost stabilizes to the value 1.4 but the efficiency fluctuates a lot. This is probably due to the branching process.
- For London and Tokyo we observe some important variations in the relative cost and the efficiency.

10.2.6 Temporal Statistics: Bursts

10.2.6.1 Dynamics

In order to get an initial feeling about the dynamics of these networks, we first estimate the simplest indicator $v = dN/dt$ which represents the number of new stations built per year. From this instantaneous velocity, we can compute the average velocity over all years. This average can however be misleading as there are many years where no stations are built and thus we describe this by the fraction of 'inactivity' time f. We provide results in Table 10.3 and some interesting facts are revealed. Note that it is clear that Shanghai and Seoul are the most recent subway networks experiencing a rapid expansion that has elevated them to amongst the largest networks in the world.

For most of these networks the average velocity is in a small range (typically $\bar{v} \in$ [1.4, 3.7] stations per year) except for Seoul and Shanghai which are more recently developed networks. This is however an average velocity and we observe that (i) for all networks, larger velocities occur at earlier stages of the network and (ii) large fluctuations occur from one year to another. Interestingly, the fraction of inactivity time (i.e. the time when no stations are built) is similar for all these networks with an average of about 58%. We also show in Fig. 10.25a, the time evolution for each city of the number of stations, using an absolute time scale. In particular, the size of the oldest networks seem to progressively reach a plateau.

Table 10.3 t_0 is the initial year considered here for the different subways networks. $\bar{v}$ is the average velocity (number of stations built per year), σ_v is the standard deviation of v, and f is the fraction of years of inactivity (no stations built)

City	t_0	$\bar{v}$	σ_v	f (%)
Beijing	1971	3.3	7.74	79
Tokyo	1927	2.8	5.47	51
Seoul	1974	11.2	14.9	20
Paris	1900	2.6	5.1	60
Mexico City	1969	3.7	5.9	55
New York City	1878	3.3	8.3	68
Chicago	1901	1.9	6.24	71
London	1863	2.3	3.8	48
Shanghai	1995	14.9	20.2	31
Moscow	1936	1.7	1.9	43
Berlin	1901	1.6	3.3	65
Madrid	1919	2.3	4.6	59
Osaka	1934	1.4	4.1	79
Barcelona	1914	1.4	4.8	78

(a)   (b)

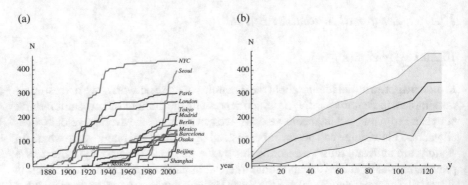

Fig. 10.25 a Evolution of the number of stations for various large world subway networks. **b** Evolution of the number of stations y years after creation, averaged over all networks (the gray area represents the standard deviation across all networks). The linear shape indicates that the relative growth in terms of percentage of new stations from a decade to another goes to zero for all these networks, signaling the possible appearance of a stationary limit

In order to compare growth across all networks, we introduce a second graph on Fig. 10.25b featuring the average, over all networks, of the number of stations after a certain number of years since network creation. This average quantity exhibits a linear increase which indicates convincingly that, overall, as these networks become large, then for a few decades thereafter new stations represent an increasingly small percentage of existing ones. This first result anticipates the fact that these large networks may reach some kind of limiting shape that we will characterize in the next section. This incremental growth of subways might reflect socio-economical concerns and pressure on the transportation networks such as diminishing return on investments as noted by various authors (see references cited in [38]).

For each subway, the number of stations constructed per year fluctuates from a year to another as we can see on the Fig. 10.26. We observe bursts of sudden activity and we can plot the corresponding histograms (see Fig. 10.27). For example in the case of Tokyo, the activity was low until the 50s and various bursts during the 70s and around 2000. These histograms show that except for Moscow where centralization seemed to have played an important role (and for which the number of new stations built is approximatively constant), there are relatively large fluctuations. For example for Paris and London the number of new stations built per year can vary from 1 to 30.

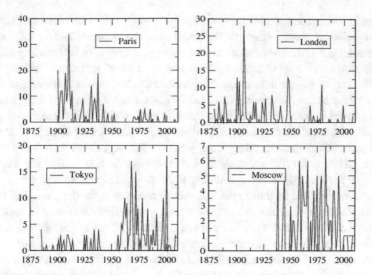

Fig. 10.26 Number of new stations versus the year for four different subways

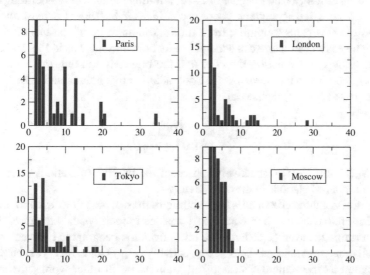

Fig. 10.27 Histograms for the number of new stations built per year for four different subways

10.2.7 Core and Branches: Measures and Model

10.2.7.1 The Universal Template

As we saw in the previous sections, there are many indicators and parameters that we can measure on a spatial time evolving network. Most of the simple measures so far suggest that subway networks converge to the same structure, despite their geographical and economical differences. It is crucial at this point to have a 'template' that can guide us about what are the most relevant quantities for characterizing these structures and their evolution. By inspection, we observe that the large subway networks consist of a set of stations delimited by a 'ring' that constitute the 'core'. From this core, quasi-one dimensional branches grow and reach out to areas of the city further and further from the core. In Fig. 10.17, we show a sample of these networks as they existed in 2009. We note here that the ring, which is defined topologically as the set of core stations which are either at the junction of branches or on the shortest geodesic path connecting these junction stations, exists or not as a subway line. For instance, for Tokyo, there is a such a circular line (called the *Yamanote* line), while for Paris the topological ring does not correspond to a single line. It is also worth noting that in those systems where the core is harder to define such as for NYC where physical constraints are strongly manifest (the east and west rivers which bound Manhattan), a pseudo core is evident where a series of lines coalesce to enable travelers to move around the core circumferentially.

More formally, branches are defined as the set of stations which are iteratively built from a terminal station, or a station of degree 1. New neighbors are added to a given branch as long as their degree is 2—continuing the line, or 3—defining a fork. In this latter case, the aggregative process continues *if and only if* at least one of the two possible new paths stemming from the fork is made up of stations of degree 2 or less. The general structure can schematically be represented as in Fig. 10.28. This template allows to distinguish the core and the branches and thus guides us to the relevant quantities to measure. We first characterize this branch and core structure with the parameter $\beta(t)$ defined as

$$\beta(t) = \frac{N_B(t)}{N_B(t) + N_B(t)} \qquad (10.13)$$

where $N_B(t)$ and $N_C(t)$ respectively represent the number of stations on branches and the number of stations in the core at time t.

We can also characterize a little further the structure of branches. Their topological properties are trivial and their complexity resides in their spatial structure. We can then determine the average distance (in kms) from the geographic barycenter of the city to all core and branches stations, respectively: $\overline{D}_C(t)$ and $\overline{D}_B(t)$ (the barycenter is computed as the center of mass of all stations, or in other words, the average location of all the stations) These distances provide information about the spatial extension of the branches and we construct the ratio $\mu(t)$

$$\mu(t) = \frac{\overline{D}_B(t)}{\overline{D}_C(t)} \qquad (10.14)$$

which gives a spatial measure of the amount of extension of the branches.

We also need information on the structure of the core. The core is usually a planar graph and can be characterized by many parameters [8]. It is important to choose those which are not simply related to each other but represent different aspects of the

Fig. 10.28 Schematic
structure of subway
networks. A large 'ring'
encircles a core of stations.
Branches radiate from the
core and reach further areas
of the urban system. The
branches are essentially
characterized by their size
(parameter $\beta(t)$, and their
spatial extension (parameter
$\mu(t)$). The core is
characterized by its average
degree ($\langle k_{core}\rangle(t)$) and the
fraction of nodes of degree 2
(f_2), its number of stations
$N_C(t)$ and its size $r_C(t)$.
Figure taken from [38]

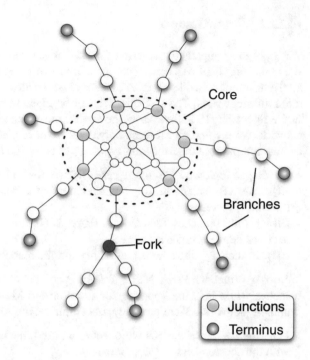

network (such as those proposed in the form of various indicators, see for example
[7, 8, 39]). At each time step t, we will thus characterize the core structure by the
following two parameters. The first parameter is simply the average degree of the
core which characterizes its 'density'

$$\langle k_{\text{core}}\rangle(t) = \frac{2E_C(t)}{N_C(t)} \tag{10.15}$$

where $N_C(t)$ is the number of core nodes and $E_C(t)$ the number of its edges. The
average degree is connected to the standard gamma index $\gamma(t) = E_C(t)/(3N_C(t) -
6)$ where the denominator is the maximum number of links admissible for a planar
network (see Chap. 2).

The average degree of the core contains a useful information about it, and there
are many other quantities (such a the standard index α, see Chap. 4) which can give
additional information. We will use another simple quantity which describes in more
detail the level of interconnections in the core and which is given by the fraction
f_2 of nodes in the core with $k = 2$. In the case of well-interconnected systems,
this fraction will tend to be small, while sparse cores with a few interconnections
will have a larger fraction of $k = 2$ nodes. Once we know the fraction f_2 and the
parameter $\mu(t)$ which characterizes the relative spatial extension of branches, we
have key information on the intertwinement of both topological and geographical
features in such "core/branch" networks.

10.2.7.2   First Measures

We apply this template to the case of Paris (for the year 2009) and obtain the result shown in Fig. 10.29. We then decompose into two parts, the core and the branches, all the different networks at different times and we monitor the number and sizes of these different parts. The size of a set will be given by the number of its stations. In the figure Fig. 10.30, we show the evolution separately for six different subway network. We see that for all these networks, we have plateaus in the time evolution and we can propose the following typology of behaviors:

- (I) 'Stagnation': we have in this case plateaus for both the core and the branches
- (II) 'Core's densification' (or 'consolidation'): the branches are not evolving and the core is increasing
- (IIbis) 'Inter branches connection': Same as (II) but for a different reason: connections between branches appear.
- (III) 'Branching': the core is plateauing and the branches only are evolving.

We observe that for Paris, NYC, Tokyo, Paris, Seoul the core is larger than the branches while it is the opposite for London and Moscow which are both very 'ramified' networks. More precisely, the empirical observation shows that

- For Moscow, since the 50s we observe a pure branching, probably going along with the development of the suburbs.

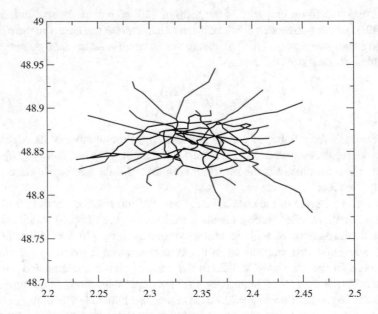

Fig. 10.29 Core and branches for the Paris subway for the year 2009

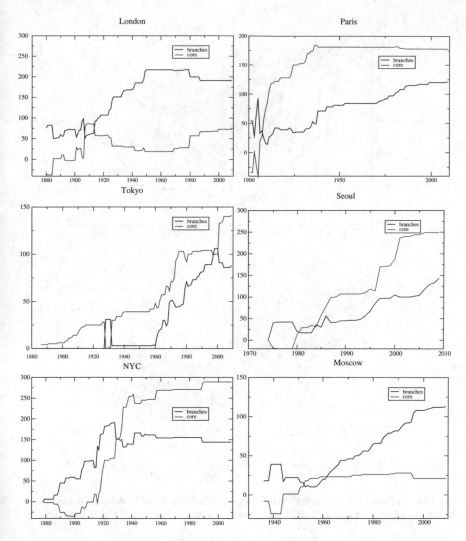

Fig. 10.30 Core and branches evolution for six different networks

- For London, we also have a ramification since the 20s but we also observe a conversion from branches to the core which probably means that some inter-branches connections have been made.
- For Seoul, the network has a dense core which seems stable so far, and there is an increase of the branches since 2005.
- For Tokyo, we observe an evolution for both the core and the branches in the period 1960−2000. Since 2000, the branches are decreasing and the core is increasing which also probably means that some inter-suburb connections have been made.

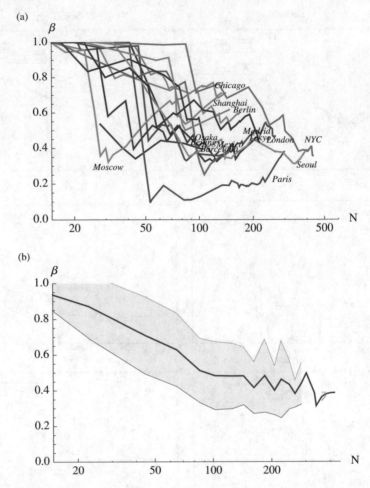

Fig. 10.31 **a** Parameter β as a function of the number of stations N for the different world subways. **b** Same as **a** but averaged over 20 bins and showing the standard deviation. Figure taken from [38]

10.2.7.3 Evolution of the Core-Branches Characteristics

In order to be able to compare the different networks across time periods and cities, we compare their evolution as described by the core and branches parameters. We first plot in Fig. 10.31a the parameter β as a function of N for the networks studied here. It is difficult to draw strong conclusions from this plot, but we can bin these data and represent the average value of β per bin and its dispersion as well (Fig. 10.31b). On this figure we may see that the average value of β seems to stabilize slowly to some value in $[0.35, 0.55]$.

We show in the Fig. 10.32 the evolution of the parameter μ with N (the data is binned). This figure shows that in the interval where we have the largest number of

Fig. 10.32 Evolution of the
ratio μ, which characterizes
the spatial extension of
branches relative to the core.
Figure taken from [38]

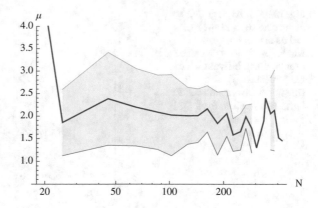

subways, the average value of μ is around 2 with relatively large fluctuations which
seem to decrease with N.

The parameters β and μ give an indication of the importance of the core but do
not say anything about its structure. A first structural indication may be given by its
average degree $\langle k_{\text{core}} \rangle$ and by the percentage f_2 of nodes in the core having a degree k
equal to 2. In particular, these two quantities shed light on how interconnections are
created in the core. We display in Fig. 10.33a the average degree of the core $\langle k_{\text{core}} \rangle$
which, even if there is a slow increase with N, displays moderate variations around
2.4 approximately. This value is relatively small and indicates that the fraction of
connecting stations (i.e. with $k > 2$) is also small and means that most core stations
belong to one single line with few that actually allow connections. More precisely,
we observe in Fig. 10.33b that on average for subways with $N < 100$ the fraction of
interconnecting stations is decreasing with N—which probably corresponds to the
growth of the subway—but that for larger subways ($N > 100$), the percentage f_2 is
increasing again, which probably corresponds to a densification process without the
creation of new interconnections.

As noted above, the number of subways with large N is smaller and the statistics
therefore less reliable. At this point and with this statistical error in mind, we observe
that the average value β and its dispersion are decreasing with N and it suggests that
β could converge to some 'limiting' value $\beta_\infty \approx 45\%$. The same remarks also apply
to μ and suggest a limiting value of order 2. Concerning the core, the dispersion of
$\langle k_{\text{core}} \rangle$ is always moderate and approximately constant showing that the fluctuations
among different networks are also moderate. We observe a slow increase of $\langle k_{\text{core}} \rangle$
pointing to a mild yet continuing densification of the core, even after a long period of
time. The fraction of connecting stations has a more complex dynamics and seems
to decrease with N for large networks. In these networks, there is an obvious cost
associated with the large value of $k > 2$ and such a decreasing fraction could be due
to the fact that a small fraction is enough to enable easy navigation in the network.

Despite non negligible fluctuations, these results suggest that large subway networks may converge to a long time limiting network largely independent of their
historical and geographical differences. So far, we can characterize the 'shape' of

Fig. 10.33 **a** Average degree
of the core $\langle k_{core} \rangle$ Eq. (10.15)
and its dispersion versus
number of stations (averaged
over 20 bins). **b** Evolution of
the percentage f_2 of $k = 2$
core nodes (averaged over 20
bins). Figure taken from [38]

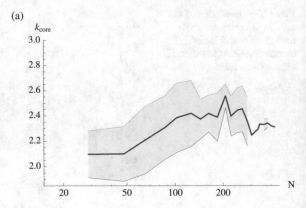

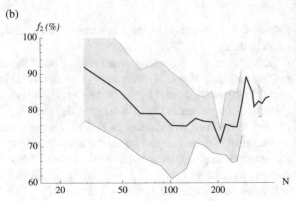

this long time limiting network with values of $\beta_\infty \approx 45\%$, $\mu_\infty \approx 2$, and a core made
of approximately 80% of non connecting stations. It will be interesting to observe
the future evolution of these networks in order to confirm (or not) these results.

10.2.7.4   Number of Branches

We now consider the number $\mathcal{N}_B$ of different branches. A naive argument would
be that the number of branches is actually proportional to the perimeter of the core
structure. This implicitly assumes that the distance between different branches is
constant. In turn, the perimeter should roughly scale as $\sqrt{N}$ as the core is a relatively dense planar graph and contains a number of nodes proportional to N. These
assumptions thus leads to

$$\mathcal{N}_B \sim \sqrt{N} \tag{10.16}$$

We display the number of branches versus the number of stations N for the various
networks considered here (and at all times) and a power law fit of the data presented

Fig. 10.34 Loglog plot of
the number of different
branches versus the number
of stations for the different
subway networks considered
here. The dashed line is a
power law fit with exponent
≈ 0.6. Figure taken from
[38]

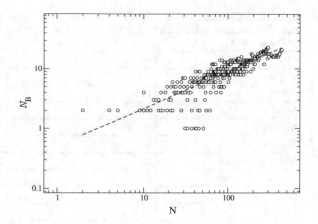

in Fig. 10.34 gives $\mathcal{N}_B \sim N^b$ with $b \approx 0.6$ ($r^2 = 0.85$) consistent with the simple
argument presented here.

10.2.8 Spatial Organization of the Core and Branches

Following earlier studies on the fractal aspects of subway networks [24], we can
inspect the spatial subway organization by considering the number of stations $N(r)$
at a distance less than or equal to r, where the origin of distances is the barycenter
of all stations considered as points. Interestingly, the barycenter of all stations is
almost motionless, except in the case of NYC where the barycenter moves from
Manhattan to Queens and thus we will exclude NYC from further study. Chicago is
a similar case: the spatial structure of the core is peculiar, mainly due to presence of
the lake which constrains the network from expanding in the other directions. We
will also exclude this network in this section. It should however be noted here that
both Chicago and NYC do follow the image of core and branches but that the main
difference with the other networks is that the core of these networks has no clear
spatial meaning due to the geographical constraints (such as the presence of a lake
for Chicago and a particular land area shape for NYC).

For the year 2009, the limiting shape made of a core and branches implies that
there is an average distance r_C which determines the core. In practice, we can measure
on the network the size N_C of the core and we then define r_C such that $N(r = r_C) =
N_C$ (which assumes implicitly an isotropic core shape, which is the case for most
networks except for the excluded cases of Chicago and NYC). For the various cities,
we can easily compute the function $N = N(r)$ from which we extract r_C and we
report the results in the Table 10.4.

Next, we rescale r by r_C and $N(r)$ by N_C and we then obtain the results shown
in the Fig. 10.35. This figure displays several interesting features. First, the short

Table 10.4 For each city, we compute the number of stations in the core (for the year 2009) and from the numerical calculation of $N(r)$ we can estimate r_C the size of the core (in kms) from $N(r = r_C) = N_C$

City	N_C	r_C (kms)
Beijing	63	4.4
Tokyo	123	5.0
Seoul	243	11.6
Mexico	90	4.7
Shanghai	57	3.7
Moscow	39	5.9
London	142	7.3
Paris	186	4.2
Madrid	113	4.4
Berlin	68	5.5
Barcelona	57	3.5
Osaka	46	3.6

distance regime $r < r_C$ is well described by a behavior of the form $N(r) \sim \rho_C \pi r^2$ consistent with a uniform density ρ_C of core stations. For very large distances, we observe for most networks a saturation of $N(r)$. The interesting regime is then for intermediate distances when r is larger than the core size but smaller than the maximum branch size r_{max}. This intermediate regime is characterized by different behaviors with r. A similar result was obtained earlier [40] where the author observed for Paris that $N(r > r_c) \sim r^{0.5}$, a result that was at that time difficult to understand in the framework of fractal geometry.

Here, we show that these regimes can be understood in terms of the core and branches model, with the additional factor that the spacing between consecutive stations on branches is increasing with r. Within this picture (and assuming isotropy), $N(r)$ is given by

$$
N(r) \sim \begin{cases} \rho_C \pi r^2 & \text{for } r < r_C \\ \rho_C \pi r_C^2 + \mathcal{N}_B \int_{r_C}^{r} \frac{dr}{\Delta(r)} & \text{for } r_C < r < r_{max} \\ N & \text{for } r > r_{max} \end{cases} \tag{10.17}
$$

where N is the total number of stations, $\mathcal{N}_B$ is the number of branches and $\Delta(r)$ is the average spacing between stations on branches at distance r from the barycenter. This equation therefore just states that above r_C the number of stations is given by the number of stations in the core plus the number of branches times each stations per branch.

In order to test Eq. (10.17)—namely $\mathcal{N}_B$, N_C, r_C, and $\Delta(r)$—we plot the resulting shape of Eq. (10.17) against the empirical data. It is easy to determine empirically the numbers $\mathcal{N}_B$, N_C, and r_C but the quantity $\Delta(r)$ is extremely noisy due to the small

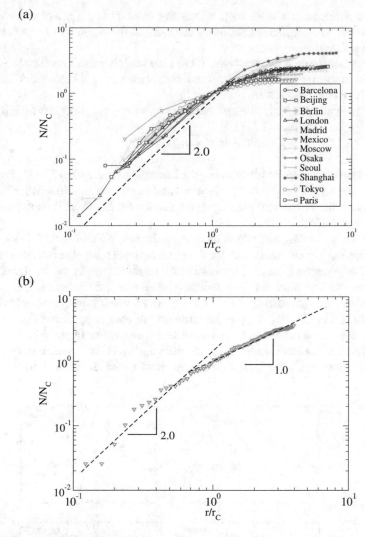

Fig. 10.35 **a** Rescaled number of stations at distance r from the barycenter as a function of the rescaled variable r/r_C where r_C is the size of the core defined as $N(r = r_C) = N_C$ (shown here in loglog). The dotted line represents a power law $\sim r^2$ and serves as a guide to the eye. **b** Case of Moscow where the two regimes ($r < r_C$ and $r > r_C$) with their different exponents are visible (the dotted lines serve here as a guide to the eye). Figure taken from [38]

number of points especially for large values of r closest to $r_{\max}$, at a distance where, often, there is no more than a handful of stations (all these numbers are determined for the year 2009).

The less noisy case is Moscow which has long branches and for which we obtain a interstation spacing roughly constant: we obtain for $r > r_C$ a behavior of the form $N(r) \sim \mathcal{N}_B r$ (see Fig. 10.35b).

More generally, the large distance behavior $r_C < r < r_{\max}$ will be of the form

$$N(r_C < r < r_{\max}) \sim r^{1-\tau} \tag{10.18}$$

where τ denotes the exponent governing the interspacing decay $\Delta(r) \sim r^\tau$. For most networks, the regime $r_C < r < r_{\max}$ is small and as already mentioned $\Delta(r)$ is very noisy. Rough fits in different cases give a behavior for Eq. (10.17) consistent with data (see Fig. 10.36).

In particular, for Moscow which has long branches, we observe a behavior consistent with $\Delta(r) \simeq$ constant while for the other networks, we observe an increasing trend but an accurate estimate of τ is difficult to obtain, given the small variation range of r—with no more than one decade of available data. For example, a fit over this decade of data gives for Paris $\tau \approx 0.5$ (with $r^2 = 0.74$) in agreement with the result obtained in [24, 40]. Despite the difficulty of obtaining accurate quantitative results, more data is needed to have a definite answer and so far we can only claim that the data are not inconsistent with the behavior Eq. (10.17), which supports this picture of a long time limit network shape made of a core and radial branches.

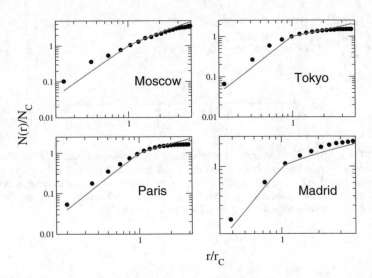

Fig. 10.36 $N(r)/N_C$ versus r/r_C for Moscow, Tokyo, Paris, and Madrid (from top to bottom and left to right). The circles represent the data and the green solid line the fit using Eq. (10.17) with parameters estimated from the empirical data. Figure taken from [38]

References

1. M. Barthelemy, *The Structure and Dynamics of Cities* (Cambridge University Press, 2016)
2. S. Hanson, G. Giuliano, *The Geography of Urban Transportation* (Guilford Press, 2004)
3. M. Batty, *The New Science of Cities* (MIT Press, 2013)
4. L.M. Bettencourt, J. Lobo, D. Helbing, C. Kühnert, G.B. West, Proc. Natl. Acad. Sci. **104**(17), 7301 (2007)
5. L. Bettencourt, G. West, Nature **467**(7318), 912 (2010)
6. D. Balcan, V. Colizza, B. Gonçalves, H. Hu, J.J. Ramasco, A. Vespignani, Proc. Natl. Acad. Sci. **106**(51), 21484 (2009)
7. P. Haggett, R.J. Chorley, *Network Analysis in Geography*, vol. 67 (Edward Arnold London, 1969)
8. M. Barthelemy, Phys. Rep. **499**(1), 1 (2011)
9. E. Strano, V. Nicosia, V. Latora, S. Porta, M. Barthelemy, Sci. Rep. **2** (2012)
10. M. Barthelemy, P. Bordin, H. Berestycki, M. Gribaudi, Sci. Rep. **3** (2013)
11. A. Masucci, D. Smith, A. Crooks, M. Batty, Eur. Phys. J. B-Condens. Matter Compl. Syst. **71**(2), 259 (2009)
12. P. Bordin, Université de Marne-la-Vallée (2006)
13. B. Costes, J. Perret, B. Bucher, M. Gribaudi, in *18th AGILE International Conference on Geographic Information Science* (2015)
14. M. Barthelemy, A. Flammini, Phys. Rev. Lett. **100**(13), 138702 (2008)
15. S. Lämmer, B. Gehlsen, D. Helbing, Phys. A **363**(1), 89 (2006)
16. I. Samuels, P. Panerai, J. Castex, J.C. Depaule, *Urban forms* (Routledge, 2012)
17. L.C. Freeman, Sociometry pp. 35–41 (1977)
18. P. Crucitti, V. Latora, S. Porta, Phys. Rev. E **73**(3), 036125 (2006)
19. A. Kirkley, H. Barbosa, M. Barthelemy, G. Ghoshal, Nat. Commun. **9**(1), 1 (2018)
20. M.P. Viana, E. Strano, P. Bordin, M. Barthelemy, Sci. Rep. **3** (2013)
21. A. Tero, S. Takaji, T. Saigusa, K. Ito, D. Bebber, M. Fricker, K. Yumiki, R. Kobayashi, T. Nakagaki, Science **327**, 439 (2010)
22. M. Ovenden, *Metro Maps of the World* (Capital Transport, 2005)
23. R. Bon, Quality & Quantity **13**(4), 307 (1979)
24. L. Benguigui, J. Phys. I **2**(4), 385 (1992)
25. V. Latora, M. Marchiori, Phys. A **314**(1), 109 (2002)
26. K.A. Seaton, L.M. Hackett, Phys. A **339**(3), 635 (2004)
27. J. Sienkiewicz, J.A. Hołyst, Phys. Rev. E **72**(4), 046127 (2005)
28. D. Gattuso, E. Miriello, Netw. Spat. Econ. **5**(4), 395 (2005)
29. P. Angeloudis, D. Fisk, Phys. A **367**, 553 (2006)
30. K. Lee, W.S. Jung, J.S. Park, M. Choi, Phys. A **387**(24), 6231 (2008)
31. C. Von Ferber, T. Holovatch, Y. Holovatch, V. Palchykov, in *Traffic and Granular Flow'07* (Springer, 2009), pp. 709–719
32. S. Derrible, C. Kennedy, Transp. Res. Rec.: J. Transp. Res. Board **2112**(-1), 17 (2009)
33. S. Derrible, C. Kennedy, Phys. A **389**(17), 3678 (2010)
34. S. Derrible, C. Kennedy, Transportation **37**(2), 275 (2010)
35. M.A. Niedzielski, E.J. Malecki, Ann. Assoc. Am. Geogr. **102**(6), 1409 (2012)
36. Un population division. http://www.unpopulation.org
37. C.F. Daganzo, Transp. Res. Part B: Methodol. **44**(4), 434 (2010)
38. C. Roth, S.M. Kang, M. Batty, M. Barthelemy, J. R. Soc. Interface **9**(75), 2540–2550 (2012)
39. F. Xie, D. Levinson, Geogr. Anal. **39**(3), 336 (2007)
40. L. Benguigui, Environ Plan A **27**(7), 1147 (1995)

Part II
Models

Families of Models of Spatial Networks

There is now a long history of modeling spatial networks and in this part of the book we adopted the following classification. First, we discuss the spatial generalizations of standard graphs such as the Erdos-Renyi, and the small-world Watts-Strogatz model. We then consider the generalization of growing models (at each time step we add a node) including in particular the preferential attachment model.

Most of the rest of this part concerns spatial networks constructed over a set of points distributed in the 2d plane. This naturally includes tessellations such as the Voronoi tessellation and its dual, the Delaunay graph. We will distinguish two large classes of models. The first one comprises proximity graphs: two nodes are connected is they are 'close' enough. Many choices are possible and we obtain a variety of graphs such as the random geometric graph, the Bluetooth, the radial graph, etc. The second large family of spatial networks that can be constructed over a set of points are excluded volume graphs. In this case, two nodes are connected if in some space between them there are no other nodes. The choice of the shape of this empty space determines the graph and we obtain various graphs such as the Beta-skeletons for example.

We then consider a specific family of graphs that are relevant for many applications although very simple. They are made of branches and a ring. This simple graph has the avantage to correspond to real-world situations (such as the subway network or other transport infrastructures) but also that analytical calculations are often amenable for this structure.

In the following chapters, we consider graphs constructed from optimal considerations (but still constructed over a set of points). The most well-known example is the minimum spanning tree that minimizes the total weight of the graph. This simple construction can however be generalized and this is what we will discuss in the chapter about optimal networks. These optimal networks are particularly important in transportation problems where the goal is to design a network that guarantees some efficiency. Finally, when nodes are added one by one and for each node we choose an optimal location (that minimizes some function) we obtain the class of greedy models discussed at the end of this part.

Chapter 11
Spatial Generalizations of Random Graphs

The simplest model of random graph where nodes are connected at random was proposed by Erdos and Renyi [1] and constitutes an archetype – or at least a benchmark – for constructing more complex random graphs. It is then natural to extend this model to the case where nodes are located in space. In this chapter we will discuss some of the possible extensions that were proposed in the literature. In particular, we will discuss the hidden variable model and the effect of space or traffic on its structure. The Waxman model that can be considered as a spatial variant of the Erdos-Renyi graph will be discussed in Chap. 15 in the framework of the random geometric model. Finally, we will end this chapter with a discussion about the geometric inhomogeneous random graph.

11.1 Spatial Version of Erdos-Renyi Graphs

11.1.1 The Erdos-Renyi Graph

We first recall here some simple facts about Erdos-Renyi (ER) graphs [1–3]. This simple model is the paradigm for random graphs and is in many cases used as a null model. One simple way to generate it is to run through all pairs of nodes and to connect them with a probability p. The average number of links is then

$$\langle E \rangle = p \frac{N(N-1)}{2} \tag{11.1}$$

giving an average degree equal to $\langle k \rangle = 2\langle E \rangle / N = p(N-1)$. This last expression implies that in order to obtain a sparse network, we have to choose a small p scaling as $p = \langle k \rangle / N$ for large N.

© The Author(s), under exclusive license to Springer Nature Switzerland AG 2022   235
M. Barthelemy, *Spatial Networks*,
https://doi.org/10.1007/978-3-030-94106-2_11

The establishment of edges are random, independent events and the degree distribution is therefore a binomial distribution

$$P(k) = \binom{N-1}{k} p^k (1-p)^{N-k+1} \tag{11.2}$$

where k is the degree of a node. This expression can be approximated by a Poisson distribution

$$P(k) \approx e^{-\langle k \rangle} \frac{\langle k \rangle^k}{k!} \tag{11.3}$$

for large N and with an average degree $pN = \langle k \rangle$ constant.

In their important paper [2], Erdos and Renyi described the behavior of these random graphs when p is varying, and more specifically the existence of a transition at $pN = 1$. They showed that if $pN < 1$, the resulting graph almost surely has no connected component larger than $\sim \log N$. For $pN = 1$, there is almost surely a largest component of size $\sim N^{2/3}$. For $Np = c > 1$ (where c is a constant), then the graph has almost surely a unique giant component. Finally, one can also show that if $pN \sim \log N$ there are no isolated vertices. We refer the reader interested in mathematical results about this random graph to the book [4].

Other classical results can be easily derived such as the clustering coefficient which can be shown to be $C = p$ and the average shortest path $\langle \ell \rangle \simeq \log N / \log \langle k \rangle$. In fact for any generalized uncorrelated random graph characterized by a probability distribution $P(k)$ it can be shown (see for example [5]) that the average clustering coefficient is

$$\langle C \rangle = \frac{1}{N} \frac{(\langle k^2 \rangle - \langle k \rangle)^2}{\langle k \rangle^3} \tag{11.4}$$

and the average shortest path is

$$\langle \ell \rangle \approx 1 + \frac{\log N / \langle k \rangle}{\log \frac{\langle k^2 \rangle - \langle k \rangle}{\langle k \rangle}} \tag{11.5}$$

These last two well-known expressions are useful in the sense that they provide a reference to which we can compare results obtained for a specific network in order to understand its features. In particular, we expect very different behavior for spatial networks signalled by different scaling with N of these quantities.

11.1.2 Planar Erdos-Renyi Graphs

The simplest idea to generate a random planar graph is to generate a set of points in the two-dimensional space and then to construct an Erdos-Renyi graph by connecting randomly the pairs of nodes. It is clear that in this way we will generate mostly non-planar graphs but we can exclude links that violate planarity. We would then obtain something like represented in Fig. 11.1 (left). It is easy to imagine even other extensions of this process and we could construct a planar BA network (see Fig. 11.1 (right)) by adding in the two-dimensional space nodes (with a random, uniformly distributed location) which will connect according to the preferential attachment. We would then keep the node if it preserves the planarity of the system. We note here that there are some visual similarities with networks obtained by random sequential adsorption of line segments [6] and it would be interesting to understand if there are deeper connections between these problems.

Instead, mathematicians studied closely related networks. If we denote by P_N the class of all simple labelled planar graphs on N vertices (a labelled graph refers to a graph where distinct labels are assigned to all vertices. The labeling thus adds configurations in the counting process of these graphs). We can draw a random planar graph R_N from this class with a uniform probability [7] and ask for some questions such as the number of vertices of a given degree, the number of faces of a given size, etc. [8, 9]. For instance, the following results have been demonstrated (see [7–9] and references therein)

- The random planar graph R_N is connected with probability at least $1/e$.
- The number $\mathcal{N}_u$ of unlabelled planar graphs scales as

$$\mathcal{N}_u \sim \gamma_u^N \tag{11.6}$$

with $9.48 < \gamma_u < 32.2$ and the number $\mathcal{N}_l$ of labelled planar graphs as

$$\mathcal{N}_l \sim \gamma_\ell (N!)^{1/N} \tag{11.7}$$

Fig. 11.1 Left: Planar Erdos-Renyi network obtained by rejecting links if they violate planarity ($N = 1000$). Right: Planar Barabasi-Albert network obtained by the same rejection method

where $27.22685 < \gamma_\ell < 27.22688$.
- The average number of edges of R_N is $\langle E \rangle \geq \frac{13}{7} N$.
- The degree distribution decreases at least as $N/\gamma_\ell^k (k+2)!$.

Other properties can be derived for this class of networks and we refer the interested reader to [7–9] and references therein for more results.

11.2   The Hidden Variable Model for Spatial Networks

In the Erdos-Renyi model, the probability p to connect two nodes is a constant. In certain situations, we could imagine that a node is described by a number of attributes (called *hidden variables* or *fitnesses*) and the connection between two nodes could depend on the respective attributes of these nodes [10, 11]. In order to give a concrete example, we assume that there is only one attribute or "fitness" η which is a real positive number distributed according to a function $\rho(\eta)$. The probability of connection for a pair of nodes (i, j) is then given by $p_{ij} = f(\eta_i, \eta_j)$ where f is a given function and $\eta_{i(j)}$ is the fitness of node $i(j)$. In the case $f = const.$ we recover the ER random graph. The average degree of a node with fitness η is given by

$$k(\eta) = N \int_0^\infty f(\eta, \eta')\rho(\eta')d\eta' \equiv N F(\eta) \tag{11.8}$$

and the degree distribution is then

$$P(k) = \int \rho(\eta)\delta(k - k(\eta))d\eta$$
$$= \rho\left[F^{-1}\left(\frac{k}{N}\right)\right]\frac{d}{dk}F^{-1}\left(\frac{k}{N}\right) \tag{11.9}$$

A surprising result appears if we choose an exponential fitness distribution ($\rho(\eta) \sim e^{-\eta}$) and for the function f a threshold function of the form

$$f(\eta_i, \eta_j) = \theta[\eta_i + \eta_j - z(N)] \tag{11.10}$$

where θ is the Heaviside function and $z(N)$ a threshold which depends in general on N. In this case, Caldarelli et al. [10] found a power law of the form $P(k) \sim k^{-2}$ showing that a scale-free network can emerge even for a peaked distribution of fitnesses.

11.2.1 Effect of Space

A spatial variant of this hidden variable model was proposed in [12] (and discussed together with other models in [13]) and where the nodes i and j are connected if the following condition is met

$$(\eta_i + \eta_j)h[d_E(i, j)] \geq \phi \qquad (11.11)$$

where $h[r]$ is a decreasing function of the distance between nodes and where ϕ is a constant threshold. For this model, large fitnesses can therefore compensate for larger distances and we will observe large-fitnesses nodes connected by long links. If the distribution of fitnesses $f(\eta)$ has a finite support or is strongly peaked around some value, we will have a typical scale $r_0 = h^{-1}(\phi/2\bar{\eta})$ above which no (or a very few) connections are possible. As a result, the average shortest path will behave as for a lattice with $\langle \ell \rangle \sim N^{1/d}$ in a d-dimensional space. For an exponential fitness distribution $f(\eta) = e^{-\lambda \eta}$ and $h(r) = r^{-\beta}$, the authors of [12] find various degree distributions according to the value of β ranging from a power-law $p(k) \sim k^{-2}$ for $\beta \to 0$ to an exponential distribution for $\beta = d$. Various other cases were also studied in [10, 12] discussing within this model when scale-free distribution could appear.

Finally, we mention a generalization to other metrics than the spatial distance [14, 15]. In particular, in [14], the probability that two individuals are connected decreases with a particular distance between these individuals. This distance is computed in a 'social' space and measures the similarity for different social attributes. This model is able to reproduce some of the important features measured in social network such as a large clustering, positive degree correlations and the existence of dense communities. Serrano et al. [16] developed the idea of hidden metric space by using the one-dimensional circle as an underlying metric space in which nodes are uniformly distributed. A degree k drawn from a law $P(k) \sim k^{-\gamma}$ for each node and each pair of nodes is then connected with a probability $r(d; k, k')$ that depends on the distance d between the nodes and also on their respective degrees k and k'. In particular, they studied the following form

$$r(d; k, k') = \left(1 + \frac{d}{d_c(k, k')}\right)^{-\alpha} \qquad (11.12)$$

where $\alpha > 1$ and where $d_c(k, k') \sim kk'$ for example. The probability that a pair of nodes is connected decreases then with distance (as $d^{-\alpha}$) and increases with the product of their degrees kk'. In this case, a long distance can be compensated by large degrees, as it is observed in various real-world networks. In this model, in agreement with other models of spatial networks, we observe a large clustering (for α large enough).

11.2.2 Effect of Traffic

In [17], the authors proposed an interesting variant of the hidden variable model, applied to spatial networks of traffic. In particular, they had in mind the airport network where nodes represent airports and there is an edge between nodes which represent the number of passengers on a direct flight (if there is any). Previous measures on this network showed that the total traffic from an airport $s_i = \sum_j w_{ij}$ and the total outreach $s_i^d = \sum_j d_{ij}$ scale with exponents β and β_d, respectively

$$\begin{cases} s_i \sim k_i^{\beta} \\ s_i^d \sim k_i^{\beta_d} \end{cases} \tag{11.13}$$

with values of order $\beta \simeq 1.5 - 1.7$ and $\beta_d \simeq 1.4$ (see [17] and references therein). Various models had difficulties to explain these values and the model of [17] provides a simple explanation. It is based on the following main assumptions:

- The nodes are described by a hidden variable x_i drawn from probability distribution $\rho(x)$.
- To each pair (i, j) of nodes correspond a weight given by $w_{ij} \sim x_i x_j$. We also assume that there is cost function c_{ij}.
- A link between i and j exists only if the expected earnings – assumed to be a function of the weight – exceeds the cost: $f(w_{ij}) > c_{ij}$.

In addition, the nodes are randomly distributed in the 2d plane according to a spatial poisson process of intensity σ. We consider a given node (with fitness x) and take it as the origin O of a polar coordinate system. We denote by r the distance between this origin and other nodes. The average degree at O is then given by

$$\langle k(x) \rangle = \int \sigma dA \int_0^{\infty} \Theta \left[f(xy) - c(r) \right] \rho(y) dy \tag{11.14}$$

where A is the area and Θ the Heaviside function. Also, we assume that the cost $c_{ij} = c(d(i, j))$ between two nodes depends on their separation distance only. Similarly, the average strength and distance strength are given by the following expressions

$$\langle s(x) \rangle = \int \sigma dA \int_0^{\infty} xy \Theta \left[f(xy) - c(r) \right] \rho(y) dy \tag{11.15}$$

$$\langle s^d(x) \rangle = \int \sigma dA \int_0^{\infty} r \Theta \left[f(xy) - c(r) \right] \rho(y) dy \tag{11.16}$$

We integrate these equations and obtain

$$\langle k(x) \rangle = \pi \sigma M_2(\rho) x^2 \qquad (11.17)$$

$$\langle s(x) \rangle = \pi \sigma M_3(\rho) x^3 \qquad (11.18)$$

$$\langle s^d(x) \rangle = \frac{2}{3} \pi \sigma M_3(\rho) x^3 \qquad (11.19)$$

where M_k ($k = 2, 3$) is the kth moment of the fitness distribution

$$M_k = \int \mathrm{d}x \rho(x) x^k \qquad (11.20)$$

These relations imply that

$$\langle s(k) \rangle \sim k^{3/2} \qquad (11.21)$$

$$\langle s^d(k) \rangle \sim k^{3/2} \qquad (11.22)$$

$$\langle w(k, q) \rangle \sim (kq)^{1/2} \qquad (11.23)$$

in very good agreement with real-world networks such as airline networks where an exponent value close to $3/2$ was measured [5, 17].

References

1. P. Erdös, A. Rényi, Publicationes Mathematicae **6**, 290 (1959)
2. P. Erdös, P. Rényi, Publ. Math. Inst. Hung. Acad. Sci. **5**, 17 (1960)
3. P. Erdös, P. Rényi, Acta. Math. Sci. Hung **12**, 261 (1961)
4. B. Bollobás, B. Béla, *Random Graphs*, vol. 73 (Cambridge university press, 2001)
5. A. Barrat, M. Barthelemy, A. Vespignani, *Dynamical Processes in Complex Networks* (Cambridge University Press, Cambridge, UK, 2008)
6. R. Ziff, R. Vigil, J. Phys. A: Math. Gen. **23**, 5103 (1990)
7. A. Denise, M. Vasconcellos, D. Welsh, Congr. Numer. **113**, 61 (1996)
8. S. Gerke, C. McDiarmid, Comb. Probab. Comput. **13**, 165 (2004)
9. C. McDiarmid, A. Steger, D. Welsh, J. Comb. Theory **93**, 187 (2005)
10. G. Caldarelli, A. Capocci, P.D.L. Rios, M.A. Munoz, Phys. Rev. Lett. **89**, 258702 (2002)
11. M. Boguñá, R. Pastor-Satorras, Phys. Rev. E **68**, 036112 (2003)
12. N. Masuda, H. Miwa, N. Konno, Phys. Rev. E **71**, 036108 (2005)
13. Y. Hayashi, I.P.S.J. Trans, Spec. Issue Netw. Ecol. **47**, 776 (2006)
14. M. Boguñá, R. Pastor-Satorras, A. Diaz-Guilera, A. Arenas, Phys. Rev. E **70**, 056122 (2004)
15. L.H. Wong, P. Pattison, G. Robins, Phys. A: Stat. Mech. Appl. **360**(1), 99 (2006)
16. M.A. Serrano, D. Krioukov, M. Boguñá, Phys. Rev. Lett. **100**, 078701 (2008)
17. M. Popović, H. Štefančić, V. Zlatić, Phys. Rev. Lett. **109**(20), 208701 (2012)

Chapter 12
Spatial Small-Worlds

A very important model in the development of the field of complex networks was proposed by Watts and Strogatz [1] and which interpolates between a lattice and the Erdos-Renyi random graph. In this way, we can produce a graph that has simultaneously a large clustering coefficient (a property of lattices) and a small diameter varying as $\log N$, a feature typical of random graphs such as in the Erdos-Renyi model. In the Watts-Strogatz model, there is an underlying lattice and it can thus be considered as a spatial network. We will discuss some of the properties of this network and end this chapter with a presentation of navigability problems and the demonstration of Kleinberg's result [2].

12.1 The Watts-Strogatz Model

Already in 1977, spatial aspects of the small-world problem were considered by geographers in the paper [3] but we had to wait until 1998 when Watts and Strogatz (WS) proposed a simple and powerful network model [1] which incorporates both a spatial component and long-range links. This model is obtained by starting from a regular lattice and by rewiring links at random with a probability p (Fig. 12.1).

The degree distribution of this network has essentially the same features as the ER random graph, but the clustering coefficient and the average shortest path depend crucially on the amount of randomness p. The average clustering coefficient has been shown to behave as [4]

$$\langle C(p) \rangle \simeq \frac{3(m-1)}{2(2m-1)}(1-p)^3 \qquad (12.1)$$

where the average degree is $\langle k \rangle = 2m$. The average shortest path has been shown to scale as [5, 6]

© The Author(s), under exclusive license to Springer Nature Switzerland AG 2022
M. Barthelemy, *Spatial Networks*,
https://doi.org/10.1007/978-3-030-94106-2_12

Fig. 12.1 Construction of the Watts-Strogatz model for $N = 8$ nodes. At $p = 0$ each node is connected to its four nearest neighbors and by increasing p an increasing number of edges is rewired. Adapted from Watts and Strogatz [1]

$$\langle \ell \rangle \sim N^* \mathscr{F} \left(\frac{N}{N^*} \right) \tag{12.2}$$

where the scaling function behaves as

$$\mathscr{F}(x) \sim \begin{cases} x & \text{for } x \ll 1 \\ \ln x & \text{for } x \gg 1 \end{cases} \tag{12.3}$$

The crossover size scales as $N^* \sim 1/p$ [4–6] which basically means that the crossover from a large-world to a small-world occurs for an average number of shortcuts of the order of one

$$N^* p \sim 1 \tag{12.4}$$

The network can thus be seen as clusters of typical size $N^*(p) \sim 1/p$ connected by shortcuts.

The interest in these networks is that they can simultaneously display some features typical of random graphs (with a small-world behavior $\ell \sim \log N$) and of clustered lattices with a large average clustering coefficient (while for the ER random graph, the clustering coefficient is small $\langle C \rangle \sim 1/N$).

12.2   Spatial Generalizations in Dimension d

One of the first variants of the Watt-Strogatz model was proposed in [2, 7, 8] and it was subsequently generalized to higher dimensions d [9]. In this variant (see Fig. 12.2), nodes are located on a regular lattice in d-dimensions with periodic boundaries. For each node, we add a shortcut with probability p which implies that on average there will be pN additional shortcuts. The length of these links are assumed to follow the distribution

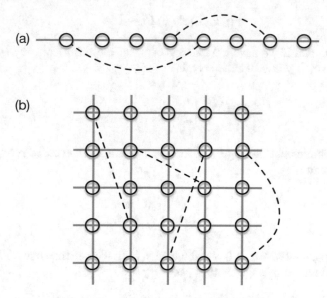

Fig. 12.2 Schematic representation of spatial small-world in **a** one dimension and **b** two dimensions. The dashed lines represent the long-range links occurring with probability $q(\ell) \sim \ell^{-\alpha}$. Figure inspired from [9]

$$q(\ell) \sim \ell^{-\alpha} \tag{12.5}$$

The main idea for justifying this choice is that if shortcuts have to be physically realized there is a cost associated with their length and thefore a probability that decreases with the length.

Concerning the average shortest path, it is clear that if α is large enough, the shortcuts will be small and the behavior of $\langle \ell \rangle$ will be 'spatial' with $\langle \ell \rangle \sim N^{1/d}$. On the other hand, if α is small enough we expect a small-world behavior $\ell \sim \log N$. In fact, various studies [7, 9, 10] discussed the existence of a threshold α_c separating the two regimes, small- and large-world. We follow here the discussion of [11] who studied carefully the behavior of the average shortest path. The probability that a shortcut is 'long' is given by

$$P_c(L) = \frac{\int_{(1-c)L}^{L} q(\ell) d\ell}{\int_{\ell_m}^{L} q(\ell) d\ell}$$
$$= \frac{L^{1-\alpha} - (1-c)^{1-\alpha} L^{1-\alpha}}{L^{1-\alpha} - \ell_m^{1-\alpha}} \tag{12.6}$$

where c is small but non-zero. This expression corresponds to the probability that a shortcut has a length in the interval $[(1-c)L, L]$. The critical fraction of shortcuts $p^*(L)$ then satisfies the following condition

$$P_c(L)p^*(L)L^d \sim 1 \tag{12.7}$$

which means that if we have a fraction $p > p^*$ of long shortcuts, the system will behave as a small-world. We then obtain

$$p^*(L) \sim \begin{cases} L^{-d} & \text{if } \alpha < 1 \\ L^{\alpha-d-1} & \text{if } \alpha > 1 \end{cases} \tag{12.8}$$

and a logarithmic behavior $\log L/L^d$ for $\alpha = 1$. For a given value of p we thus have one length scale

$$L^*(p) \sim \begin{cases} p^{-1/d} & \text{if } \alpha < 1 \\ p^{1/(\alpha-d-1)} & \text{if } \alpha > 1 \end{cases} \tag{12.9}$$

which in the special case $\alpha = 0$ was obtained in [12]. We will then have the following scaling form for the average shortest path

$$\langle \ell \rangle = L^* \mathcal{F}_\alpha \left(\frac{L}{L^*} \right) \tag{12.10}$$

where the scaling function varies as

$$\mathcal{F}_\alpha(x) \sim \begin{cases} x & \text{if } x \ll 1 \\ \ln x & \text{if } x \gg 1 \end{cases} \tag{12.11}$$

(or even a function of the form $(\ln x)^{\sigma(\alpha)}$ with $\sigma(\alpha) > 0$ for $x \gg 1$). The characteristic length for $\alpha > 1$ thus scales as

$$L^*(p) \sim p^{1/(\alpha-d-1)} \tag{12.12}$$

and displays a threshold value $\alpha_c = d + 1$, a value already obtained with the average clustering coefficient for $d = 1$ in [9]. For $\alpha > \alpha_c$, the length scale $L^*(p)$ is essentially finite and less than 1, which means that for all values of L the system has a large-world behavior with $\langle \ell \rangle \sim L$. In other words, the links in this case cannot be long enough and the graph can always be coarse-grained to reproduce an almost regular lattice. In the opposite case $\alpha < \alpha_c$, the length $L^*(p)$ diverges for $p \to 0$ and there will always be a regime such that $L^*(p) \gg L$ implying a small-world logarithmic behavior.

Finally, we mention a numerical study [13] of this model which seems to show that for $\alpha > d$ there are two regimes. First, for $d < \alpha < 2d$

$$\langle \ell \rangle \sim (\log N)^{\sigma(\alpha)} \tag{12.13}$$

with

$$\sigma = \begin{cases} \frac{1/\alpha}{2-\alpha} & \text{for } d = 1 \\ \frac{4/\alpha}{4-\alpha} & \text{for } d = 2 \end{cases} \qquad (12.14)$$

The second regime is obtained for $\alpha > 2d$ where the 'spatial' regime $\langle \ell \rangle \sim N^{1/d}$ is recovered. We note that numerically, the scaling prediction of [11] with two regimes only and the result of [13] are however difficult to distinguish. For $d = 1$ there are no discrepancies ($\sigma = 1$ for $\alpha = d = 1$) and for $d = 2$, the results for $\alpha = 2$ and $\alpha = 4$ are consistent with the analysis of [11]. A problem thus subsists here for $d = 2$ and $\alpha = 3$ for which $\sigma = 4/3$, a value probably difficult to distinguish numerically from corrections obtained at $\alpha = \alpha_c = d + 1$.

12.3   Navigability in the Kleinberg Model

12.3.1   Searchable Networks

The original 1967 experiment of Milgram [14] showing that the average shortest path in North-America is around 6 raises a number of questions. The first one is about the structure of the social network and it is now relatively clear that enough shortcuts will modify the scaling of $\langle \ell \rangle$ and induce a logarithmic dependence on the number of nodes. Another question raised by Kleinberg [2] is actually how a node can find a target efficiently with a local knowledge of the network only (the answer being trivial if you know the whole network). It thus seems that in some way the social network is search-efficient—or is a *searchable* or *navigable* network—meaning that the shortcuts are easy to find, even by having access to local information only. In these cases, one speaks of navigability or searchability when the greedy search is efficient.

   This problem goes beyond social networks as decentralized searches, where nodes only possess local information (such as the degree or the location of their neighbors for example) in complex networks have many applications ranging from sensor data in wireless sensor networks, locating data files in peer-to-peer networks, finding information in distributed databases (see for example [15] and references therein). It is thus important to understand the efficiency of local search routines and the effect of the network structure on such decentralized algorithms.

   In the case of social networks, it seems that there is a local spatial component comprising nodes that belong to the spatial neighborhood (such as in a regular lattice) and a purely social component, not correlated with space and which can connect regions which are geographically very far apart. In a search process, it is thus natural to try these links which open the way to very different parts of the world (in Milgram's experiment it is indeed interesting to note that individuals were passing the message only according geography or proximity in the space of professional activities [16],

Fig. 12.3 Lower bounds of the exponent governing the behavior of the average delivery time $\overline{T}$ as a function of the exponent α controlling the distribution of shortcuts. After [2]

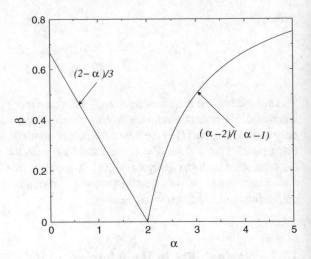

as it was known to them that the target individual was a lawyer). In order to quantify this, Kleinberg [2] constructed a d-dimensional Watts-Strogatz model where each node i of the lattice

- (i) is connected to all neighbors such that their lattice distance is less than p (with $p \geq 1$).
- (ii) has q shortcuts to node j with a probability decreasing with the distance

$$p(i \rightarrow j) \sim d_E(i, j)^{-\alpha} \qquad (12.15)$$

where α is a tunable parameter.

The greedy search process is the following one: a message needs to be sent to a target node t whose geographical position is known and a node i which receives the message forwards it to one of its neighbor j that is the closest (geographically) to t. This is the simplest decentralized algorithm that we can construct (and which requires only geographical information). The most important figure of merit for this type of algorithm is the delivery time T (or its average $\overline{T}$ which is easier to estimate analytically) and its scaling with the number of nodes N. Kleinberg found bounds on the exponent of $\overline{T}$ (see Fig. 12.3) and the important result is that the delivery time is optimal for $\alpha = d$ for which it scales as $\log^2 N$ while for $\alpha \neq d$, it scales faster (as a power of N). This behavior can be intuitively understood: for $\alpha > d$, long links are rare and the network looks essentially like a lattice (with a renormalized spacing). In the opposite case $\alpha < d$, shortcuts are all long and not necessarily useful. The best case is obtained when the shortcuts explore all spatial scales, which is obtained for $\alpha = d$ (This result was extended in [17] to the case of small-world networks constructed by adding shortcuts to a fractal set of dimension d_f).

12.3.2   Sketch of Kleinberg's Proof

Inspired from Kleinberg's original rigorous derivation [18] of the bounds shown in Fig. 12.3, we can give the following hand-waving arguments in order to grasp some intuition about the effects of the link distribution on the average number of steps to reach a target in a decentralized algorithm (we will discuss in detail here the case for $d = 2$ but the extension to a generic d is trivial). For the interested reader, we also note that a detailed study of the $d = 1$ case is done in [19], that exact asymptotic results were obtained in [20], and that the 'greedy' paths connecting a source to the target were studied in [21] by defining a greedy connectivity.

For $\alpha = 2$ (and we assume here that $p = q = 1$), the probability to jump from node u to node v is given by

$$P(u \rightarrow v) = \frac{1}{Z} \frac{1}{d_E(u, v)^2} \tag{12.16}$$

where Z is the normalization constant given by

$$Z = \sum_{v \neq u} \frac{1}{d_E(u, v)^2} \simeq 2\pi \int_1^{N/2} \frac{r\,dr}{r^2} \tag{12.17}$$

$$\simeq 2\pi \ln N \tag{12.18}$$

implying that $P(u \rightarrow v) \sim 1/\ln N d_E(u, v)^2$ (here and in the following we will use continuous approximation and neglect irrelevant prefactors; for rigorous bounds we refer the reader to [18]). Following Kleinberg [18], we say that the execution of the algorithm is in phase j when the lattice distance d from the current node (which is holding the message) is such that $2^j \leq d < 2^{j+1}$. The largest phase is then $\ln N$ and the smallest 0 when the message reaches the target node. The goal at this point is to compute the average number of steps $\overline{T}$ to reach the target. For this we decompose the problem in computing the average time duration $\overline{T_j}$ that the message stays in phase j. We thus have to compute the probability that the message leaves the phase j and jumps in the domain B_j defined as the set of nodes within a distance 2^j to the target node t (see Fig. 12.4). The size of this set B_j is $|B_j| \sim 2^{2j}$ and the distance between u and any node of B_j is $d(u, v \in B_j) \leq 2^{j+2}$. The probability to get out of phase j by using a long-range link is thus

$$P_{out} \sim \frac{|B_j|}{(2^{j+2})^2 \ln N} \sim \frac{1}{\ln N} \tag{12.19}$$

(the actual exact bound found by Kleinberg is $P_{out} \geq 1/(136 \ln N)$. We then have $P(T_j = i) = [1 - P_{out}]^i P_{out}$ from which we obtain $\overline{T_j} \sim \ln N$. The average time to reach the target is then

Fig. 12.4 The message goes
from the phase j to the
domain B_j and we have to
compute the corresponding
probability. The domain B_j
is defined as the set of nodes
within a distance 2^j to the
target node t

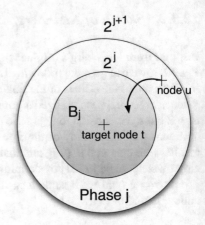

$$\overline{T} = \sum_{j=0}^{\ln N} \overline{T_j} \sim \ln^2 N \tag{12.20}$$

which is the minimum time obtained for a decentralized algorithm for $\alpha = 2$ (and $\alpha = d$ for the general d-dimensional case).

We now consider the minimum scaling of $\overline{T}$ in the case $\alpha < 2$ (and general p and q). The normalization constant behaves then as

$$Z = \sum_{v \neq u} d_E(u, v)^{-\alpha} \simeq 2\pi \int_1^{N/2} r^{-\alpha} r \, dr \tag{12.21}$$

$$\sim \left(\frac{N}{2}\right)^{2-\alpha} \tag{12.22}$$

We assume now that the minimum number of steps to reach the target scales as N^δ. In this case, there is necessarily a last step along a long-range link leading to a node which is different from the target node and which is in the region U centered at t and of size $\sim pN^\delta$ (Fig. 12.5). The probability P_i that this long-range link leading to U at step i is given by

$$P_i \leq \frac{q|U|}{Z} \sim N^{2\delta - 2 + \alpha} \tag{12.23}$$

and the probability that it happens at any step less than N^δ is

$$P = \sum_{i \leq N^\delta} P_i \leq N^{3\delta - 2 + \alpha} \tag{12.24}$$

Fig. 12.5 One dimensional representation of Kleinberg's theorem in the case $\alpha < 2$

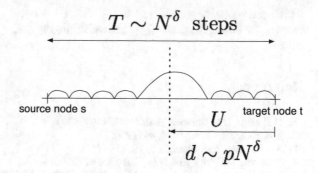

This probability is non-zero only if $3\delta - 2 + \alpha \geq 0$ leading to the minimum possible value for δ such that $\overline{T} \sim N^\delta$

$$\delta_{min} = \frac{2 - \alpha}{3} \tag{12.25}$$

(the d-dimensional generalization would give $\delta_{min} = (d - \alpha)/(d + 1)$).

In the last case $\alpha > 2$, we will have mostly short links and the probability to have a link larger than m is given by

$$P(d_E(u, v) > m) \sim \int_m^N \frac{r\,dr}{r^\alpha} \tag{12.26}$$

$$\sim m^{2-\alpha} \tag{12.27}$$

In the following we will use the notation $\epsilon = \alpha - 2$. The probability to have a jump larger than N^γ for $T < N^\beta$ is then given by

$$P(N^\gamma, T < N^\beta) \sim q N^\beta (N^\gamma)^{-\epsilon} \tag{12.28}$$

$$\sim q N^{\beta - \gamma\epsilon} \tag{12.29}$$

This probability will be non-zero for

$$\beta - \gamma\epsilon \geq 0 \tag{12.30}$$

Also, if at every step during a time $T \sim N^\beta$, we perform a jump of size N^γ the traveled distance must be of order N which implies that $N^\beta N^\gamma \sim N$ leading to the condition $\beta + \gamma = 1$. This last condition together with Eq. (12.30) leads to the minimum value of β

$$\beta_{min} = \frac{\alpha - 2}{\alpha - 1} \tag{12.31}$$

which can be easily generalized to $(\alpha - d)/(\alpha - d + 1)$ in d-dimensions. We thus recover the bounds δ_{min} and β_{min} shown in the Fig. 12.3.

References

1. D.J. Watts, D.H. Strogatz, Nature **393**, 440 (1998)
2. J.M. Kleinberg, Nature **406**, 845 (2000)
3. A. Stoneham, Env. Plann. A **9**, 185 (1977)
4. A. Barrat, M. Weigt, Eur. Phys. J. B **13**, 547 (2000)
5. M. Barthelemy, L.A.N. Amaral, Phys. Rev. Lett. **82**, 3180 (1999)
6. M. Barthelemy, L.A.N. Amaral, Phys. Rev. Lett. **82**, 5180 (1999)
7. S. Jespersen, A. Blumen, Phys. Rev. E **62**, 6270 (2000)
8. P. Sen, B. Chakrabarti, J. Phys. A **34**, 7749 (2001)
9. P. Sen, K. Banerjee, T. Biswas, Phys. Rev. E **66**, 037102 (2002)
10. C. Moukarzel, M.A. de Menezes, Phys. Rev. E **65**, 056709 (2002)
11. T. Petermann, P. De Los Rios, arXiv:cond-mat/0501 (2005)
12. M.E.J. Newman, D.J. Watts, Phys. Rev. E **60**, 7332 (1999)
13. K. Kosmidis, S. Havlin, A. Bunde, Europhys. Lett. **82**, 48005 (2008)
14. S. Milgram, Psychol. Today **2**, 60 (1967)
15. H. Thadakamalia, R. Albert, S. Kumara, New J. Phys. **9**, 190 (2007)
16. P. Killworth, H. Bernard, Soc. Netw. **1**, 159 (1978)
17. M. Roberson, D. ben Avraham, Phys. Rev. E **74**, 017101 (2006)
18. J. Kleinberg, The small-world phenomenon: an algorithmic perspective (1999). Technical report 99-1776, Cornell Computer Science
19. H. Zhu, Z.X. Huang, Phys. Rev. E **70**, 021101 (2004)
20. C.C. Cartozo, P. De Los Rios, Phys. Rev. Lett. **102**, 238703 (2009)
21. J. Sun, D. ben Avraham, Phys. Rev. E **82**, 016109 (2010)

Chapter 13
Growing Spatial Networks

Most networks, including spatial graphs, evolve and grow in time. Understanding the main processes governing this growth and the resulting structure is therefore crucial in many disciplines ranging from urban planning to the study of neural networks. There are essentially three ingredients for growing a network:

- At each time step, one node (or more) is added to the network
- New nodes are located according to a distribution that depends in general on the structure of the existing network.
- Once located in space, the new nodes connect to the existing network according to a certain rule.

In this chapter, we will first study the case where nodes are located at random, and where the connection rule is governed by preferential attachment. In a second part, we will discuss the case where nodes are located according to a potential that depends on the structure of the network. In order to introduce the different ingredients we will use the language of road networks in cities, although the model presented here is more general and could be applied to other spatial networks. We note here that according to the general definition given here, greedy models discussed in the Chap. 20 are also network growth model, but for the sake of clarity, we divided models where the connection rule is probabilistic (which will be the case here) and models with a deterministic rule such as local optimization.

13.1 Preferential Attachment and Space

We will review models of growing networks which essentially elaborate on the preferential attachment model [1, 2] described by a propensity to connect a new node to an already well-connected one, which is probably an important ingredient in the formation of various real-world networks.

© The Author(s), under exclusive license to Springer Nature Switzerland AG 2022
M. Barthelemy, *Spatial Networks*,
https://doi.org/10.1007/978-3-030-94106-2_13

In the absence of space, the process to generate such a Barabasi-Albert (BA) network is extremely simple. Starting from a small 'seed' network, we introduce a new node n at each time step which is allowed to make m connections towards nodes i with a probability

$$\Pi_{n\to i} = \frac{k_i(t-1)}{\sum_{j\in G_{t-1}} k_j(t-1)} \tag{13.1}$$

where G_t is the network at time t and where $k_i(t)$ is the degree of node i at time t. We refer the interested reader to the various books and reviews that describe in detail this model [3–8] . In particular, the degree distribution behaves as a power-law with exponent

$$P(k) \sim k^{-\gamma} \tag{13.2}$$

with $\gamma = 3$, the average shortest path behaves at the dominant order as $\log N$, and the average clustering coefficient is given by

$$\langle C \rangle = \frac{m}{8N}(\ln N)^2 \tag{13.3}$$

while $C(k) \sim 1/k$.

In many networks such as transportation or communication networks, distance is however a relevant parameter and real-world examples suggest that when long-range links are existing, they usually connect to hubs-the well-connected nodes. Many variants of the BA model were proposed and a few of them were concerned with space. The growth process is the same as for the BA model, but in addition one has to specify the location of the new node. In most models, the location is taken at random and uniformly distributed in space. The attachment probability is then written as

$$\Pi_{n\to i} \propto k_i F[d_e(n, i)] \tag{13.4}$$

where F is a function of the euclidean distance $d_e(n, i)$ from the node n to the node i. When F is a decreasing function of distance (as in most cases), this form Eq. (13.4) implies that new links preferentially connect to hubs, unless the hub is too far in which case it is better to connect to a less connected node but closer in space. In order to have long links, the target node must have a large degree in order to compensate for a small $F(d)$ such that $kF(d) \sim 1$. This is for instance the case for airlines: Short connections go to small airports while long connections point preferably to big airports, i.e. well-connected nodes.

13.1.1 *Preferential Attachment and Distance Selection*

Several models including distance were proposed [9–17] and we review here the main results obtained in these studies. The N nodes of the network are supposed to be in a d-dimensional space of linear size L and we assume that they are distributed randomly in space with uniform density ρ (one could use other distributions: for instance in cities the density usually decreases from the center [18]). The case of randomly distributed points is interesting since it preserves on average natural symmetries such as translational and rotational invariance, in contrast with regular lattices.

The influence of space is encoded in the function $F(d)$ and we can distinguish essentially two different categories that were considered in the literature.

13.1.1.1 Finite Range Case.

In this case, the function $F(d)$ decreases sharply with distance, typically as an exponential [17]

$$F(d) = e^{-d/r_c} \tag{13.5}$$

and thus introduces a new scale in the system, the interaction range r_c. When the interaction range is of the order of the system size (or larger), the distance is irrelevant and the obtained network will be scale-free. In contrast, when the interaction range is small compared to the system size, we expect new properties and a crossover between these two regimes.

The important dimensionless parameters are here the average number n of points in a sphere of radius r_c

$$n = \rho r_c^d \frac{\pi^{d/2}}{\Gamma(1 + \frac{d}{2})} \tag{13.6}$$

and the ratio which controls the importance of spatial effects

$$\eta = \frac{r_c}{L} \tag{13.7}$$

(where L is the system size). It can then be shown [17] that the degree distribution follows the scaling form valid only for $\eta \ll 1$

$$P(k) \sim k^{-\gamma} f(k/k_c) \tag{13.8}$$

with $\gamma = 3$ and where the cut-off k_c behaves as

$$k_c \sim n^\beta \tag{13.9}$$

where $\beta \simeq 0.13$. The distance effect thus limits the choice of available connections and thereby limits the degree distribution for large values.

Also, when the distance effect is important we expect a large value of the average clustering coefficient. In the limit of small η, we expect the result of random geometric graphs (see Chap. 15 and [12]) to hold

$$C_0 \equiv \langle C \rangle (\eta = 0) = 1 - 3\sqrt{3}/4\pi \simeq 0.59 \qquad (13.10)$$

(for $d = 2$). We expect to recover this limit for $\eta \to 0$ and for an average connectivity $\langle k \rangle = 6$ which is a well-known result in random geometry. If η is not too small, the preferential attachment is important and induces some dependence of the clustering coefficient on n. In addition, we expect that $\langle C \rangle (\eta)$ will be lower than $\langle C \rangle (0)$ since the longer links will not connect to the nearest neighbors. Numerical results show that there is a good collapse (see Fig. 13.1a) when $\langle C \rangle (\eta)$ is expressed in terms of n and is a decreasing function: when n increases the number of neighbors of the node will increase thus decreasing the probability that two of them will be connected.

The average shortest path is described by a scaling ansatz which governs the crossover from a spatial to a scale-free network

$$\langle \ell(N, \eta) \rangle = [N^*(\eta)]^\alpha \Phi_\alpha \left[\frac{N}{N^*(\eta)} \right] \qquad (13.11)$$

with

$$\Phi_\alpha(x) \sim \begin{cases} x^\alpha & \text{for } x \ll 1 \\ \ln x & \text{for } x \gg 1 \end{cases} \qquad (13.12)$$

The crossover size N^* depends on η and we can find its behavior in two extreme cases. For $\eta \gg 1$, space is irrelevant and

$$N^*(\eta \gg 1) \sim N_0 \qquad (13.13)$$

where N_0 is a finite constant. When $\eta \ll 1$, the existence of long-range links will determine the behavior of $\langle \ell \rangle$. If we denote by $a = 1/\rho^{1/d}$ the typical inter-node distance, the transition from a large to a small-world will be observed for $r_c \sim a(N^*)$ which leads to

$$N^*(\eta \ll 1) \sim \frac{1}{\eta^d} \qquad (13.14)$$

In Fig. 13.1b, the ansatz Eq. (13.11) together with the results Eqs. (13.13), (13.14) is shown. This data collapse is obtained for $\alpha \simeq 0.3$ and $N_0 \simeq 180$ (for $d = 2$). For $N > N^*$, the network is a small-world: the diameter is growing with the number of points as $\langle \ell \rangle \sim \log N$. In the opposite case of the spatial network with a small interaction range, the network is much larger: To go from a point A to a point B, we

Fig. 13.1 **a** Clustering
coefficient versus the mean
number $n = \rho \pi r_c^2$ of points
in the disk of radius r_c
(plotted in Log-Lin). The
dashed line corresponds to
the theoretical value C_0
computed when a vertex
connects to its adjacent
neighbors without
preferential attachment. **b**
Data collapse for the average
shortest path obtained. The
first part can be fitted by a
power-law with exponent
≈ 0.3 followed by a
logarithmic regime for
$N > N^*$. Both figures are
taken from [17]

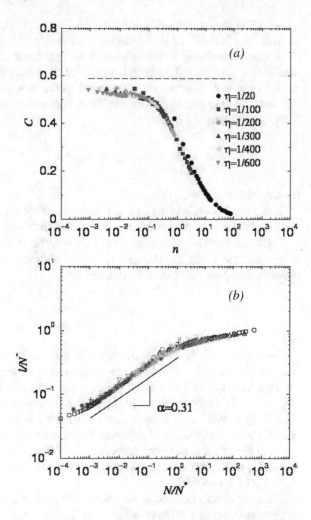

essentially have to pass through most of the points in between and the behavior of
this network is much that of a lattice with $\langle \ell \rangle \sim N^\alpha$, although the average shortest
path is here smaller probably due to the existence of some rare longer links (in the
case of a lattice we expect $\alpha = 1/d$). Probably larger networks and better statistics
are needed here.

This model was extended [19] in the case of weighted growing networks in a
two-dimensional geometrical space. The model considered consists of growth and
the probability that a new site connects to a node i is given by

$$\Pi_{n \to i} = \frac{s_i^w e^{-d_{ni}/r_c}}{\sum_j s_j^w e^{-d_{nj}/r_c}}, \tag{13.15}$$

where s_i^w is the strength of node i ($s_i^w = \sum_{j \in \Gamma(i)} w_{ij}$), and where r_c is a typical scale and d_{ni} is the Euclidean distance between n and i. This rule of *strength driven preferential attachment with spatial selection*, generalizes the preferential attachment mechanism driven by the strength to spatial networks. Here, new vertices connect more likely to vertices which correspond to the best interplay between Euclidean distance and strength. In this model, the weights are also updated according to the following rule studied in [20]

$$w_{ij} \rightarrow w_{ij} + \delta \frac{w_{ij}}{s_i^w}. \qquad (13.16)$$

for all neighbors $j \in \Gamma(i)$ of i.

This model contains thus two relevant parameters: the ratio between the typical scale and the size of the system $\eta = r_c/L$, and the ability to redistribute weights, δ.

The most important results concerning the traffic are the following. The exponents β_w characterizes how the strength s_i^w scales with the degree, and the exponant β_d characterizes how the 'outreach' defined by

$$s_i^d = \sum_{j \in \Gamma(i)} d_E(i, j) \qquad (13.17)$$

()where $d_E(i, j)$ is the euclidean distance between i and j) scales with the degree. The correlations appearing between traffic and topology of the network are largely affected by space as the value of the exponents β_w and β_d depend on η (for β_d see Fig. 13.2). Strikingly, the effect of the spatial constraint is to increase both exponents β_w and β_d to values larger than 1 and although the redistribution of the weights is linear, non-linear relations $s^w(k)$ and $s^d(k)$ as a function of k appear. For the weight strength the effect is not very pronounced with an exponent of order $\beta_w \approx 1.1$ for $\eta = 0.01$, while for the distance strength the non-linearity has an exponent of order $\beta_d \approx 1.27$ for $\eta = 0.02$.

The nonlinearity induced by the spatial structure can be explained by the following mechanism affecting the network growth. The increase of spatial constraints affects the trend to form global hubs, since long distance connections are less probable, and drives the topology towards the existence of 'regional' hubs of smaller degree. The total traffic however is not changed with respect to the case $\eta = \infty$, and is in fact directed towards these regional hubs. These medium-large degree vertices therefore carry a much larger traffic than they would do if global hubs were available, leading to a faster increase of the traffic as a function of the degree, eventually resulting in a super-linear behavior. Moreover, as previously mentioned, the increase in distance costs implies that long range connections can be established only towards the hubs of the system: this effect naturally leads to a super-linear accumulation of $s^d(k)$ at larger degree values.

The spatial constraints act at both the local and global levels of the network structure by introducing a distance cost in the establishment of connections. It is therefore important to look at the effect of space in global topological quantities such as the betweenness centrality. Hubs are natural crossroads for paths and it is

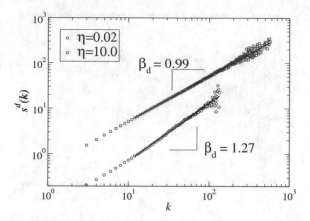

Fig. 13.2 Distance strength versus k for two different cases ($\eta = 0.02$ and $\eta = 10.0$). When η is not small, space is irrelevant and there are no correlations between degree and space as signalled by $\beta_d \approx 1$. When spatial effects are important ($\eta = 0.02 \ll 1$), non-linear correlations appear and $\beta_d > 1$. We observe a crossover for $k \simeq 10 - 20$ to a power-law behavior and the power-law fit over this range of values of k is shown (full lines). Figure taken from [19]

natural to observe a correlation between g and k as expressed in the general relation $g(k) \sim k^{\mu}$ where the exponent μ depends on the characteristics of the network. We expect this relation to be altered when spatial constraints become important. In particular, the betweenness centrality displays relative fluctuations which increase as η decreases and become quite large. This can be understood by noticing that the probability to establish far-reaching short-cuts decreases exponentially in Eq. (13.15) and only the large traffic of hubs can compensate this decay. Far-away geographical regions can thus only be linked by edges connected to large degree vertices, which implies a more central role for these hubs. The existence of fluctuations means that nodes with small degree may have a relatively large betweenness centrality (or the opposite), as observed in the air-transportation network (see [21]). This model defines an intermediate situation in that we have a random network with space constraints that introduces a local structure since short distance connections are favored. Shortcuts and long distance hops are present along with a spatial local structure that clusters spatially neighboring vertices. In Fig. 13.3 we plot the average distance $d(G, C)$ between the barycenter G and the 10 most central nodes. As expected, as spatial constraints become more important, the most central nodes get closer to the spatial barycenter of the network.

13.1.1.2 Power Law Decay of $F(d)$

In this case, the function F in Eq. (13.4) is varying as

$$F(d) = d^{\alpha} \tag{13.18}$$

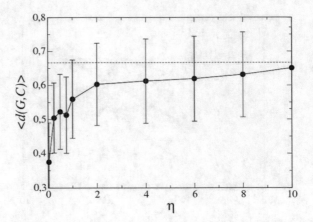

Fig. 13.3 Average Euclidean distance between the barycenter G of all nodes and the 10 most central nodes (C) versus the parameter η (Here $\delta = 0$, $N = 5,000$ and the results are averaged over 50 configurations). When space is important (i.e. small η), the central nodes are closer to the gravity center. For large η, space is irrelevant and the average distance tends to the value corresponding to a uniform distribution $\langle r \rangle_{unif} = 2/3$ (dotted line). Figure from [19]

and this problem was considered in [9, 15, 16]. The numerical study presented in [16] shows that in the two-dimensional case, for all values of α the average shortest path behaves as $\log N$. The degree distribution is however different for $\alpha > -1$ where it is broad, while for $\alpha < -1$, it is decreasing much faster (the numerical results in [16] suggest according to a stretched exponential).

In [15], Manna and Sen studied the same model for various dimensions and for values of α going from $-\infty$ to $+\infty$ where the node connects to the closest and the farthest node, respectively (Fig. 13.4). These authors indeed found that if $\alpha > \alpha_c$ the network is scale-free and in agreement with [16] that $\alpha_c(d = 2) = -1$ while for large dimensions α_c decreases with d (the natural guess $\alpha_c = 1 - d$ is not fully supported by their simulations). This study was complemented by another one by the same authors [22] in the $d = 1$ case and where the probability to connect to a node i is given by (which was already proposed in [9])

$$\Pi_{n \to i} \sim k_i^\beta d_E(n, i)^\alpha \tag{13.19}$$

For $\alpha > \alpha_c = -0.5$ the network is scale-free at $\beta = 1$ with an exponent $\gamma = 3$. They found a scale-free network for a line in the $\alpha - \beta$ plane and also for $\beta > 1$ and $\alpha < -0.5$. The degree-dependent clustering coefficient $C(k)$ behaves as

$$C(k) \sim k^{-b} \tag{13.20}$$

where the authors found numerically that b varies from 0 to 1 (which is the value obtained in the BA case).

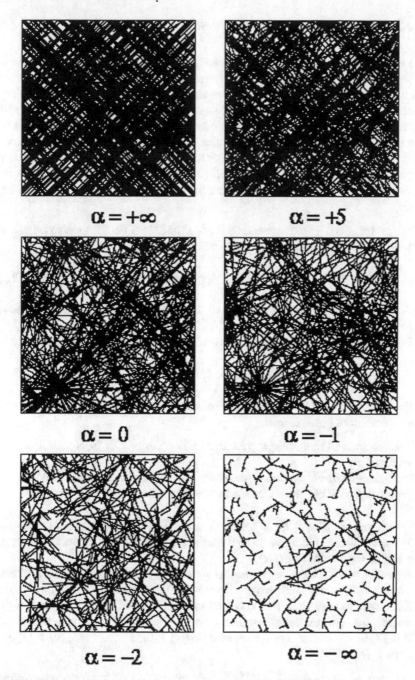

Fig. 13.4 Various networks obtained with the rule $F(d) = d^\alpha$. Figure taken from [15]

13.1.2   Searching in Spatial Scale-Free Networks

We saw in Chap. 12 that Kleinberg showed how a simple greedy search passing the message to the neighbor which is the nearest to the target is able in the best case to find its target in a squared logarithmic time. The lattice considered in Kleinberg's paper is a variant of the Watts-Strogatz model and has a low degree heterogeneity. When a large degree heterogeneity is present it is not clear that the greedy search used in [23] will work well, as it might be best to jump to a hub even if there is neighbor closer to the target node. In order to understand the effect of heterogeneity in spatial networks, Thadakamalla et al. [24] studied decentralized searches in a family of spatial scale-free network where the nodes are located in a d-dimensional space:

- With probability $1 - p$ a new node n is added and is connected to an existing node i with a preferential attachment probability weighted by the distance

$$\Pi_{n \to i} \propto k_i F(d_E(i, j)) \tag{13.21}$$

where $F(d)$ is a decreasing function of distance (and can be chosen as a power law $d^{-\sigma}$ or as an exponential $\exp(-d/d_0)$).
- With probability p, a new edge is connecting existing nodes with probability

$$\Pi_{i \leftrightarrow j} \propto k_i k_j F(d_E(i, j)) \tag{13.22}$$

The authors of [24] investigated the following search algorithms which cover a broad spectrum of possibilities:

1. *Random walk:* The message goes from a node to one of its randomly chosen neighbor.
2. *High-degree search:* The node passes the message to the neighbor which has the largest degree. This algorithm is already very efficient for non-spatial network [25].
3. *Greedy search:* This is the algorithm used in Kleinberg's study [23] and where the node i passes the message to the neighbor which is the closest to the target (i.e. with the smallest $d_E(i, t)$).
4. *Algorithms 4–8:* The node passes the message to the neighbor which minimizes a function $\mathcal{F}[\|\rangle, \lceil \varepsilon(), \sqcup)]$ which depends both on the degree of the node and its distance to the target. The function $\mathcal{F}$ considered here are: (i) $F_1[k, d] = d - f(k)$ where $f(k)$ is the expected maximum length of an edge from a node with degree k; (ii) $F_2[k, d] \propto d^k$; (iii) $F_3[k, d] = d/k$; (iv) $F_4[k, d] \propto d^{\ln k + 1}$; (v) $F_5[k, d] = d/(\ln k + 1)$.

The main result obtained in [24] is that algorithms $(4 - 8)$ perform very well and are able to find a path between the source and the target whose length is at most one hop more than the average shortest path. This result is surprising: the calculation of the shortest average path requires the knowledge of the whole network, while the algorithms used here have only local information. This success can be attributed to the fact that scale-free networks have hubs which allows to find efficiently the target. It should also be noted that the greedy search performance is not too bad but with the severe drawback that in some cases it doesn't find the target and stays stuck in a loop, which never happens with algorithms $(4 - 8)$. Similar results were obtained for different values of p and σ. These results allow the authors to claim that the class of spatial networks considered here belong to the class of searchable networks. The authors checked with these different algorithms that it is also the case for the US airline network.

Finally, we mention Hajra and Sen [26] who studied the effect of the transition scale-free/homogeneous network ([15]) on the navigability for three different search algorithms. In particular, they showed that the effect of the transition on navigability is marginal and is the most pronounced on the highest degree-based search strategy which is less efficient in the power-law regime.

13.2 Attraction Potential Models

In previous sections of this chapter, nodes are added to the system one at the time and connect to the existing network. We will consider another type of models where both the choice of location and the attachment rule are different. Although the primary interest was on roads' networks, it is worth mentioning that transportations networks appear in variety of different fields including plant/leaves morphology [27], river networks [28], mammalian circulatory systems [29, 30], networks for commodities delivery [31], and technological networks [32] and that the common purpose of these structures is to convey energy or matter from one point to another.

As we saw in various chapters in this book, empirical studies have shown that roads' networks, despite the peculiar geographical, historical, social-economical mechanisms that have shaped distinct urban areas in different ways, exhibit unexpected quantitative similarities, suggesting the possibility to model these systems through quite general mechanisms.

In the simplified model presented in this section, we represent cities as a collection of points scattered on a two dimensional area (a square throughout this study), and connected by a roads' network. The description of the street network adopted here, therefore, consists of a graph whose links represent roads, and vertices represent roads' intersections.

We will discuss the mechanisms of road formation and location choice separately. In particular, we explicitly consider the shape of the network and model its evolution as the result of a local cost-optimization principle [33], akin to greedy models developed in Chap. 20. We will first consider a connection rule such that a new node will connect in an economical way to the network. In a second part, we will consider the case where a new node will be located in x with a probability of the form

$$P(x) \sim e^{-W(x)} \tag{13.23}$$

which generalizes the case considered so far where nodes are distributed uniformly in the plane. Here, the 'potential' $W(x)$ depends in general on the state of the network that exists before the new node enters the system. In [33] this potential depends on the centrality and the density, and in [34] it depends on the network only.

13.2.1 The Connection Rule

When new nodes (or 'centers' such as new homes or businesses in the case of street networks) appear, they need to connect to the existing road network. We are here in a particular situation where (i) the network is still not very dense, or (ii) the new center is important enough to trigger the evolution of the transportation network. In the opposite situation where small centers (such as homes) appear, they usually connect to an existing network without triggering the evolution of the network. If only one new center is present, it is reasonable to assume that it will connect to the nearest point of the existing road network. When two or more new centers (as in Fig. 13.5) want to connect to the same point in the network, we assume that economic considerations impose that a single road—from the chosen network's point—is built to serve both of them. In this Fig. 13.5, the nearest point in the network to both new nodes A and B is M. We grow a single new portion of road MM' of fixed length dx in order to grant the maximum reduction of the cumulative distance of A and B from the network. This translates in the requirement that

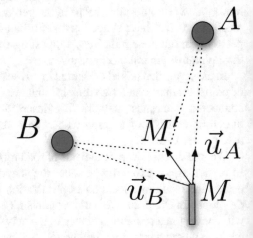

Fig. 13.5 The nearest road to the centers A and B is M. The road will grow to point M'. The proposed minimum expenditure principle suggests that the next point M' will be such that the variation of the total distance to the two points A and B is maximal. Figure taken from [35]

$$\delta d = []d(M', A) + d(M', B)] - [d(M, A) + d(M, B)] \tag{13.24}$$

is maximal (dx being fixed). A simple calculation shows the maximization of δd leads to

$$\overrightarrow{dMM'} \propto \overrightarrow{u}_A + \overrightarrow{u}_B \tag{13.25}$$

where $\overrightarrow{u}_A$ ($\overrightarrow{u}_B$) is the unitary vector from M to A (B). The procedure described above is iterated until the road from M reaches the the line connecting A and B, where a singularity occurs: $\overrightarrow{dMM'} = 0$. From there two independent roads to A and B need to be build to grant connection to the two new centers. The rule in Eq. (13.25) can be easily generalized to the general case of n new centers, and, interestingly, was proposed in [27] in the context of visualization of leaves' venation patterns. The present mechanism obviously provides nodes with a single connection to the network, but it can be extended, through the notion of relative neighborhood, to connect new nodes to more than one node in the existing network, therefore creating loops. We refer the interested reader to [33, 35] for a detailed exposition of the algorithm and of its extension to the multiple connection case.

13.2.2 Uniform Distribution of Nodes

We first assume that the nodes/centers are distributed uniformly in the plane and that there are no correlations between the evolution of the network and the location choice for new centers.

We show in Fig. 13.6 examples of patterns obtained for different times. The model also gives information about the time evolution of the road network. At earlier times, the density is low and the typical inter-distance between nodes is large (see Fig. 13.6). As time passes, the density increases and the typical length to connect a center to the existing road network becomes shorter. Since the number of points grows with time, the simple assumption that the typical road length is given by $1/\sqrt{\rho}$ leads to $\ell_1 \sim 1/\sqrt{t}$ which is indeed what we observe in this model.

Visual similarities are however not enough and we now compare the emergent properties of this model with empirical findings. The ratio $e = E/N$ is initially close to 1 (and the corresponding network for low density is tree-like) and increases very fast with e reaching a value of order 1.25 which is in the ballpark of empirical findings (see Chap. 4 and [36]). We also computed the cumulative length of the roads produced by the model (Fig. 13.7a) and found a behavior of the form $a\sqrt{N}$ with $a \approx 1.90$ in reasonable agreement with the empirical findings as well as the shape factor distribution (Fig. 13.7b): we find an average value $\phi = 0.6$ and values essentially in the $[0.3, 0.7]$ in agreement with the results obtained in [37] for 20 German cities.

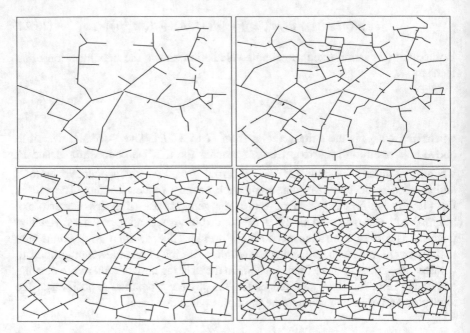

Fig. 13.6 Snapshots of the network at different times of its evolution: for **a** t =1,000, **b** t =2,000, **c** t =3,000, **d** t =4,000. At short times, we have a tree structure and loops appear for larger density values obtained at larger times. For t =4,000, we have approximately 1,700 nodes connected by 2,000 roads. Figure taken from [33]

13.2.3   Exponential Distribution of Centers

An important feature of street networks seem to be a large diversity of cell shapes as well as a broad distribution of cell areas. So far, we used the assumption that centers are distributed uniformly across the plane. Within this assumption, the model predicts a cell area distribution following an exponential (with a large cut-off however) as shown in Fig. 13.8.

The distribution of centers in real cities is however not accurately described by an uniform distribution but, as shown by previous studies [39, 40], display a decrease from the center and which is often fitted by an exponential (see [41] and references therein). We thus use such an exponential distribution $P(r) = \exp(-|r|/r_c)$ for the node spatial repartition and measure the areas formed by the resulting network. Although most quantities (such as the average degree and the total road length) are not sensitive to the node distribution, the impact on the area distribution is drastic. In Fig. 13.8, a power law with exponent equal to 2.0 is found, in remarkable agreement with the empirical facts reported by [37] (see Chap. 4) for the city of Dresden. This agreement confirms the fact that the simple local optimization principle is a good candidate for the main process driving the evolution of city street patterns and demonstrates that the spatial distribution of nodes is also crucial.

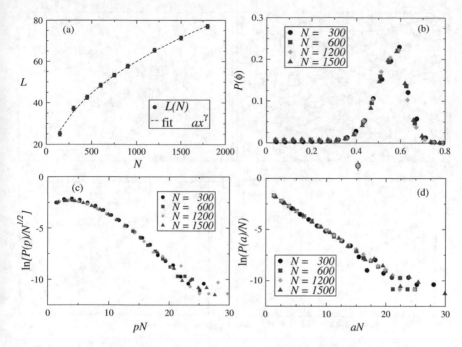

Fig. 13.7 Results of the model (averaged over 1000 configurations). **a** Total length of roads versus the number of nodes. The dotted line is a square root fit. **b** Shape factor distribution showing a good agreement with the empirical results of [37]. **c–d** Rescaled distributions of the perimeter (**c**) and of the areas **d** of the cells displaying an exponential behavior. Figure taken from [38]

The optimization process described above has several interesting consequences on the global arrangement of street networks when geographical constraints are imposed, as illustrated by the following example. We simulate the presence of a river assuming that new centers cannot appear on a stripe of given width and they are otherwise uniformly distributed, the resulting pattern is shown in Fig. 13.9. The local optimization principle naturally creates a small number of bridges that are roughly equally spaced along the river and organizes the road network to provide the most efficient connectivity given the constraint.

13.2.4   Effect of Centrality and Density

In the simple version of the model presented above, the location of centers is independent of the existing road network topology. In real urban systems, this is however unlikely to happen. There is an extensive spatial economics literature [42] that focuses on the several factors that may potentially influence the choice location for new businesses, homes, factories, or offices (see also [43] and references therein).

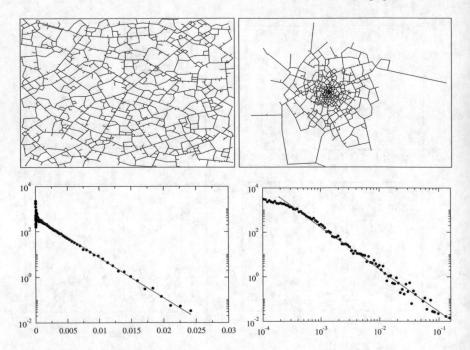

Fig. 13.8 Upper left plot: Uniform distribution of points (1000 centers, 100 configurations). In this case, the area distribution is exponentially distributed (bottom left). Upper right plot: Exponential distribution of centers (5000 centers, 100 configurations, exponential cut-off $r_c = 0.1$). In this case, we observe a power law (bottom right). The line is a power law fit which gives an exponent ≈ 2.0. Figure taken from [33]

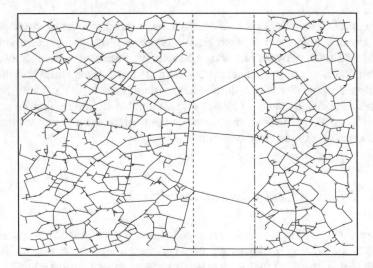

Fig. 13.9 In the presence of an obstacle (here a 'river' delimited by the two dotted line) in which the centers are not allowed to be located, the local optimization principle leads to a natural solution with a small number of bridges. Figure taken from [33]

Our interest here is to discuss, based on very simple and reasonable assumptions, the coupled evolution of the road network and the population density. More specifically, we focus on three main features (in the language of street networks):

- the local density of nodes,
- the network,
- the traffic on the network.

Density, network and traffic constitute three different faces of the same system, and their evolution is in general strongly correlated. In the case of cities, the road network evolves to better serve the changing density of population. In turn, the road network influences the patterns of traffic, making different zones more or less 'central' and therefore more or less attractive for further growth of population density. Further, 'attractiveness'—leading to a very high demand—triggers a self-inhibitory mechanism through the increase of prices that limits the unbounded growth of most desirable areas. It is the mutual interaction between these three aspects that we aim to model here.

Although there are many other economical mechanisms (type of centers, income variations, etc.) which enter the choice of a home, we limit ourselves to the two antagonist mechanisms of centrality and attractiveness. The goal here is not to as realistic as possible, but to capture the essence of the coupled evolution of the network system and the node density, and to understand the interplay between fundamental mechanisms and their impact on the structure of the network.

13.2.4.1 Effect of Density

The two main factors that affect in opposite directions location's attractiveness are the 'price' and the 'centrality'. It is reasonable to assume that the rent price for example is an increasing function of the local density. For the sake of simplicity, we will assume here that the price is directly proportional to the local density of population (which can be seen as the first term of an expansion of the price as a function of the density). This is implemented in the model by dividing the city in square sectors of area S, and computing the local density as $\rho(i) = N(i)/S$ where $N(i)$ is the number of centers in the sector (i). The price or rent cost is then given by

$$C_R(i) = A\rho(i) \qquad (13.26)$$

where A is some positive prefactor corresponding to the price per density.

13.2.4.2 Network Effect

The second important factor is the transport accessibility. Locations that are easily accessible and that allow to reach easily arbitrary destinations are obviously more attractive, all other parameters being equal. Also, for a new commercial activity, high

traffic areas can strongly enhance profit opportunities. In terms of the existing network, the best locations are therefore the most central, where the notion of centrality
has to be understood not simply in geographical terms, but also in terms of traffic.
The simplest choice to model traffic, given the road network, is to assume that it is
proportional to betweenness centrality (Chap. 5). For each sector S_i of the grid, the
average betweenness centrality is computed

$$\overline{g}(i) = \frac{1}{N(i)} \sum_{v \in S_i} g(v) \tag{13.27}$$

where $g(v)$ is the BC of node v (and $N(i)$ the number of nodes in S_i). If the sector is
empty, we use the centrality of the closest center. The transportation cost is a decreasing function of the betweenness centrality and we assume that the transportation cost
$C_T(i)$ for a center in sector (i) is given by

$$C_T(i) = B[g_m - \overline{g}(i)] \tag{13.28}$$

where B and g_m are positive constants.

13.2.4.3   Combining These Effects

Finally, we assume that all new centers have the same income $Y(c) = Y$. The net
income of the center c in a sector (i) is then

$$K(c, i) = Y(c) - C_R(i) - C_T(i) \tag{13.29}$$

The higher the net income $K(c, i)$ and the more likely the location i will be chosen
for the implantation of a new business, home, etc. In urban economics the location
is usually chosen by minimizing costs, and we relax this assumption by defining the
probability that a new center will choose the sector i as its new location. Absorbing
irrelevant factors into the parameters β and λ, this probability is given by

$$\begin{aligned}
P(i) &= \frac{e^{\beta K(c,i)}}{\sum_j e^{\beta K(c,j)}} \\
&= \frac{e^{\beta(\lambda \overline{g}(i) - \rho(i))}}{\sum_j e^{\beta(\lambda \overline{g}(j) - \rho(j))}}
\end{aligned} \tag{13.30}$$

where the local density is normalized by the global density N/L^2 in order to have
both the density and centrality contributions defined in the same interval $[0, 1]$. Their
relative weights is modeled by the parameter λ. The parameter β implicitly describes
in an 'effective' way all the factors that have not been explicitly taken into account,
and that may potentially influence the choice of location. If $\beta \approx 0$, costs are irrelevant

and new centers will appear uniformly distributed across the different sectors:

$$P(i) \sim \frac{1}{N(i)} \tag{13.31}$$

In the opposite case, $\beta \to \infty$, the location with the minimal cost will be chosen deterministically.

$$P(i) = 1 \quad \text{for i such that } K(c, i) \text{ is minimum}$$
$$P(i) = 0 \quad \text{for all other sectors} \tag{13.32}$$

The parameter β can thus be used in order to adjust the importance of the cost relative to that of other factors not explicitly included in the model.

13.2.4.4 One-Dimensional Model

Before discussing the full model, it is worth to study a 1–dimensional toy model where the network plays no role, because a single path exists between each couple of points. Despite the simplicity of the setting, it is possible to draw some general conclusion.

We assume that the centers are located on a one-dimensional segment $[-L, L]$. Since there exists a single path between any two points, the calculation of centrality is trivial. In the continuous limit, and for a generic location x it can be written as the product of the number of points that lie at the right and left of the given location:

$$g(x) = \int_{-L}^{x} \rho(y, t) \left[N - \int_{-L}^{x} \rho(y, t) dy \right] dy \tag{13.33}$$

where $\rho(x, t)$ is the density at x. The equation for the density therefore reads:

$$\partial_t \rho(x, t) = e^{\beta \left[\lambda \frac{\int_{-L}^{x} \rho(y,t)dy}{N} \frac{(N - \int_{-L}^{x} \rho(y,t)dy)}{N} - \frac{\rho(x,t)}{N} \right]} \tag{13.34}$$

where $N = \int_{-L}^{L} \rho(y, t) dy$. The long term behavior can be obtained by separation of variables, i.e. $\rho(x, t) = f(x) g(t)$. Without loss of generality one can set $\int_{-L}^{L} f(x) dx = 1$, which implies that in the long time limit $g(t \gg 1) \simeq N$. We thus obtain

$$\alpha f(x) = e^{\beta [\lambda \int_{-L}^{x} f(y)dy (1 - \int_{-L}^{x} f(y)dy) - f(x)]} \tag{13.35}$$

where α is an integration constant to be determined. An explicit solution for the inverse $x(f)$ can be achieved via the Lambert function (the Lambert's function is the principal branch of the inverse of $z = w \exp^w$), but the expression is not particularly

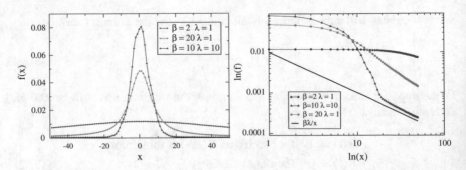

Fig. 13.10 The stationary growth rate for different values of the parameters. **a** Large values of β and λ implies larger degree of centralization and a faster decay of density from center to periphery. **b** At large values of λ the decay of density becomes algebraic for location away from the center. The exponent approaches -1 and $f(x)$ is approximated in that region by $\beta\lambda/x$. Figure taken from [38]

illuminating and it is therefore not presented here. Several facts can however be understood using a numerical simulation of the process:

- At large times, population in different areas grows with a rate $f(x)$ that is constant in time, but depends on the area (separation of variables).
- Although β models the 'noise' in the choice of location and λ the relative importance of centrality as compared to density, they have similar effects on the expected density in a given area. To an increase in β and λ corresponds a concentration of density in the areas of large centrality and a steeper decay of density towards the periphery, as shown in Fig. 13.10a.
- Fluctuations from the expected value increases as noise increases (it corresponds to a decrease of β).
- As λ increases the decay of density assumes a power law form whose exponent depends on β and λ and approaches -1 as λ gets very large. This can be explained assuming $f(x) \approx \gamma x^{-r}$, and using Eq. (13.35), its derivative (both computed in L, and the fact that $\int_{-L}^{L} f(x)dx = 1$. This leads to $r = 1$ and $\gamma = 1/\beta\lambda$ (see Fig. 13.10b).
- as β and λ approaches 0, the density $f(x)$ becomes flat.

This simple one-dimensional model allowed us to understand some basic features of this model and we will turn in the next section to the study of the full model.

13.2.4.5   Two-Dimensional Case: A Sharp Localization Transition

We now apply the probability Eq. (13.30) to the growth model described in the first part of this chapter. The process starts with a 'seed' population settlement (few centers distributed over a small area) and a small network of roads that connects them. At any stage, the density and the betweenness centrality of all different subareas are computed, and few new centers are introduced. Their location in the existing subareas is determined according to the probability defined in Eq. (13.30). Then, roads are grown for as many steps as needed to connect the centers that just entered the scene to the existing network. The above process is iterated until the desired number of centers has been introduced and connected. In the two panels of Fig. 13.11 we show the emergent pattern of roads that is obtained when λ is small and very large, respectively.

When λ is small the density plays the dominant role in determining the location of new centers. New centers appear preferably where density is small, smoothing out the eventual fluctuations in density that may occur by chance. The resulting density is then uniform. When, on the other hand, λ is very large, the key role is played by centrality, leading to a city where all centers are located in the same small area. The centrality has thus an effect opposite to that of density and tends to favor concentration. For intermediate values of λ centrality and density compete leading to a large density region connected to a distinct, low-density, 'sub-urban' area. The low density regions are attractive from the point of view of rent but are usually inconvenient from the point of view of centrality (i.e. they have large transportation cost, both in terms of time and money). The difference and the transition between the two scenarios described above is amenable to be described quantitatively.

We compute, in the two cases, the following quantity (previously introduced in a different context [44, 45]):

$$Y_2 = \sum_i \left[\frac{N(i)}{N} \right]^2 \tag{13.36}$$

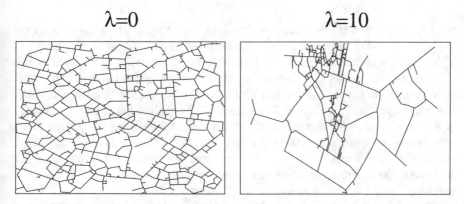

Fig. 13.11 Networks obtained for different values of λ (and for $N = 500$ and $\beta = 10$). On the left, $\lambda = 0$ and only the density plays a role and we obtain a uniform distribution of centers. On the right, we show the network obtained for $\lambda = 10$. In this case, the centrality is the most important factor leading to a dominant area with high density. Figure taken from [38]

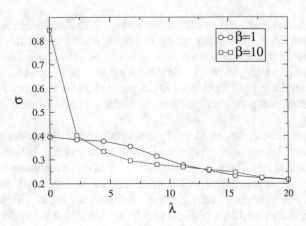

Fig. 13.12 Fraction of dominating sectors (obtained for 500 centers and averaged over 100 configurations). When λ is small, the center distribution is uniform and σ is large (close to 100%). When λ increases, we see the appearance of a few sectors dominating and concentrating most of the population. This effect is smoothen out for smaller values of β corresponding to the possibility of choice. Figure taken from [33]

where the sum runs over all N_S sectors. In the uniform case, all the $N(i)$ are approximately equal and one obtains $Y_2 \sim 1/N_S$, which is usually small. In contrast, when most population concentrates in just few sectors, $Y_2 \sim 1/n$ where n is of the same order of magnitude of the number of highly-populated sectors. The quantity

$$\sigma = \frac{1}{Y_2 N_S} \tag{13.37}$$

gives therefore the fraction of dense grid sectors. The behavior of σ verses λ is shown in Fig. 13.12. We observe that σ decreases very fast when λ increases, signaling that a phenomenon of 'concentration' sets in as soon as transportation costs are involved.

We conclude this section discussing the role played by the parameter β. Analogously to what happens in the one dimensional case, the concentration effect is weakened by a small values of β. The parameter β describes the overall importance of the cost-factors with respect to other factors that have not been explicitly taken into account, or, equivalently, the possibility of choice. Indeed, when β is very large, the location which maximizes the cost is chosen. In contrast, when the parameter β is small, the cost differences are smoothen out and a broader range of choices is available for new settlements. Figure 13.12 illustrates the importance of choice. In particular, the appearance of large-density zones (controlled by the importance of transportation accessibility) is counterbalanced by the possibility of choice and the resulting pattern is more uniform.

13.2.5   The Appearance of Core Districts

In this last part, we describe the effect of the interplay of transportation and rent costs on the decay of population density from the city center. In the following, the core district is identified with the sector with the largest density. The whole plane is then divided in concentric shells with internal radius r and width dr. The density profile $\rho(r)$ is given by the ratio of the number δn of centers in a shell to its surface $\delta S(r)$

$$\rho(r) = \frac{\delta n}{\delta S} \tag{13.38}$$

For small λ, the density is uniform, as expected. In Fig. 13.13 we show the density profile $\rho(r)$ in the case of λ large, where we a fast exponential decay is observed of the form $\exp -r/r_c$, in agreement with empirical observations [39, 41]. This behavior is the signature of the appearance of a well-defined core district of typical size r_c, whose typical size r_c decreases with λ. This simplified model predicts, therefore, the existence of a highly populated central area whose size can be estimated in terms of the relative importance of transport and rent costs.

This basic model thus describes the impact of economical mechanisms on population density and on the topology of the road network. The interplay between rent costs and demand for accessibility leads to a sharp transition in population density. When transportation costs are moderate, the density is approximately uniform and the road network is a typical planar network that does not show any strong heterogeneity. In contrast, if transportation costs are higher, we observe the appearance of a very densely populated area around which the density decays exponentially.

Fig. 13.13 Density profiles for $\lambda = 10$ ($N = 125$, averaged over 1000 configurations). The decay of the density profile is well fitted by an exponential, signalling the appearance of a well-defined core district. Figure taken from [33]

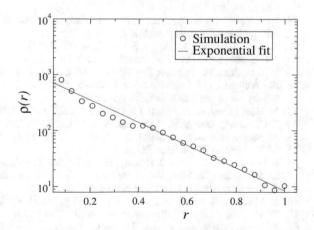

References

1. H.A. Simon, Biometrika **42**, 425 (1955)
2. R. Albert, H. Jeong, A.L. Barabasi, Nature **401**, 130 (1999)
3. R. Albert, A.L. Barabasi, Rev. Mod. Phys. **74**, 47 (2002)
4. S.N. Dorogovtsev, J.F.F. Mendes, Adv. Phys. **51**, 1079 (2002)
5. R. Pastor-Satorras, A. Vespignani, *Evolution and Structure of the Internet: A Statistical Physics Approach* (Cambridge University Press, Cambridge, 2003)
6. M. Newman, SIAM Rev. **45**, 167 (2003)
7. G. Caldarelli, *Scale-free networks* (Oxford University Press, Oxford, 2007)
8. R. Cohen, S. Havlin, *Complex Networks: Structure, Robustness and Function* (Cambridge University Press, Cambridge, 2010)
9. S.H. Yook, H. Jeong, A.L. Barabasi, Proc. Natl. Acad. Sci. (USA) **99**, 13382 (2002)
10. A.F. Rozenfeld, R. Cohen, D. ben Avraham, S. Havlin, Phys. Rev. Lett. **89**, 218701 (2002)
11. C.P. Warren, L.M. Sander, I.M. Sokolov, Phys. Rev. E **66**, 056105 (2002)
12. J. Dall, M. Christensen, Phys. Rev. E **66**(1), 016121 (2002)
13. P. Sen, K. Banerjee, T. Biswas, Phys. Rev. E **66**, 037102 (2002)
14. J. Jost, M. Joy, Phys. Rev. E **66**, 036126 (2002)
15. S.S. Manna, P. Sen, Phys. Rev. E **66**, 066114 (2002)
16. R. Xulvi-Brunet, I.M. Sokolov, Phys. Rev. E **66**, 026118 (2002)
17. M. Barthelemy, Europhys. Lett. **63**, 915 (2003)
18. C. Clark, J.R. Stat. Soc. (Series A) **114**, 490 (1951)
19. A. Barrat, M. Barthelemy, A. Vespignani, J. Stat. Mech. p. P05003 (2005)
20. A. Barrat, M. Barthelemy, A. Vespignani, Phys. Rev. Lett. **92**, 228701 (2004)
21. R. Guimerá, L.A.N. Amaral, Eur. Phys. J. B **38**, 381 (2004)
22. P. Sen, S. Dasgupta, A. Chatterjee, P. Sreeram, G. Mukherjee, S. Manna, Phys. Rev. E **67**(3), 036106 (2003)
23. J.M. Kleinberg, Nature **406**, 845 (2000)
24. H. Thadakamalla, R. Albert, S. Kumara, New J. Phys. **9**, 190 (2007)
25. L.A. Adamic, R.M. Lukose, A.R. Puniyani, B.A. Huberman, Phys. Rev. E **64**, 046135 (2001)
26. K. Hajra, P. Sen, J. Stat. Mech. **2007**, P06015 (2007)
27. A. Runions, A.M. Fuhrer, P.F.B. Lane, A.G. Rolland-Lagan, P. Prusinkiewicz, ACM Trans. Graph. **24**, 702 (2005)
28. I. Rodriguez-Iturbe, A. Rinaldo, *Fractal River Basins: Chance and Self-Organization* (Cambridge University Press, Cambridge, 1997)
29. G. West, J. Brown, B. Enquist, Science **276**, 122 (1997)
30. J.R. Banavar, A. Maritan, A. Rinaldo, Nature **399**(6732), 130 (1999)
31. M.T. Gastner, M. Newman, Phys. Rev. E **74**(1), 016117 (2006)
32. M. Schwartz, *Telecommunication Networks: Protocols, Modelling and Analysis* (Addison-Wesley Longman Publishing, Boston, 1986)
33. M. Barthelemy, A. Flammini, Phys. Rev. Lett. **100**(13), 138702 (2008)
34. T. Courtat, C. Gloaguen, S. Douady, Phys. Rev. E **83**(3), 036106 (2011)
35. M. Barthelemy, A. Flammini, J. Stat. Mech. p. L07002 (2006)
36. M. Barthelemy, Phys. Rep. **499**(1), 1 (2011)
37. S. Lämmer, B. Gehlsen, D. Helbing, Physica A **363**(1), 89 (2006)
38. M. Barthelemy, A. Flammini, Netw. Spat. Econ. **9**(3), 401 (2009)
39. H. Makse, S. Havlin, H. Stanley, Nature **377**, 608 (1995)
40. H. Makse, J. Andrade, M. Batty, S. Havlin, Phys. Rev. E **58**, 7054 (1998)
41. M. Guérois, D. Pumain, Environ Plan A **40**(9), 2186 (2008)
42. M. Fujita, P. Krugman, A. Venables, *Spatial Economy: Cities, Regions, and International Trade* (MIT Press, Cambridge, 1999)
43. P. Jensen, Phys. Rev. E **74**(3), 035101 (2006)
44. B. Derrida, H. Flyvbjerg, J. Phys. A **20**, 5273 (1987)
45. M. Barthelemy, Phys. Rev. Lett. **91**, 189803 (2003)

Chapter 14
Tessellations of the Plane

Another way to think of planar networks is by focusing on their faces and how to generate them. The system of edges forming the faces then produces a spatial, planar network. An example is given by tessellations of a plane that are divisions of the plane into polygons. There is a vast literature on tessellations (see [1] and references therein) and in this chapter we will discuss selected examples only. Many tessellations are constructed from a random set of points and we will first focus in this chapter on the well-known Voronoi tessellation and its dual, the Delaunay graph. We then discuss briefly another class of tessellation obtained by growing lines from points (with the Gilbert model) or obtained by adding random lines (STIT models) and we end this chapter with a discussion on planar fragmentation.

14.1 The Voronoi Tessellation

The Voronoi tessellation is a standard tool in the analysis of random point distribution and we discuss here some of its properties and refer the interested reader to the very complete book [2]. In the Voronoi tessellation, we start from N points (sometimes also called seeds or generators) randomly distributed in the plane with position x_i ($i = 1, \ldots, N$). The Voronoi tessellation is obtained by partitioning the plane into regions – called the Voronoi cells – based on the distance to the points. More precisely the Voronoi cell associated to the seed i is the set of points defined by

$$V(i) = \{ x \mid d(x, x_i) < d(x, x_j), \forall j \neq i \} \tag{14.1}$$

where $d(x, y)$ is the euclidean distance here but we can generalize this tessellation to any distance. The edges of the Voronoi cells constitute the Voronoi diagram or,

© The Author(s), under exclusive license to Springer Nature Switzerland AG 2022
M. Barthelemy, *Spatial Networks*,
https://doi.org/10.1007/978-3-030-94106-2_14

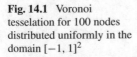

Fig. 14.1 Voronoi
tesselation for 100 nodes
distributed uniformly in the
domain $[-1, 1]^2$

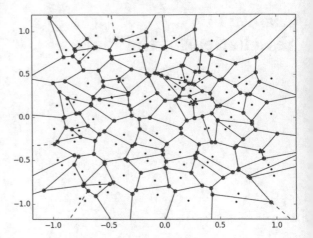

in other words, the edges of the Voronoi diagram are the mediatrices between the
generators. We show in Fig. 14.1 an example of a Voronoi tesselation computed for
100 nodes distributed uniformly in the plane.

It is interesting to note that the Voronoi tessellation can be seen as a growth
model where the points/seeds act as nuclei which start to grow at the same time and
isotropically. The initial growing surfaces are spheres (disks in 2d) and when two
spheres meet the growth stops at the contact point, leading to the emergence of a
facet (or a straight line in 2d). By definition, this contact line lies at the same distance
from the two nuclei they originally come from, implying that each polygon in this
construction are Voronoi cells.

Obviously the properties of the Voronoi tessellation will strongly depend on the
point distribution. The simplest case corresponds to a uniform distribution of points
in the plane – a Poisson process with density ρ – and is usually referred to as the
Poisson-Voronoi tessellation [2]. Most results concern this case and much less is
known when the point distribution is not uniform.

We end this section by noting that the graph generated by the Voronoi construction
has vertices given by the intersection of the mediatrices between points. The so-called
'general position assumption' often used in mathematics states that in general we
do not have 4 co-circular points. In other words, given 3 points we always will find
an equidistant node, but that doesn't happen – in general – for 4 points. When we
construct the Voronoi tessellation we start first with two nodes, say s_1 and s_2 (see
Fig. 14.2) and the next step consist in finding the nearest node such that the three
nodes s_1, s_2, and s_3 lie on the same circle. The center of this circle corresponds
also to the intersection of the three mediatrices M_{12}, M_{23}, M_{13} between the pairs
of points and therefore represents a vertex in the Voronoi graph. It is very unlikely
that there is also a node s_4 that lies on this circle and this is precisely what states the
general position assumption. This is an important fact that implies that the dual of the
Voronoi -the Delaunay graph - is a triangulation (see Chap. 16 for more details about
the Delaunay graph). Exceptions to this assumption will happen in very specific cases

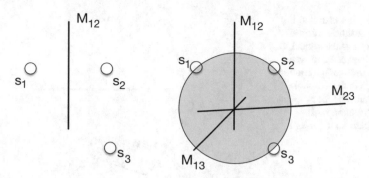

Fig. 14.2 For two nodes s_1 and s_2 we construct the mediatrice. We then look for the nearest node s_3 and we can construct a circle which contains the three points. The center of this circle is also the intersection of the three mediatrices and is a vertex in the Voronoi tessellation

such as aligned points on a line or points in regular lattices for example. However, for random points this assumption is almost always satisfied.

14.1.1 Average Properties of the Poisson-Voronoi Tessellation

When the nodes are uniformly distributed in the plane (according to a Poisson point process), the Voronoi construction on this set is called the Poisson-Voronoi tessellation. It is a very simple tessellation, and represents a good benchmark or null model that we can compare with empirical properties and assess the effect of interactions. There are obviously many studies about this object and we will discuss the main ones only. We refer the interested reader to the excellent and complete book [2].

We denote by $\rho = N/A$ the intensity of the Poisson process. In other words $\rho = N/A$ is the density of the N generators (or seeds) that are randomly distributed in the plane of surface area A (we won't consider here the generalization to d dimensions, see [1] and references therein).

The Poisson-Voronoi tessellation is the simplest one but calculations can rapidly become involved and non trivial. Simple average properties were computed some time ago in [3] and we first discuss the scaling obtained by simple arguments. In dimension $d = 2$, the typical distance between two points is then given by

$$\langle \ell \rangle \sim \frac{1}{\sqrt{\rho}} \tag{14.2}$$

where the brackets denote the average over the Poisson process. This result implies that the average area $\langle a \rangle$ of Voronoi cells scales as (see Fig. 14.3)

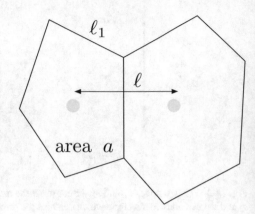

$$\langle a \rangle \sim \langle \ell \rangle^2 \sim \frac{1}{\rho} \tag{14.3}$$

The average perimeter $\langle p \rangle$ also will be proportional to $\langle \ell \rangle$ and the general position
assumption (which leads to the fact that the Delaunay graph is a triangulation, see
above) implies that on average the Voronoi cells are hexagons. This implies that

$$\langle p \rangle \sim 6\ell_1 \tag{14.4}$$

where ℓ_1 is the average length of the edges in the Voronoi graph. Finally the total
length of the Voronoi graph is given by $L = N\langle p \rangle/2$ and we obtain $L/A \sim 2\sqrt{\rho}$.

Theses scalings are correct and the difficult thing is to compute exactly the pref-
actors which is done in [3] and the results read

$$\langle p \rangle = \frac{4}{\sqrt{\rho}} \tag{14.5}$$

$$\langle \ell_1 \rangle = \frac{2}{3} \frac{1}{\sqrt{\rho}} \tag{14.6}$$

$$\langle a \rangle = \frac{1}{\rho} \tag{14.7}$$

It is interesting to note that on average the Poisson-Voronoi tessellation is made
of hexagons, and for this figure the edge length is given by

$$\ell_1^{hexa} = \sqrt{2/3\sqrt{3}}/\sqrt{\rho} \approx 0.62/\sqrt{\rho} \tag{14.8}$$

which is not far from the exact result (a prefactor equal to $2/3$) obtained for $\langle \ell_1 \rangle$ but
nonetheless different.

The second moment of the area was computed in [4] and is given by

$$\langle a^2 \rangle \simeq 1.28/\rho^2 \tag{14.9}$$

Discussions about higher moments can for example be found in [1].

The average length of the Voronoi and the Delaunay graphs (see Chap. 16) have been computed in the limit $N \to \infty$ and the results are (see [5] and references therein)

$$\lim_{N \to \infty} \frac{\overline{L_N}(vor)}{N^{1/2}} = 2 \tag{14.10}$$

$$\lim_{N \to \infty} \frac{\overline{L_N}(del)}{N^{1/2}} = \frac{32}{3\pi} \tag{14.11}$$

where *vor* stands for the Voronoi tessellation and *del* for the Delaunay graph. These results indicate in particular that the Delaunay graph is much longer (by a factor almost equal to 2) than the Voronoi graph.

14.1.2 Statistical Properties

14.1.2.1 Cell Area

Despite a large number of studies, distributions of various quantities are unknown, even in the simplest case of the Poisson-Voronoi tessellation. In particular, the cell area distribution $P(a)$ has been approximated using the Gamma or the lognormal distributions with appropriate parameters. A rough argument about why the Gamma function is a good candidate is presented in [6] and we reproduce it here.

For a Poisson point process of intensity ρ, the probability to find k points in a surface a is given by

$$P(k) = \frac{p^k}{k!} e^{-p} \tag{14.12}$$

where $p = \rho a$ is the average number of points inside the area a. This implies that the probability to have $k = 0$ points in a disk of radius r around a given point is given by

$$P(0) = e^{-\rho \pi r^2} \tag{14.13}$$

which also implies that the probability $P_>(r)$ that the distance to a nearest neighbor is larger than r is given by

$$P_>(r) = e^{-\rho \pi r^2} \tag{14.14}$$

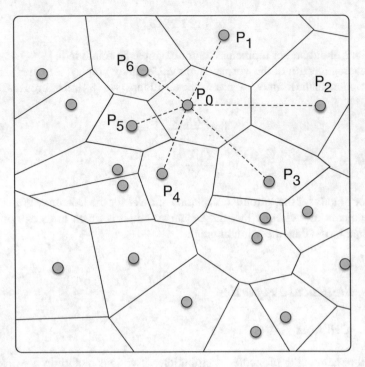

Fig. 14.4 An example of a Voronoi cell with 6 neighbors. The dotted lines represent the distance r_i to the ith neighbor

The probability to have a nearest neighbor at a distance r is then given by $P(r) = -dP_>(r)/dr$ and reads

$$P(r) = 2\pi\rho r e^{-\rho\pi r^2} \tag{14.15}$$

We now consider a typical cell, centered at point P_0, with $n = 6$ neighbors which divide the plane from a given point in 6 sectors of angle $2\pi/n$. The point P_0 has thus six neighbors P_i ($i = 1, \ldots, 6$) as show in Fig. 14.4. The distribution $P_{sector}(r)$ that the nearest neighbor in this sector is at distance r, is then given by

$$P_{sector}(r) = \frac{2\pi\rho r}{n} e^{-\rho\pi r^2/n} \tag{14.16}$$

The average distance

$$\bar{r} = \frac{1}{n}\sum_1^n r_i \tag{14.17}$$

where $r_i = d(P_0, P_i)$ is the distance to the nearest neighbor in sector i, and has thus a distribution which is the convolution of P_{sector} with itself 5 times. This distribution P_{sector} is the derivative of a Gaussian and in Fourier of the form ke^{-k^2} which implies that the convolution will still have a Gaussian component. The inverse Fourier transform will then lead to a Gaussian behavior times a polynomial function of r^2. The dominant behavior will thus be of the form $(r^2)^{\alpha}e^{-\pi\rho r^2}$. The area of an hexagon with apothem $\bar{r}$ is given by $A = \sqrt{3}\bar{r}^2/2$ and we thus obtain

$$P(A) \sim A^{\alpha}e^{-\nu\rho A} \tag{14.18}$$

where $\nu = 2\pi/\sqrt{3} \approx 3.63$ and where α is unknown. If we impose the constraint that $\int P(A)dA = 1/\rho$ we obtain the following result

$$\alpha = \nu - 1 \approx 2.63 \tag{14.19}$$

This argument thus suggests that the area distribution could be a Gamma distribution with exponent close to 3.6. In the article [6], numerics are in agreement with this result and more recent simulations [7] found $\nu \approx 3.31$ in agreement with previous results found in [8].

On the basis of the one dimension result which is known exactly [9]

$$P_{d=1}(A) = 4\rho Ae^{-2\rho A} \tag{14.20}$$

and numerical simulation for $d = 2$ and $d = 3$ the authors of [9] proposed the ansatz

$$P_d(A) = C_d(\rho A)^{(3d-1)/2}e^{-\frac{3d+1}{2}\rho A} \tag{14.21}$$

where

$$C_d = \left(\frac{3d+1}{2}\right)^{(3d+1)/2} \frac{1}{\Gamma((3d+1)/2)} \tag{14.22}$$

This result is however not exact but, as claimed by the authors, could be a useful ansatz for practical applications.

14.1.2.2 The Number of Sides and the Perimeter

The probability distribution $p(n)$ to have a Voronoi cell with n sides has a peak at $n = 6$ and decreases quickly with n [10]. This quantity has been discussed in many studies (see [11] and references therein) and its analytical expression is unknown. Many studies focused on the asymptotic behavior for $n \to \infty$ [10, 11] and a key role is played by the angles ξ and η defined in Fig. 14.5. These angles satisfy the sum rules

Fig. 14.5 A Voronoi cell
showing the angles ξ and η
defined and used in [11] and
in the text

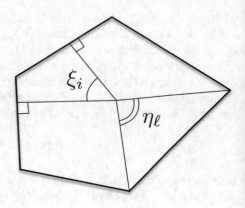

$$\sum_{m=1}^{n} \xi_m = 2\pi \qquad\qquad (14.23)$$

$$\sum_{\ell=1}^{n} \eta_\ell = 2\pi \qquad\qquad (14.24)$$

These constraints make this problem very difficult but become irrelevant for $n \to \infty$ where these angles ξ and η become $2n$ independent variables and distributed according to [11]

$$P(x = n\xi/2\pi) = 4xe^{-2x} \qquad\qquad (14.25)$$
$$P(y = n\eta/2\pi) = e^{-y} \qquad\qquad (14.26)$$

The independence of the angles implies that for $n \to \infty$ the shape of the n-sided cell converges to a circle with probability 1 and of radius $R = \sqrt{N/4\pi\rho}$.

The quantity $p(n)$ can be written in terms of ξ and η and for large n the following expansion has been obtained [10]

$$P(n) \sim \frac{C}{4\pi^2} \frac{(8\pi^2)^n}{(2n)!} \left[1 + \mathcal{O}\left(\frac{1}{\sqrt{n}} \right) \right] \qquad\qquad (14.27)$$

where C can be computed exactly and is $C = 0.344\ldots$ The average value of n is $\langle n \rangle = 6$ which corresponds indeed to the intuitive idea that most Voronoi cells are hexagons, but there are obviously other polygons with a number of sides distributed according to $P(n)$.

We end this section with a note on Aboav's law [12] which is reasonably verified in experiments (see [11] and references therein) and which states that the neighbor of a cell with n sides has on average a number $m(n)$ of sides given by

$$m(n) = a + \frac{b}{n} \qquad (14.28)$$

where a and b are positive constants and found in general to be $a \simeq 5$ and $5.7 \le b \le 8.5$. This means that a cell with a large number of neighbors tend to have neighbors with a smaller number of sides. In other words large grains tend to be surrounded by small ones and vice-versa [13]. This law is however not exact for Poisson-Voronoi tessellation and Hilhorst showed that for $n \gg 1$ [14]

$$m(n) = 4 + 3\sqrt{\frac{\pi}{n}} + \ldots \qquad (14.29)$$

which is in agreement with Monte-Carlo data [11].

14.1.3 Central Limit Theorem for the Total Length

The total length of the graph G is defined as

$$L(G) = \sum_{e \in G} l(e) \qquad (14.30)$$

where $l(e)$ is the length of the link e. Although this quantity $L(G)$ is defined here as a sum over random variables, the existence of central limit theorems (CLT) for this class of problems is not obvious as these links are not independent from each other. The existence of correlations could induce dramatic differences with the usual CLT. Similarly to results obtained for combinatorial optimization problems, Avram and Bertsimas [5] proved the existence of central limit theorems (CLTs) and estimated the rate of convergence, for various graphs defined over a set of N points distributed in $[0, 1]^2$ according to a Poisson process. In particular, they considered the total length for three important models of graphs:

- The nearest graph $N(k, N)$ where each point is connected to its kth nearest neighbors. In particular $N(1, N)$ is the nearest neighbor graph (see Chap. 15).
- The Delaunay triangulation connecting nodes if their Voronoi cells are neighbors
- The Voronoi graph constructed by the Voronoi cells

Avram and Bertsimas [5] were able to show that these correlations are somewhat 'local' and a CLT indeed applies. In other words, they could show that the impact of a local configuration of links has a finite cut-off distance, allowing to cluster correlated nodes and to apply a central limit theorem to these groups of nodes. More precisely they showed that the length $L_N = L(G)$ (for N nodes) satisfies

$$\lim_{N \to \infty} \text{Prob}\left(\frac{L_N - \overline{L_N}}{\sqrt{\text{Var}(L_N)}} \le x \right) = \Phi(x) \qquad (14.31)$$

where $\overline{L_N} \simeq \beta N^{1/2}$ [15] is the average over many realizations and Var is the variance of the length L_N (see Sect. 14.1.1 for behavior of the average length for the Voronoi and Delaunay graphs, and Chap. 18 for the minimum spanning tree). The function $\Phi(x) = 1/2 + \mathrm{erf}(x/\sqrt{2})/2$ (erf is the error function) corresponds to the cumulative of the normal distribution $\mathcal{N}(0, 1)$. These authors were also able to estimate the rate of convergence and found that

$$\left| \mathrm{Prob}\left(\frac{L_N - \overline{L_N}}{\sqrt{\mathrm{Var}(L_N)}} \leq x \right) - \Phi(x) \right| \sim \mathcal{O}\left(\frac{(\ln N)^{7/4}}{N^{1/4}} \right) \tag{14.32}$$

indicating a relatively slow convergence to the normal distribution.

14.2   Effect of the Density of Points

We considered so far the properties of the Voronoi tessellation by assuming implicitly that space is homogeneous and the density of points inside each cell is uniform. In real-world applications there are in general two different densities. First, the density $D(x)$ of the seeds of the Voronoi tessellation, which can represent for example facilities such as airports, post offices, hospitals, etc. Second, the local population density $\rho(x)$ around these seeds (the population is distributed in some volume $\mathcal{D}$). Both these distributions can be measured and compared to the optimal case given by a tessellation, as discussed below.

We will assume that the average distance from an individual to the nearest facility is minimized [16] and we follow here the derivation given in [17]. We examine the distribution of N facilities such that their locations $x_1, x_2, \ldots, x_n$ minimize the total distance F to reach them and which is given by the following integral

$$F(x_1, x_2, \ldots, x_N) = \int \rho(x) \min_{i=1,\ldots,N} |x - x_i| \mathrm{d}^2 x. \tag{14.33}$$

These N facilities partition the domain $\mathcal{D}$ under consideration into Voronoi cells. Since the Voronoi cell for a facility i is the set of points that are closer to this facility than others. We denote by $a(x)$ the area of the Voronoi cell to which the point x belongs to. The distance to the facility for these individuals will then be of the order of $g\sqrt{a(x)}$ where g is a geometrical factor of order 1 that depends on the shape of the Voronoi cell. The function to be minimized can then be rewritten as

$$F[a(x)] \sim \int \rho(x)\sqrt{a(x)}\mathrm{d}^2 x \tag{14.34}$$

such that $$\int_{\mathcal{D}} \frac{1}{a(x)} \mathrm{d}^2 x = N \tag{14.35}$$

where this constraint expresses the fact that we have a total of N facilities. The minimization equation using functional derivative then reads

$$\frac{\delta}{\delta a(x)} \left[\int \rho(x)\sqrt{a(x)} d^2x + \lambda \left(\int \frac{1}{a(x)} d^2x - N \right) \right] = 0 \qquad (14.36)$$

where λ is a Lagrange multiplier. This equation leads to

$$\frac{\rho(x)}{2\sqrt{a(x)}} = \frac{\lambda}{a(x)^2} \qquad (14.37)$$

which gives $a(x) \sim (\lambda/\rho(x))^{2/3}$. Inserting this result in the constraint allows us to extract λ and we finally get

$$D(x) = \frac{1}{a(x)} = C\rho(x)^{2/3} \qquad (14.38)$$

where the prefactor is $C = N/\int \rho(x)^{2/3} d^2x$.

As noted in [17], this simple argument has a number of consequences on the statistics of the Poisson-Voronoi tessellation. In particular, we have the following properties:

- The area of the Voronoi cells scales as

$$a(x) \sim \rho(x)^{-2/3} \qquad (14.39)$$

  which implies that

$$D(x) \sim \frac{1}{a(x)} \sim \rho(x)^{2/3} \qquad (14.40)$$

- The average radius of Voronoi cells scales as

$$r(x) \sim \rho(x)^{-1/3} \qquad (14.41)$$

- Finally, the number of individuals in each Voronoi cell is given by

$$n(x) \sim \rho(x)a(x) \sim \rho(x)^{1/3} \qquad (14.42)$$

The result Eq. (14.38) was tested by the authors of [16] who plotted the facility density $D(x)$ versus the population density for 5,000 facilities (hospitals, airports, or malls) for the 48 lower states of the US and their result is shown in Fig. 14.6. The density of facilities is thus scaling with the density of individuals raised to a power consistent with 2/3. This result suggests that (at least for the US), the spatial distribution of some facilities is indeed given by an optimality argument. However, we note that in [18] the empirical analysis led to different exponents according to the type of facilities considered, suggesting the existence of other mechanisms [18].

Fig. 14.6 Density of
facilities $D(x)$ versus the
population density $\rho(x)$
shown here in loglog. The
power law fit gives a slope of
0.663 ± 0.002 in agreement
with an optimality argument.
Figure taken from [16]

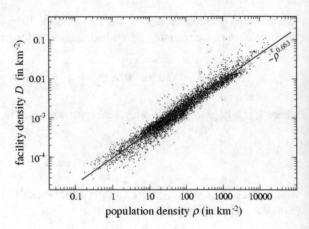

14.3   Crack and STIT Tessellations

Fracture and crack processes in materials also lead to tessellations of the plane [1].
In contrast with the Voronoi tessellation, the network is generated in these cases by
the growth of lines and not by the construction of cells.

For cracks that appear step by step, there is hierarchical ordering from longer
cracks to shorter ones. An important model for this class of process is the Gilbert
model [19] which is obtained from a set of points with a mark which is here an angle
drawn from a given distribution (typically the uniform distribution in $[0, 2\pi]$). The
process is then the following one: one chooses a point at random and edges are grown
from the point in the two opposite directions at random angle θ and stop when other
edges are hit. This process is described on a simple example in Fig. 14.7 and the
resulting tessellation is shown in Fig. 14.8. We note that this process implies that all

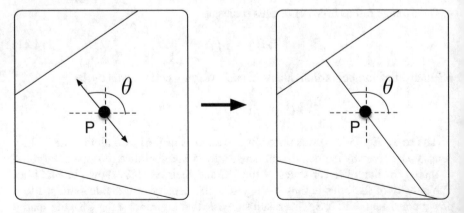

Fig. 14.7 Construction of the Gilbert tessellation: the point P is chosen and has the attribute angle
θ. We then grow a line in both direction along this angle and stop when the lines meet other edges

Fig. 14.8 Example of a
Gilbert tessellation with free
angles. Figure created by
Claudio Rocchini (licensed
under CC 3.0)

nodes of the resulting graph have a degree 3 leading to 'T-shaped' intersections since
they result from a line stopping at another one. This model is extremely difficult to
analyze mathematically and one reason is that the tessellation depends strongly on
the order in which the points are chosen.

The STIT (acronym for stability under iteration) tessellation is another process
introduced by Nagel and Weiss [20]. In its simplest version this model consists in
splitting a polygon with random lines at random times (for an example see Fig. 14.9).
This model therefore requires to generate random sequences of time intervals and
random lines. If the time t is larger than a random variable (constructed as the sum of
random time intervals), the polygon under consideration is split into two parts with
the help of a random line. This process is iterated and applied to all polygons and
the time t controls the density of edges. This model has attracted a lot of attention in
stochastic geometry studies as a tractable model for hierarchical face splitting and for
describing natural processes such as crack formation (see [21] or [1] and references
therein). For more details we refer the interested readers to [1, 20, 21].

14.4   Planar Fragmentation

Planar fragmentation is a particular case of the general class of fragmentation processes, where we take into account the geometry (for an excellent account of fragmentation processes from a statistical physics point of view, see the book [22]). For
example, in the simple rectangular fragmentation process, a point and an orientation
(vertical or horizontal) are taken at random and a line is constructed in the same spirit

Fig. 14.9 Example of a
STIT tessellation. Figure
taken from [21]

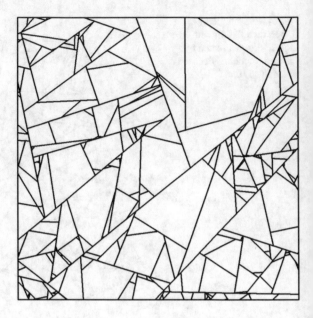

as in the Gilbert model and stops when it meets other edges (see Fig. 14.10 for a few
steps for this process).

This simple rectangular fragmentation model can be explored analytically and
we follow here the derivations given in [22]. We denote by $\rho(x_1, x_2, t)dx_1dx_2$ the
probability to have a rectangle of length x_1 and height x_2 and the equation governing
its time evolution is

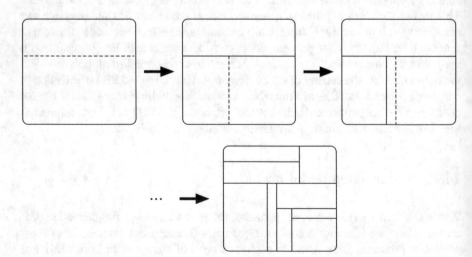

Fig. 14.10 A few steps in a rectangular fragmentation process. Figure inspired from [22]

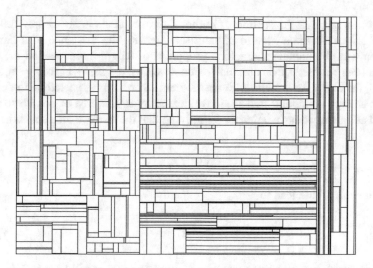

Fig. 14.11 Rectangular fragmentation model obtained for a probability $p = 1/2$ to draw an horizontal line (here $t = 500$ lines)

$$\frac{\partial \rho(x_1, x_2, t)}{\partial t} = -x_1 x_2 \rho(x_1, x_2, t) + x_2 \int_{x_1}^{1} \rho(y_1, x_2, t) dy_1 + x_1 \int_{x_2}^{1} \rho(x_1, y_2, t) dy_2$$

$$(14.43)$$

The first term of the right handside accounts for the decay of rectangle of size (x_1, x_2) while the second term corresponds to the breaking of a rectangle of size (y_1, x_2) integrated over all $y_1 > x_1$ (and the third term corresponds to the fragmentation of a rectangle of size (x_1, y_2) integrated over $y_2 > x_2$). We refer the reader interested in details of the calculation to [22] and the essential result is that the general moment scales as

$$\langle x_1^n x_2^m \rangle \sim t^{-\alpha(n,m)-2} \tag{14.44}$$

where α is the function

$$\alpha(n, m) = \frac{n + m}{2} - 1 - \sqrt{\frac{(n - m)^2}{4} + 1} \tag{14.45}$$

This result shows in particular that the moment of the width of faces (or equivalently the height) scale as

$$\langle x \rangle \sim t^{-(3-\sqrt{5})/2} \sim t^{-0.38...} \tag{14.46}$$

$$\langle x^2 \rangle \sim t^{-(2-\sqrt{2})/2} \sim t^{-0.29...} \tag{14.47}$$

First, the fact that the scaling of higher order moments is not simply related to low-order ones is the signature of multiscaling with the existence of a family of exponents. It also signals anomalous properties of length moments and the aspect ratio that can be observed on the small example shown in Fig. 14.11 for $t = 500$ lines. Second, the decay for $\langle x \rangle$ and $\langle x^2 \rangle$ are slower than what is expected on the basis of a naive scaling argument. Indeed, if we denote by a_i the area of the ith rectangle created at time i, we have $\sum_i^t a_i = 1$ implying that the average area scales as

$$\langle a \rangle \sim \frac{1}{t} \tag{14.48}$$

which naively would lead to

$$\langle x \rangle \sim \sqrt{\langle a \rangle} \sim \frac{1}{t^{1/2}} \tag{14.49}$$

which decays faster than the exact result. Also, the aspect ratio $\langle x_1/x_2 \rangle$ (for $x_1 > x_2$) diverges for large time as $t^{\sqrt{2}-1}$ showing that typical fragments become very elongated in time.

Finally, for this model it is also possible to compute for large times the scaling of the area distribution $P(a, t)$ (i.e. the density of rectangles of area a) and it is [22]

$$P(a, t) \sim t^2 e^{-at} \tag{14.50}$$

showing that indeed the typical area scales as $1/t$.

References

1. S.N. Chiu, D. Stoyan, W.S. Kendall, J. Mecke, *Stochastic Geometry and Its Applications* (Wiley, 2013)
2. A. Okabe, B. Boots, K. Sugihara, S.N. Chiu, *Spatial Tessellations: Concepts and Applications of Voronoi Diagrams*, vol. 501 (Wiley, 2009)
3. J. Meijering, Philips Res. Rep. **8**(4), 270 (1953)
4. E. Gilbert, Ann. Math. Stat. **33**(3), 958 (1962)
5. F. Avram, D. Bertsimas, Ann. Appl. Probab. pp. 1033–1046 (1993)
6. D. Weaire, J. Kermode, J. Wejchert, Philos. Mag. B **53**(5), L101 (1986)
7. M. Tanemura, FORMA-TOKYO **18**(4), 221 (2003)
8. A. Hinde, R. Miles, J. Stat. Comput. Simul. **10**(3–4), 205 (1980)
9. J.S. Ferenc, Z. Néda, Phys. A: Stat. Mech. Appl. **385**(2), 518 (2007)
10. H.J. Hilhorst, J. Stat. Mech. Theory Exp. **2005**(09), P09005 (2005)
11. H. Hilhorst, Eur. Phys. J. B-Condens. Matter Complex Syst. **64**(3), 437 (2008)
12. D. Aboav, Metallography **3**(4), 383 (1970)
13. D. Weaire, Metallography **7**(2), 157 (1974)
14. H.J. Hilhorst, J. Phys. A: Math. Gen. **39**(23), 7227 (2006)
15. R.E. Miles, Math. Biosci. **6**, 85 (1970)
16. M.T. Gastner, M. Newman, Phys. Rev. E **74**(1), 016117 (2006)
17. S.M. Gusein-Zade, Geogr. Anal. **14**(3), 246 (1982)

18. J. Um, S.W. Son, S.I. Lee, H. Jeong, B.J. Kim, Proc. Natl Acad. Sci., **106**(34), 14236–14240 (2009)
19. E. Gilbert, *Applications of Undergraduate Mathematics in Engineering* (1967)
20. W. Nagel, V. Weiss, Adv. Appl. Probab. **37**(04), 859 (2005)
21. W. Nagel, J. Mecke, J. Ohser, V. Weiss, Image Anal. Stereol. **27**(2), 73 (2011)
22. P.L. Krapivsky, S. Redner, E. Ben-Naim, *A Kinetic View of Statistical Physics* (Cambridge University Press, 2010)

Chapter 15
Proximity Graphs

The main idea for constructing these graphs is that two nodes have to be sufficiently near in order to be connected which justifies the name 'proximity' graphs.

The simplest case, the random geometric graph (also called the unit disk graph) is obtained by applying a proximity rule on randomly distributed points in the plane, which states that nodes only within a certain distance are connected. There is an extensive mathematical litterature (see the book [1] and references therein) on these graphs and they were also studied by physicists in the context of continuum percolation (see for example [2, 3]). This process extends usual percolation theory to continous space where shapes are randomly positioned in continuous space and can overlap [4].

Random geometric graphs are probably the simplest models of spatial networks and are also a good model for a number of applications, in particular when a finite range is in play. This is the case of wireless networks, smart-grids, disaster relief, etc. (see [5] and references therein). In particular, in the context of wireless ad-hoc networks, the nodes represent devices that can communicate with each other and an edge between two nodes signals the possibility of a communication between the corresponding devices. The percolation threshold and other quantities have then a direct interest in applications. We will first discuss the 'hard' case where the range is fixed in contrast with the 'soft' case where there is probability of a communication (that usually decreases with the distance between nodes). This includes in particular the Waxman model used as a model for the topology of the intra-domain internet [6].

We will then consider variants of the random geometric graphs, obtained by removing randomly some edges (which gives the Bluetooth graph) or by allowing nodes to move. Another graph obtained by a simple rule is the k-nearest neighbour model, where we connect a node to its k nearest neighbours.

We will focus in this chapter on graphs constructed with the euclidean distance, but one could use any other distance. For instance, radial graphs [7] are constructed over a graph G with its set of nodes, and two vertices are adjacent if the distance between them is equal to the radius of the graph $r(G)$ (see Chap. 4).

© The Author(s), under exclusive license to Springer Nature Switzerland AG 2022   295
M. Barthelemy, *Spatial Networks*,
https://doi.org/10.1007/978-3-030-94106-2_15

15.1   Random Geometric Graphs

The usual random geometric graph is defined in [8] and was introduced by Gilbert [9] who assumes that points are randomly located in the plane and have each a communication range r. Two nodes are connected by an edge if they are separated by a distance less than r. This corresponds to the 'hard' case. The 'soft' random geometric graph [10–12] is formed on a subset of points by adding an edge between distinct pairs with probability $H(|x - y|)$ where $H : \mathbb{R}^+ \to [0, 1]$ is called the *connection function* ($|x - y|$ is the Euclidean distance). We focus on the case of *Rayleigh fading* where, with $\gamma > 0$ a parameter and $\eta > 0$ the path loss exponent, the connection function, with $|x - y| > 0$, is given by

$$H(|x - y|) = \exp\left(-\gamma |x - y|^{\eta}\right) \tag{15.1}$$

and is otherwise zero (this choice is for example discussed in [13]).

15.1.1   The Hard Case

This model was introduced by Gilbert [9] who assumed that N points are located randomly in the plane and have each a communication range R. This also could be seen as a system of disks (or spheres in dimension d) of radius r. Two nodes are connected by an edge if they are separated by a distance less than R (or $2r$ for the distance between the centers of the disks). We show an example of such a network in Fig. 15.1.

If we denote by $\rho = N/V$ the density of nodes in the volume V (or area in the $d = 2$ case), the average degree is given by

$$\langle k \rangle = \rho \pi R^2 \tag{15.2}$$

Similarly to the Erdos-Renyi random graph, there are different quantities that we can compute and we will discuss some of them in the following. In particular, there is a percolation transition for a critical density and another transition to full connectivity.

We note that most studies are conducted in the limit $N \to \infty$ which can be achieved in different ways. A first way is to consider a finite total volume V, an increasing density $\rho = N/V$ but a range R that decreases with N such that $\langle k \rangle$ is fixed. Another way—mostly considered by mathematicians (see for example [1]) is to study this limit by considering a fixed R and a fixed density (usually taken equal to 1) but a total volume that varies as N/ρ.

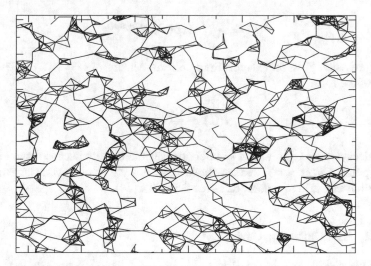

Fig. 15.1 Example of a 2d random geometric graph obtained with a density and radius such that the average degree is $\langle k \rangle \approx 6$

15.1.1.1 Degree Distribution

The degree distribution can be computed analytically for any distribution of points following for example [14]. We assume that the points are distributed according to a distribution $\rho(x)$ and the condition for connecting to nodes i and j located at positions x_i and x_j, respectively, is $d_E(i, j) \leq R$. We denote by $B_R(x)$ the ball of radius R and centered at x, and the probability $q_R(x)$ that a given node is located in $B_R(x)$ is

$$q_R(x) = \int_{B_R(x)} \mathrm{d}x' \rho(x') \tag{15.3}$$

The degree distribution for a node located at x is thus given by the binomial distribution

$$P(k; x, R) = \binom{N-1}{k} q_R(x)^k [1 - q_R(x)]^{N-1-k} \tag{15.4}$$

In the limit $N \to \infty$ and $R \to 0$, we obtain that the degree distribution for a node located at x is Poissonian and reads

$$P(k; x, \alpha) = \frac{1}{k!} \alpha^k \rho(x)^k e^{-\alpha \rho(x)} \tag{15.5}$$

where $\alpha = \langle k \rangle / \int \mathrm{d}x \rho^2(x)$ fixes the scale of the average degree. Averaging over x we then obtain the degree distribution under the general form

$$P(k) = \frac{\alpha^k}{k!} \int dx \rho(x)^{k+1} e^{-\alpha\rho(x)} \tag{15.6}$$

This expresssion gives for example for a uniform density $\rho(x) = \rho_0$ a degree distribution of the form

$$P(k) \sim \frac{(\alpha\rho_0)^k}{k!k^d} \tag{15.7}$$

which decays very rapidly with k. In contrast if the density decays slowly from a central point as $\rho(r) \sim r^{-\beta}$ we then obtain $P(k) \sim k^{-d/\beta}$ showing that large density fluctuations can lead to spatial scale-free networks [14].

15.1.1.2 The Clustering Coefficient

One of the first studies done by physicists on the random geometric graph can be found in [15]. In particular, these authors computed the average clustering coefficient and we reproduce their argument here. If two vertices i and j are connected to a vertex k it means that they are both in the excluded volume of k. In turn, these vertices i and j are connected only if j is in the excluded volume of i, and vice-versa. Putting all pieces together, the probability to have two connected neighbors i and j of a node k is given by the fraction of the excluded volume of i which lies within the excluded volume of k. By averaging over all points i in the excluded volume of k we then obtain the average clustering coefficient. We thus have to compute the volume overlap ρ_d of two spheres which for spherical symmetry reasons depends only on the distance between the two spheres. In terms of this function, the clustering coefficient is simply given by

$$\langle C_d \rangle = \frac{1}{V_e} \int_{V_e} \rho_d(r) dV \tag{15.8}$$

For $d = 1$, we consider a node i which defines the interval $[-R, R]$ of length $2R$ where the neighbors have to be located (see Fig. 15.2).

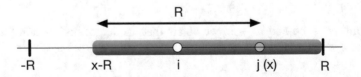

Fig. 15.2 Calculation of the clustering coefficient for a random geometric graph in $d = 1$. The node j is located at x and all the points that are located in the grey area are both neighbors of j and i

Fig. 15.3 The overlap between the two disks (area comprised within the bold line) gives the quantity $\rho_2(r)$, which by integration gives the average clustering coefficient. Figure taken from [15]

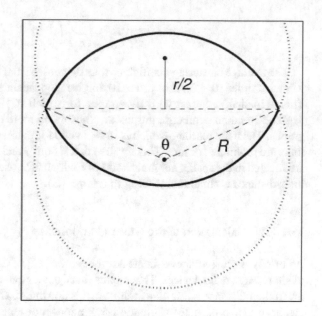

Two neighbors j and k of this node i must then be in the interval $[x - R, R]$ where $x = d(i, j)$. The probability that j and a node k chosen at random are connected is then given by

$$\rho_1(x) = (2R - x)/2R = 1 - x/2R \tag{15.9}$$

which gives on average

$$\langle C_1 \rangle = \frac{1}{2R} \int_{-R}^{R} dx \rho_1(x) \tag{15.10}$$

$$= 3/4 \tag{15.11}$$

For $d = 2$, we have to determine the area overlapping in the Fig. 15.3 which gives

$$\rho_2(r) = (\theta(r) - \sin(\theta(r)))/\pi \tag{15.12}$$

with $\theta(r) = 2 \arccos(r/2R)$ and leads to

$$\langle C_2 \rangle = 1 - 3\sqrt{3}/4\pi \approx 0.58650 \tag{15.13}$$

Similarly, an expression can be derived in d dimension [15] which for large d reduces to

$$\langle C_d \rangle \sim 3\sqrt{\frac{2}{\pi d}}\left(\frac{3}{4}\right)^{\frac{d+1}{2}} \tag{15.14}$$

The average clustering coefficient thus decreases from the value $3/4$ for $d = 1$ to values of order 10^{-1} for d of order 10 and is independent from the number of nodes. This is in sharp contrast with ER graphs for which $\langle C \rangle \sim 1/N$ (for fixed average degree). Random geometric graphs are thus much more clustered than random ER graphs. The main reason—which is in fact valid for most spatial graphs – is that long links are prohibited or rare. This implies that if both j and k are connected to i, there are located in the spatial neighborhood of i which increases the probability that their inter-distance is small too, leading to a large $\langle C \rangle$.

15.1.1.3   Calculation of the Giant Component.

As briefly discussed above, in ad-hoc networks [16], users communicate by means of short range radio devices. This means that a device can communicate with another one if their distance is less than their transmission range. The set of connected devices can be used to propagate information over a longer distance by going from the source to the destination hoping through intermediate nodes. If there is a large density of nodes, alternate routes are even available which allows to split the information into separate flows. Usually, the users are mobile and the network evolves in time and it is important to understand the condition for the existence of a giant cluster.

It has been demonstrated (see [1] and references therein) that for large N there is a critical density (at fixed R and volume given by N/ρ) below which we have small components of typical size $\sim \log N$ and over which a giant cluster of size $\sim N$. Gilbert [9] discussed already the probability to belong to an infinite cluster P_∞ and found a critical radius R_c above which there is a giant cluster. Using 1-independent percolation Ballister, Bollobas, and Walters [17] showed that with confidence 99.99% we have the bounds on the critical average degree

$$4.508 \leq k_c \leq 4.515 \tag{15.15}$$

which corresponds to $R_c \simeq 1.19 \pm 0.001$ and which is consistent with the non-rigourous bounds [3]

$$4.51218 \leq k_c \leq 4.51228 \tag{15.16}$$

Computing exactly this critical value for various dimensions led Dall and Christensen [15] to propose the following fit

$$k_c(d) = k_c(\infty) + \frac{A}{d^\gamma} \tag{15.17}$$

where $k_c(\infty) = 1$, $A = 11.78$ and $\gamma = 1.74$. This result indicates that random geometric graphs in infinite dimensions behave (in terms of the giant component) as an Erdos-Renyi random graph with $k_c = 1$. Computing exactly this critical value is an open problem and some bounds can be found (see for example [18]).

We can however estimate the scaling of the critical radius above which the graph is completely connected (which corresponds to the so-called full connectivity) with the following simple argument (that we will encounter again in the case of soft random geometric graphs). In a system of total area 1, the probability that a vertex is isolated is given by

$$p_0 = (1 - \pi R^2)^{N-1} \tag{15.18}$$

which means that all the other $N - 1$ nodes are not in the disk of radius R. The average number of isolated vertices is then given by $N_0 = N(1 - \pi R^2)^{N-1} \sim N e^{-N\pi R^2}$. From this simple calculation, we deduce that if $N_0 \approx 0$, the RGG is certainly connected. In constrast, if $N_0 \to \infty$, the RGG is almost surely disconnected. The transition happens then for $N_0 \sim \mathcal{O}(1)$ implying that the critical radius for full connectivity scales as

$$R_c \sim \left(\frac{\ln N}{N}\right)^{1/2} \tag{15.19}$$

(and a similar expression in dimension d). In addition, it can be shown (see [19]) that when $N_0 \sim \mathcal{O}(1)$ the RGG consists of one giant component of size larger than $N/2$ together with a number of isolated vertices distributed according to a Poisson distribution of average N_0. Another way to see the connectivity problem in the random geometric graph relies on the longest link in the MST constructed over the same set of point [1] (and Chap. 18). More precisely. A sufficient condition for the RGG to be connected is that the range R should be larger than the longest link R^* of the MST. This length scales as $R^* \sim \sqrt{\ln N/N}$ and we recover the previous result. Besides these hand-waving arguments, this sharp transition at $\sqrt{\ln N/N}$ can be proven more rigorously and we refer the interested reader to [20] and references therein.

15.1.2 Soft Random Geometric Graphs

Motivated by applications to wireless networks, random geometric graphs have been extended [1, 5] to the 'soft' case for which pairs of nodes are connected with a probability $H(r)$ which depends in general on the distance r betweeen the nodes (we note that this model was also used for social networks [21]). These independent probabilities $H(r)$ typically decrease as r increases (see for example Fig. 15.4). The 'hard' random geometric graph is recovered with the choice $H(r) = \Theta(r_0 - r)$ (where $\Theta(x)$ is the Heaviside function) and is therefore deterministic in this case.

Fig. 15.4 Example of the probability function $H(r)$ for soft random geometric graph. The 'hard' case corresponds to $H(r)$ given by a Heaviside function $H(r) = \Theta(r_0 - r)$

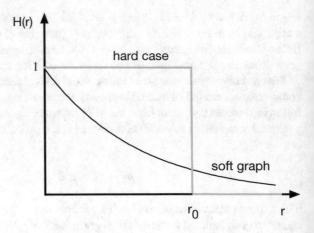

In contrast, in the soft case, there are two sources of randomness: the nodes and also the links. The clustering coefficient was discussed in [16] and recently in [22] in the case of $H(r)$ decreasing as a power law, and the problem of full connectivty was considered in [5] (and references therein) and we reproduce here the main results of this study.

15.1.3 The Full Connectivity Probability

At low density (or at low R), the graph is disconnected and comprises many small clusters. At the percolation threshold, a giant cluster which contains a macroscopic fraction of nodes appears. If the density is increased beyond the percolation threshold the graph forms a single cluster and this transition is described by the probability P_{fc} of full connectivity. This quantity is of interest in practical applications where it is important that all devices (described by nodes) are connected.

We will follow [10] where the authors used a cluster expansion approach in order to derive a systematic perturbative method for computing the full connection probability P_{fc} as a function of the density of nodes $\rho = N/V$ (where V is the total volume of the system).

For N nodes there are $N(N-1)/2$ pairs which are either connected or not and the number of possible graphs is then $2^{N(N-1)/2}$. We therefore obtain the trivial equality

$$1 = \sum_{g \in G_N} P(g) \tag{15.20}$$

where $P(g)$ is the probability to obtain the graph g among the $2^{N(N-1)/2}$ possible graphs with N nodes. We can then classify these graphs according to their number n of isolated nodes

$$1 = \sum_{g \in G_{n,N}} P_n(g) \tag{15.21}$$

where $G_{n,N}$ is the family of graphs with N total nodes and n isolated nodes. We denote by P_n the probability that a graph has n isolated nodes. This identity allows us to express the full connection probability $P_{fc} \equiv P_{n=0}$

$$P_{fc} = 1 - P_1(g) - P_2(g) - \ldots \tag{15.22}$$

At first order we consider graphs with one single isolated nodes only. Denoting by H_{ij} the probability to connect nodes i and j we obtain for the probability that a given node, say 1, is isolated reads

$$P_1(1) = \prod_{j=2}^{N}(1 - H_{1j}) \tag{15.23}$$

Averaging over the position r_1 of the node 1, we obtain

$$P_{fc} = 1 - \rho \int dr_1 \left[1 - \frac{M(r_1)}{V}\right]^{N-1} \tag{15.24}$$

where the probability to be connected can be written in the continuous limit

$$\frac{M(r_1)}{V} = \frac{1}{V} \int dr_2 H(r_2 - r_1) \tag{15.25}$$

In the limit of large N we then obtain from Eq. (15.24)

$$P_{fc} \simeq 1 - \rho \int dr_1 e^{-\rho M(r_1)} \tag{15.26}$$

This important result was demonstrated rigorously by Penrose [11] and we reproduce here the main steps of his argument (and refer the reader interested in the technical—difficult—details to [11]). More precisely, if we denote by $\mathcal{K}$ the set of fully connected graphs, the probability P_{fc} is given by

$$P_{fc} = P(G \in \mathcal{K}) \sim P(N_0(G) = 0) \tag{15.27}$$

where $N_0(G)$ is the number of isolated nodes in the graph G. The important result is that the number of isolated node N_0 is distributed according to a Poisson law

$$P(N_0 = k) = \frac{m^k}{k!} e^{-m} \tag{15.28}$$

where $m = \langle N_0 \rangle$ is the average value of N_0. This result in particular implies that the probability that there are no isolated nodes is

$$P_{fc} \sim e^{-\langle N_0 \rangle} \tag{15.29}$$

We can estimate (the rigorous derivation can be found in [11]) this average $\langle N_0 \rangle$ by noting that the probability that a node at location r is isolated reads

$$P_0(r) = \prod_{r'}(1 - H(r - r')) \tag{15.30}$$

which implies

$$\langle N_0 \rangle = \rho \int dr \prod_{r'}(1 - H(r - r')) \tag{15.31}$$

$$\simeq \rho \int dr e^{-\rho \int dr' H(r-r')} \tag{15.32}$$

It is expected that this result holds for a large class of probability functions $H(r)$ (for $d \geq 2$). We then finally obtain for the full connectivity probability

$$P_{fc} \sim e^{-\rho \int dr e^{-\rho \int dr' H(r-r')}} \tag{15.33}$$

which is the result Eq. (15.26).

In particular in the hard case with a range $R = 1$, a volume $V = N/\rho$ (in the limit $N \to \infty$), the probability is given by $H(r) = \Theta(r)$ and the Eq. (15.33) predicts

$$P_{fc} \sim e^{-Ne^{-\rho\pi}} \tag{15.34}$$

Defining $t = \pi\rho - \log N$ we obtain

$$P_{fc} \sim e^{-e^{-t}} \tag{15.35}$$

which allows to express this quantity P_{fc} as a function of ρ and N.

15.1.4 The Waxman Model

The Waxman model [23] is a widely accepted model for the topology of intra-domain internet network and used to test routing performances for example (see [6] and references therein). It can be seen as a spatial variant of the Erdos-Renyi model (see Chap. 11) with a probability to connect two nodes given by

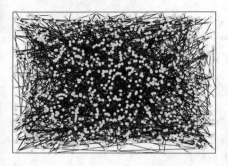 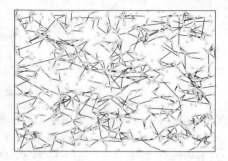

Fig. 15.5 Left: Erdos–Renyi graph with $E = 926$ links between $N = 1000$ nodes. Right: Waxman graph obtained for $N = 1000$ nodes and the same number of links E ($\beta = 0.5$, $\alpha = 0.05$). As expected there is no spatial structure in the ER graph, while in the Waxman graph we observe a spatial organization with a local clustering effect

$$p_{ij} = h(r_i, r_j) \tag{15.36}$$

where $r_{i(j)}$ denotes the position of node $i(j)$. The Waxman model is recovered as a particular case with $h(u, v) \propto e^{-\alpha|u-v|}$.

Another way to see the Waxman model is as a soft random geometric graph which is why we included it in this part. More precisely In this model, the nodes are uniformly distributed in the plane and edges are added with probabilities that depend on the distance between the nodes

$$P_{ij} = \beta e^{-d_E(i,j)/d_0} \tag{15.37}$$

The quantity d_0 determines the typical length of the links and β controls the total density of links. In the original model d_0 is written as αL where L is the maximum distance allowed. In terms of hidden variables the attribute is here the spatial location of the node and the pair connection probability depends on the distance between the nodes. For $d_0 \to \infty$, length is irrelevant and we recover the ER model while for $d_0 \sim 1/\sqrt{\rho}$ (where ρ is the average density of nodes in the plane) long links are prohibited and we are in the limit of a lattice-like graph (Fig. 15.5).

Even if this model is very simple, it served as a first step towards the elaboration of more sophisticated models of the Internet [24]. Also, despite its simplicity, this Waxman model can be used in order to understand the importance of space in different processes taking place on this network. We can cite for example navigation or congestion problems in communication systems.

Some quantities can be calculated for this model such as the average number of links which is given by

$$\mathscr{L}(\{r\}) = \sum_{i=1}^{N} \sum_{j=i+1}^{N} h(r_i, r_j) \tag{15.38}$$

$$= \sum_{i=1}^{N} \sum_{j=i+1}^{N} e^{-\alpha|r_i - r_j|} \tag{15.39}$$

which depends on the quenched positions of the nodes. The average over the disorder is then obtained by integrating over the distribution of the positions (assumed to be described by a Poisson process), one obtains [25]

$$\langle \mathscr{L} \rangle = \frac{N(N-1)}{2} \int \frac{du\,dv}{V^2} h(u, v) \tag{15.40}$$

This integral cannot be computed in general but in the Waxman case in a square of linear size L an explicit calculation is possible and the result depends only on the variable αL [25]. The exact result can be found in [25] and the limiting behavior for $\alpha \to \infty$ is

$$\frac{\langle \mathscr{L} \rangle}{\mathscr{L}_{max}} \sim \frac{2\pi}{\alpha^2} + \mathcal{O}\left(\frac{1}{\alpha^3}\right) \tag{15.41}$$

(where $\mathscr{L}_{max} = N(N-1)/2$).

Van Mieghem [25] also estimated the average number of paths of a given length and we reproduce his argument here. For a complete graph, the number χ_j of paths of length j between two nodes A and B is maximum and is given by

$$\chi_j(A, B) = \sum_{k_1 \notin \{A,B\}} \sum_{k_2 \notin \{A,k_1,B\}} \cdots \sum_{k_{j-1} \notin \{A,k_1,\ldots,k_{j-2},B\}} 1$$

$$= (N-2)(N-3)\ldots(N-j)$$

$$= \frac{(N-2)!}{(N-j-1)!} \tag{15.42}$$

For a ER graph, the probability of a path of length j is p^j and the average number of paths is then given by

$$\langle \chi_j \rangle = \frac{N-2)!}{(N-j-1)!} p^j \tag{15.43}$$

The total number of paths between two nodes is then

$$\chi = \sum_{j=1}^{N-1} \chi_j \tag{15.44}$$

and the following dominant behavior can be found

$$\chi \sim (N - 2)! p^{N-1} e^{1/p} \tag{15.45}$$

When $p = 1/N$ we obtain in the large N limit

$$\chi \sim \frac{1}{\sqrt{N}} \tag{15.46}$$

which is explained by the fact that there are no paths between arbitrary nodes for large N. The condition $\chi < 1$ gives therefore a 'total disconnectivity threshold' p_d which can be estimated as [25]

$$p_d = \frac{1}{N} + \mathcal{O}(\sqrt{\log N/N}) \tag{15.47}$$

and if $p > p_d$ all the pairs of nodes are connected.

In the case of the Waxman graph, Van Mieghem [25] showed that the number of paths of length j is given by the (expected) expression

$$\langle \chi \rangle = \frac{(N - 2)!}{(N - j - 1)!} F_j \tag{15.48}$$

where

$$F_j = \int_V \prod_{i=1}^{j} \frac{dr_i}{V} h(r_1 - r_2) h(r_2 - r_3) \cdots h(r_{j-1} - r_j) \tag{15.49}$$

In the exponential case $h(r) = e^{-\alpha|r|}$ and using the Jensen inequality

$$\langle e^{-X} \rangle \geq e^{-\langle X \rangle} \tag{15.50}$$

we obtain the following lower bound

$$\frac{(N - 2)!}{(N - j - 1)!} F^j \leq \chi \tag{15.51}$$

with

$$\begin{cases} F & = e^{-\langle r \rangle} \\ \langle r \rangle & = \int_V \frac{dr}{V} r \end{cases} \tag{15.52}$$

We end this section by noting that a growth model close to the Waxman model was proposed in [26] where at each time step a new node u is added in the plane and is connected to existing nodes v with a probability

$$p_{uv} = \beta e^{-\alpha d_E(u,v)} \tag{15.53}$$

as in the Waxman model, but if it fails to connect the node is discarded (a probability decreasing as a power law $P(u, v) \sim d_E(u, v)^{-\tau}$ was also studied). In this way, 'surviving' nodes are necessarily in the vicinity of the existing network and this model could be more suited to describe the growth of biological or artificial systems. Networks generated with this algorithm have a large clustering coefficient (as expected) and probably a large diameter (although there is no quantitative prediction in [26]).

15.1.5   Random Geometric Graphs in Hyperbolic Space

Motivated by studies on the Internet, a model of a random geometric graph in hyperbolic space was proposed in [27] (see also [28, 29] for more details). In these studies, Boguñá, Krioukov and Serrano considered the two-dimensional hyperbolic space $\mathbb{H}^2$ of constant negative curvature equal to $K = -\zeta^2 = -1$ and used a polar representation (r, θ) for the nodes. They placed N points distributed uniformly in a disk of radius R and in the Euclidean disk projection this implies that the nodes have a uniform angle distribution in $\theta \in [0, 2\pi]$ and that the radial coordinate is distributed according to

$$\rho(r) = \frac{\sinh r}{\cosh R - 1} \approx e^r \tag{15.54}$$

They considered then the usual geometric graph rule and connect two nodes if their hyperbolic distance d is less than R which can be written in terms of the connection probability as

$$H(d) = \Theta(R - d) \tag{15.55}$$

where Θ is the Heaviside function. The hyperbolic distance d between two nodes (r, θ) and (r', θ') is defined by

$$\cosh \zeta d = \cosh \zeta r \cosh \zeta r' - \sinh \zeta r \sinh \zeta r' \cos(\theta' - \theta) \tag{15.56}$$

In line with random geometric graphs in euclidean space, there is also a strong clustering for these graphs constructed on hyperbolic space (unfortunately the clustering coefficient cannot be computed analytically).

Finally, the authors of [28] extended their model and introduce an inverse 'temperature' β akin to usual statistical mechanics in the connection probability

$$H(x) = \left(1 + e^{\beta(\zeta/2)(x-R)}\right)^{-1} \tag{15.57}$$

and showed that this $\mathbb{H}^2$ model reproduces well the Internet measurements for $P(k)$, the assortativity $k_{nn}(k)$, and the clustering coefficient $C(k)$ for $\alpha = 0.55$, $\zeta = 1$ and $\beta = 2$.

15.2 Bluetooth Graph and Sparsification

The Bluetooth graph [30] is an example of a sparsification of a graph where one extracts a subgraph from a known graph. A simple way to achieve this is to consider a graph and to keep a number c of allowed edges for each node. The simplest example of sparsification is starting from the complete graph $G = K_n$ and in [31] the constant case $c = const.$ has been considered. It was proven that the resulting graph is connected with a large probability (i.e. the probability tends to 1 in the infinite limit of the number of nodes). After this complete graph example, the natural candidate that was the object of many studies (see [30] and references therein) is the sparsification of the random geometric graph where each node uses effectively a number c of allowed edges. There are therefore two parameters here and which correspond to the two sources of randomness in this model: the range R that allows to construct the random geometric graph and c. In the paper [30], the authors characterize the connectivity of this graph in the (R, c) space. In particular, we saw above that a condition for a random geometric graph to be connected is that the range is larger than a threshold R^* that corresponds to the longest link in the MST (see for example [1] and Chap. 18).

$$R \geq R^* \sim \left(\frac{\log N}{N} \right)^{1/d} \qquad (15.58)$$

in dimension d. Obviously, the interesting case is obtained for values of R that are just above the threshold, $R \sim R^*$. If R is much less than the threshold, the graph cannot be connected. When c is sufficiently large, the bluetooth graph will be connected, and in [32] it was shown that $c \sim \log N$ is enough to ensure connectivity. This bound was improved in [30] where it was shown that there is a threshold value for c when $R \sim (\log N/N)^{1/d}$ and which is given by

$$c^* \sim \sqrt{ \frac{\log N}{\log \log N} } \qquad (15.59)$$

and which is independent of the dimension d. This relation gives the minimum number of neighbors needed to ensure connectivity (among the $\log N$ neighbours in the ball of volume $\log N/N$).

15.3   The $k-$nearest Neighbour Model

15.3.1   Definition and Connectivity Properties

This model is very similar to the random geometric graph and starts from a distribution of points in space, but instead of joining nodes whose distance is less than a certain range, we connect a node to its k first nearest neighbours (sorted according to their euclidean distance to it). We note that this creates a priori a directed graph with k outgoing links but all directions can be removed and we get an undirected graph.

For this graph, we connect points to their $k \in \mathbb{N}$ nearest neighbours. When $k = 1$, we obtain the nearest neighbour graph (1-NNG) (see for example the Chap. 8 of [33]). The model is notably different from the RGG because local fluctuations in the density of nodes do not lead to local fluctuations in the degrees. The typical degree is much lower than the RGG when connected [33]. For large enough k, the graph contains the RGG as a subgraph. The average degree is always then between k and $2k$ and it can be shown that the maximum degree is bounded by $6k$ (see [34] and references therein).

Percolation in this model can also be considered and the threshold k_c is defined as the minimum value of k such that the probability P_∞ to belong to the giant cluster is strictly positive. Numerical simulations show that for $d = 2$

$$k_c = 3 \tag{15.60}$$

but this result is not yet rigorously demonstrated (although there are theoretical bounds such as $k_c \leq 11$ for example, see [34]). For larger dimensions, numerical simulations seem to suggest that (see [34])

$$k_c(d \geq 3) = 2 \tag{15.61}$$

The full connectivity problem is here more difficult. We need to estimate how large k has to be in order to have a single giant cluster which connects everyone. In particular, there are no isolated nodes by definition (the minimum degree is k) and the previous argument exposed above cannot be applied. A simple argument [34] however shows that $k \sim \log N$ is enough to ensure connectivity. Indeed, assume that we tessellate space (of total area N so that the density is $\rho = 1$) in squares of area of order $\log N$, the probability that there are no nodes in a square is of order $e^{-\log N} \sim 1/N$. This means that there is a large probability to have at least one node per square. Also, if $k \sim \log N$ then the probability that the number of points K in the square is larger than k is given by

$$P(K > k) = e^{-\rho \log N} \sum_{\ell \geq k+1} \frac{(\rho \log N)^\ell}{\ell!} \tag{15.62}$$

and using the inequality $r! > (r/e)^r$, we obtain

$$P(K > k) < e^{-\rho \log N} \left(1 + \frac{1}{e} + \frac{1}{e^2} + \cdots \right) \sim \frac{1}{N} \qquad (15.63)$$

and is therefore very small. This means that all points in the square are connected to each other, but also connected to points in adjacent squares. This is obviously a sufficient—but not necessary—condition for getting a fully connected graph. However, in [35], the authors found the rigorous bounds $0.3043 \log N < k_c < 0.5139 \log N$ which justifies a posteriori the argument described above.

15.3.2 A Scale-Free Network on a Lattice

As we saw previously, geometric graphs constructed on uniformly distributed points naturally lead to networks with degrees distributed according to a Poisson distribution. It is however interesting to generate spatial graphs with a broad distribution (scale-free network) in order to understand the effect of strong heterogeneity on spatial networks. In [36], Rozenfeld et al. proposed a simple method to construct a scale-free network on a lattice (and in [37] another variant is suggested). This model is defined on a d-dimensional lattice with periodic boundary conditions. A random degree k is assigned to each node on this lattice according to the probability distribution

$$P(k) = Ck^{-\lambda} \qquad (15.64)$$

The idea is then simple: we connect a randomly chosen node i to all its nearest neighbors until its degree reaches its assigned value k_i. The larger k and the larger the region contains connected neighbors (we note here that outdegree is considered only). The size of this region is such that $\rho r^d \sim k$ which implies that

$$r(k) \sim \left(\frac{k}{\rho} \right)^{1/d} \qquad (15.65)$$

where ρ is the density of points. We show in Fig. 15.6 the obtained networks for two different values of λ.

The larger λ and the shorter the links (large degrees k and therefore long links are very rare) and the closer we are to a regular lattice. We define a chemical shell ℓ as consisting of all sites at shortest distance ℓ from a given site. For large λ these chemical shells are essentially concentric (as in the case of a regular lattice) while for smaller λ the presence of long links destroys this order. Scaling arguments proposed in [36] suggest that the minimal length exponent is given by

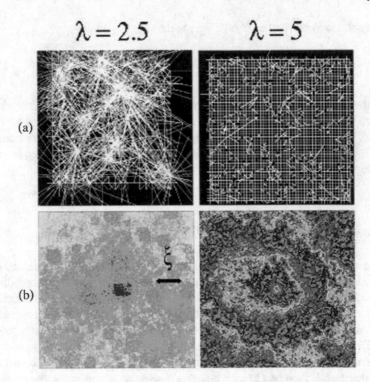

Fig. 15.6 Spatial structure of the network obtained by the method proposed in [36] for different values of the exponent λ of degree distribution. In **a** the networks are shown and in **b** the corresponding chemical shells of equidistant sites form the central node. Figure taken from [36]

$$d_{min} = \frac{\lambda - 2}{\lambda - 1 - 1/d} \qquad (15.66)$$

This network has then the curious property to have a fractal dimension which stays identical to the euclidean dimension, but with a minimal length exponent $d_{min} < 1$ for all λ and $d > 1$.

The authors of [37] studied the percolation properties of such a model and found that for these spatial scale-free networks with a degree distribution $P(k) \sim k^{-\lambda}$, the percolation threshold in the limit of infinite networks does not go towards zero (for λ > 2), in sharp contrast with non-spatial scale-free network which have $p_c = 0$ for λ < 3 (in the limit $N \to \infty$). In fact, for λ > 2, most of the links are short and the network behaves as an almost regular lattice with a finite percolation threshold.

In a non-spatial scale-free network, there are many short paths between the different hubs of the system easing the percolation. In contrast, for spatial (scale-free) networks there is a high local clustering due to the limited range of links which naturally lengthen the distance between hubs. This negative assortativity makes it thus more difficult to achieve percolation in such a system hence the existence of a non-zero threshold. As noted in [37], the spread of a disease too would be easier to control than on a scale-free network which is expected as spatial containment is usually easier to set up.

15.4 A Dynamical Proximity Model

A model for contact networks based on mobile agents was proposed in [39–41]. In this model, individuals are described by disks with the same radius and are moving and colliding in a two-dimensional space. The contact network is built by keeping track of the collisions: a link connects two nodes if they have collided at a previous time. A collision takes place whenever two agents are at a certain distance and this model can thus be seen as a dynamical version of random geometric graphs. In particular, Gonzalez, Lind, Herrmann [41] used this framework to model the sexual interaction network. Interestingly enough, they found that in some cases and for a simple collision rule such that the velocity grows with the number of previous collisions, a scale-free network with exponent $\gamma = 3$ emerges.

15.4.1 The Model

The model proposed in [39] is composed of N mobile agents described by disks of radius r and moving in a two-dimensional square-shaped space of linear size L and with periodic boundary conditions. The density $\rho \equiv N/L^2$ is supposed to be low at all times. A link is formed whenever two agents collide and after each collision, the agents move with updated velocity in random directions.

More precisely, the agents are placed at random locations with the same initial speed v_0 and with random directions. At each time step Δt, the position $\mathbf{x}_i$ of agent i is updated according to

$$\mathbf{x}_i(t + 1) = \mathbf{x}_i(t) + \mathbf{v}_i(t)\Delta t. \tag{15.67}$$

After a collision with agent i at time t, its speed is updated according to

$$|\mathbf{v}_i(t + 1)| = v_0 + \bar{v}k_i(t), \tag{15.68}$$

where $\bar{v}$ is a constant having unit of velocity and v_0 is the initial velocity of the agents, and where $k_i(t)$ is the number of connections of node i at time t. There is a characteristic time between collisions given by

$$\tau_o \equiv \frac{1}{2\sqrt{2\pi}r\rho v_0} \tag{15.69}$$

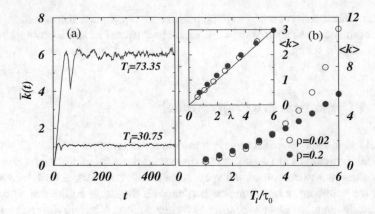

Fig. 15.7 (Color online) **a** Average degree $\bar{k}$ per agent as function of time t, illustrating the convergence towards a quasi-stationary state ($N = 4096$). **b** Average degree $\langle k \rangle$ versus T_l/τ_0 for $N = 10^4$, averaged over 100 realizations. Inset: linear dependence between $\langle k \rangle$ and n_C (see text); the solid line indicates $\langle k \rangle = n_C/2$. In all cases $v_0 = \sqrt{2}$ and $\bar{v} = 1$. Figure taken from [40]

In addition, agents arrive and depart after a time of residence which implies that the total number of agents is roughly constant in time, and the system can in this way reach a stationary state. The 'age' of each agent is denoted by $\mathscr{A}_i$ and is initially chosen randomly in the interval $[0, T_l]$. It is then updated as

$$\mathscr{A}_i(t + 1) = \mathscr{A}_i(t) + \Delta t. \tag{15.70}$$

When the age of the agent is equal to the maximum lifetime $\mathscr{A}_i = T_l$, agent i leaves the system, all its links are removed, and a new agent replaces its position with initial conditions (with speed v_0, random lifetime). Each agent thus stays a time $T_l - \mathscr{A}_i(0)$ in the system before leaving.

15.4.2   Stationary State

The authors of [39, 40] showed numerically that the system reaches a stationary state characterized by constant features that depends on two parameters, the density ρ and the average number of collisions T_l/τ_o that each agent experiences. This convergence is shown in Fig. 15.7a for the average degree $\bar{k}(t)$ per agent.

The stationary value of the average degree is shown in Fig. 15.7b versus vs. T_l/τ_o and displays a non-linear increase that depends on the density. Also, the average degree depends on the average number n_C of collisions during the average residence time $T_l - \langle \mathscr{A} \rangle$, and defined as

Table 15.1 Critical exponents related to the emergence of the giant cluster for the network of mobile agents, compared to the mean-field and $2D$ percolation values. Table taken from [40]

	Mean-field	$2D$ percolation	Mobile agents
ν	0.5	$4/3 \sim 1.33$	1.3 ± 0.1
γ	1	$43/18 \sim 2.39$	2.4 ± 0.1
β	1	$5/36 \sim 0.139$	0.13 ± 0.01
σ	0.5	$36/91 \sim 0.397$	0.40 ± 0.01

$$n_C \equiv \frac{1}{v_o \tau_o} \langle v \rangle (T_l - \langle \mathscr{A} \rangle). \tag{15.71}$$

The result, shown on the inset of Fig. 15.7b, suggests that $\langle k \rangle = n_C/2$ for all values of the density.

15.4.3 Percolation Properties

In this model, the authors [40] find a percolation transition with a critical value $n_C = 2.04$, beyond which a giant cluster of connected nodes emerges. We show on the Table 15.1 the numerical values obtained for this model and the standard values for the 2d percolation, together with the mean-field results for the critical exponents.

This table shows clearly that the model is in same universality class of 2d percolation on lattices. This can be understood as the agents move in the 2d plane during a finite time and will be able to create connection with a limited number of other agents that are in a restricted vicinity. This corresponds effectively to a short-range connectivity—even if the clusters are not quenched—and therefore to the standard 2d percolation.

15.4.4 Degree Distribution

The degree distribution $P(k)$ depends on the collision rule and on the value $\bar{v}$ used for updating the speed (Eq. (15.68)). For $\bar{v} = 0$, the degree distribution is well fitted by a Poisson distribution

$$P_p(k) = \frac{\langle k \rangle^k}{k!} e^{-\langle k \rangle} \tag{15.72}$$

In contrast for $\bar{v} = 1$, the distribution is closer to an exponential of the form

$$P_e(k) = \frac{1}{\langle k \rangle - 1} e^{-(k-1)/(\langle k \rangle - 1)} \tag{15.73}$$

However, while for small $\langle k \rangle$ the degree distribution of the giant cluster is exponential of the form of $P_e(k)$, for larger $\langle k \rangle$ it deviates from this shape. So far, these results are essentially numerical and theoretical derivations are missing.

The degree correlations were also characterized in [39] using the average degree of the nearest neighbors of a vertex of degree k, and the mixing is found to be assortative, as expected for spatial networks in general and for social networks in particular (see Chap. 4).

15.5 Apollonian Networks

These networks were originally proposed as a model for spatial scale-free graphs. They were introduced by Andrade et al. [42] who constructed a scale-free network (Fig. 15.8) from a space-filling packing of spheres and by connecting the centers of touching spheres by lines. In this respect, we can consider that they belong to the family of proximity graphs.

These networks are simultaneously planar, scale-free with exponent $\gamma = 1 + \ln 3 / \ln 2$, small-world—in fact 'ultra-small'—with an average shortest path varying as

$$\langle \ell \rangle \sim (\log N)^{3/4} \tag{15.74}$$

and a clustering coefficient larger than 0.8 for large N. Various quantities for the Apollonian network and one of its variant are also computed and discussed in [43, 44]. Due to all these simultaneous properties, Apollonian networks provide an interesting playground to test theoretical ideas.

In [45], Auto et al. studied percolation on Apollonian networks (see [42]) using real-space renormalization. For this two-dimensional spatial, scale-free, and planar network, the percolation threshold goes to zero in the thermodynamic limit in agreement with general results for scale-free networks with $\gamma < 3$. The mass of the percolating cluster however behaves as $M \sim e^{-\lambda/p}$ (where λ is a constant), a result reminiscent of the marginal case $\gamma = 3$.

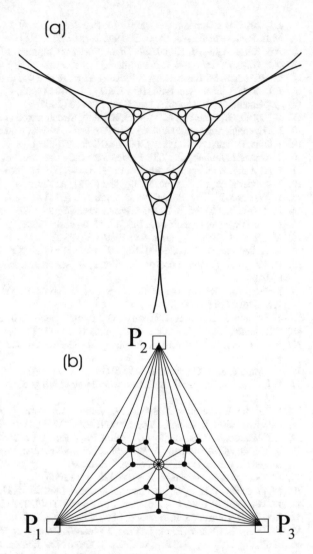

Fig. 15.8 Top: Classical Apollonian packing. Bottom: Apollonian network (showing the first, second and third generation with circles, squares, dots symbols respectively). Figure taken from [42]

References

1. M. Penrose, *Random Geometric Graphs*, vol. 5 (Oxford University Press, 2003)
2. I. Balberg, Phys. Rev. B **31**, 4053 (1985)
3. J. Quantanilla, S. Torquato, R. Ziff, Phys. A **33**, L399 (2000)
4. R. Meester, R. Roy, *Continuum Percolation*, vol. 119 (Cambridge University Press, 1996)
5. C.P. Dettmann, O. Georgiou, Phys. Rev. E **93**(3), 032313 (2016)
6. M. Naldi, Comput. Commun. **29**(1), 24 (2005)
7. K. Kathiresan, G. Marimuthu, Ars Combin. **96**, 353 (2010)
8. M. Penrose, et al., *Random Geometric Graphs*, vol. 5 (Oxford university press, 2003)
9. E.N. Gilbert, J. Soc. Ind. Appl. Math. **9**(4), 533 (1961)

10. J. Coon, C.P. Dettmann, O. Georgiou, J. Stat. Phys. **147**(4), 758 (2012)
11. M.D. Penrose et al., Ann. Appl. Probab. **26**(2), 986 (2016)
12. A.P. Kartun-Giles, S. Kim, IEEE Trans. Wireless Commun. **17**(5), 3201 (2018)
13. A.P. Giles, O. Georgiou, C.P. Dettmann, J. Stat. Phys. **162**(4), 1068 (2016)
14. C. Herrmann, M. Barthelemy, P. Provero, Phys. Rev. E **68**, 026128 (2003)
15. J. Dall, M. Christensen, Phys. Rev. E **66**(1), 016121 (2002)
16. G. Nemeth, G. Vattay, Phys. Rev. E **67**, 036110 (2003)
17. P. Balister, B. Bollobás, M. Walters, Random Struct. Algorithms **26**(4), 392 (2005)
18. M. Haenggi, *Stochastic Geometry for Wireless Networks* (Cambridge University Press, 2012)
19. J. Diaz, D. Mitsche, X. Perez (2007). arXiv:cs/0702074
20. G. Ganesan, *Annales de l'IHP Probabilités et Statistiques*, vol. 49 (2013), pp. 1130–1140
21. L.H. Wong, P. Pattison, G. Robins, Phys. A **360**(1), 99 (2006)
22. J. van der Kolk, M. Serrano, M. Boguñá (2021). arXiv:2106.08030
23. B. Waxman, IEEE. J. Select. Areas. Commun. **6**, 1617 (1988)
24. E. Zegura, K. Calvert, S. Bhattacharjee, *Proceedings of IEEE INFOCOM* (1996), pp. 594–602
25. P. Van Mieghem, Probab. Eng. Inf. Sci. **15**(04), 535 (2001)
26. M. Kaiser, C. Hilgetag, Phys. Rev. E **69**, 036103 (2004)
27. M.A. Serrano, D. Krioukov, M. Boguñá, Phys. Rev. Lett. **100**, 078701 (2008)
28. D. Krioukov, F. Papadopoulos, M. Kitsak, A. Vahdat, M. Boguña, Phys. Rev. E **82**, 036106 (2010)
29. M. Boguna, I. Bonamassa, M. De Domenico, S. Havlin, D. Krioukov, M.Á. Serrano, Nat. Rev. Phys. **3**(2), 114 (2021)
30. N. Broutin, L. Devroye, N. Fraiman, G. Lugosi, Random Struct. Algorithms **44**(1), 45 (2014)
31. T.I. Fenner, A.M. Frieze, Combinatorica **2**(4), 347 (1982)
32. P. Crescenzi, C. Nocentini, A. Pietracaprina, G. Pucci, Concurr. Comput.: Pract. Exp. **21**(7), 875 (2009)
33. M. Walters, Surv. Comb. **392**, 365 (2011)
34. P. Balister, A. Sarkar, B. Bollobás, *Handbook of Large-Scale Random Networks* (2008), pp. 117–142
35. P. Balister, B. Bollobás, A. Sarkar, M. Walters, Adv. Appl. Probab. **37**(01), 1 (2005)
36. A.F. Rozenfeld, R. Cohen, D. ben Avraham, S. Havlin, Phys. Rev. Lett. **89**, 218701 (2002)
37. C.P. Warren, L.M. Sander, I.M. Sokolov, Phys. Rev. E **66**, 056105 (2002)
38. A. Barrat, M. Barthelemy, A. Vespignani, *Dynamical Processes in Complex Networks* (Cambridge University Press, Cambridge, UK, 2008)
39. M. Gonzalez, H. Herrmann, Phys. A **340**, 741 (2004)
40. M. Gonzalez, P. Lind, H. Herrmann, Phys. Rev. Lett. **96**, 088702 (2006)
41. M. Gonzalez, P. Lind, H. Herrmann, Eur. Phys. J. B **49**, 371 (2006)
42. J.A. Jr., H. Herrmann, R. Andrade, L. da Silva, Phys. Rev. Lett. **94**, 018702 (2005)
43. Z. Zhang, L. Chen, S. Zhou, L. Fang, J. Guan, T. Zou, Phys. Rev. E **77**, 017102 (2008)
44. Z. Zhang, J. Guang, B. Ding, L. Chen, S. Zhou, New J. Phys. **11**, 083007 (2009)
45. D. Auto, A. Moreira, H. Herrmann, J.A. Jr, Phys. Rev. E **78**, 066112 (2008)

Chapter 16
Excluded Volume Graphs

Assume that we have a set of points distributed in the 2d plane. These points will be the nodes of the graph and the main idea for excluded volume graphs is that two nodes are connected if some region between them is empty. Obviously there is a large number of possibilities for chosing the shape of the empty region and depending on this choice we get different graphs. Among these graphs, we have the Delaunay triangulation (DT), the Gabriel graph (GG), or the relative neighborhood graph (RNG). More generally, if we use the concept of a lune which is the intersection of two disks of equal radius, we obtain the $\beta-$skeleton graphs.

16.1 Delaunay Graph

We first discuss the construction of the dual G^* of a planar graph G. We assign to each face F of G a node f, and there is a link between nodes f and f' of the dual, if the corresponding faces F and F' are adjacent (i.e. have an edge in common). We represent in Fig. 16.1 an example of such a construction. Note that the external face is represented by a single vertex.

Now that we have the construction of the dual of a graph we can apply it to the Voronoi tessellation and the resulting graph is called the Delaunay graph (or Delaunay triangulation). According to the general position assumption (see Chap. 14), the degree of Voronoi vertices is 3 which implies that the Delaunay construction is actually a triangulation. A triangulation is a division of the plane into triangles, which is also the face with the smallest number of edges possible. For a given set of points there are many triangulations and the Delaunay one favors a relative homogeneity in the aspect ratio of triangles, leading to a visually appealing triangulation.

The Delaunay triangulation has many interesting properties and in particular contains the Euclidean minimum spanning tree (see Chap. 18). For example, if the points are distributed uniformly in the plane (Poisson process), then each node has on aver-

© The Author(s), under exclusive license to Springer Nature Switzerland AG 2022
M. Barthelemy, *Spatial Networks*,
https://doi.org/10.1007/978-3-030-94106-2_16

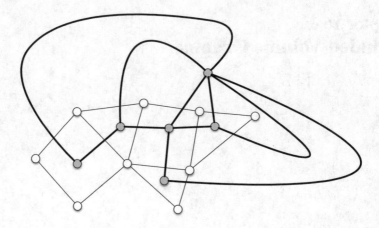

Fig. 16.1 Starting from a planar graph G represented here by white vertices, we construct the dual G^* where the nodes (in grey) correspond to the faces of G. If two faces share a common edge, the correspond nodes in G^* are connected

age six surrounding triangles. Also, this procedure leads to a set of triangles that are not too elongated: the minimum angle is maximized (but note that this triangulation does not necessarily minimize the maximum angle). This triangulation can therefore be used as a simple model for terrain or other objects.

The Delaunay triangulation does not minimize the length of the edges, but it is a t-spanner [1] (see Chap. 18), such that the shortest path between two points of the plane along edges of the triangulation is not larger than $t < 1.998$ times the Euclidean separation [2].

There is another way to construct the Delaunay triangulation which implies that it belongs to the class of excluded volume graphs: if we construct the graph such that no point is inside the circumcircle of any triangle, we then obtain the Delaunay graph (Fig. 16.2).

Also, if n is the closest neighbor of a point i, the edge ni belongs to the Delaunay triangulation, since the nearest neighbor graph is included in the Delaunay triangulation. The Delaunay graph also includes the Gabriel graph (see below) and this implies that if $i - j$ is an edge of the triangulation, the disk of diameter $d(i, j)$ cannot contain any other point.

16.2   Gabriel Graph

In a discussion in spatial statistics, Gabriel and Sokal [3] developed statistical methods for categorizing sets of populations sampled from different localities. In this work they introduced a graph constructed over a set of points which is now known as the Gabriel graph. The connection rule in this case is particularly simple: for two

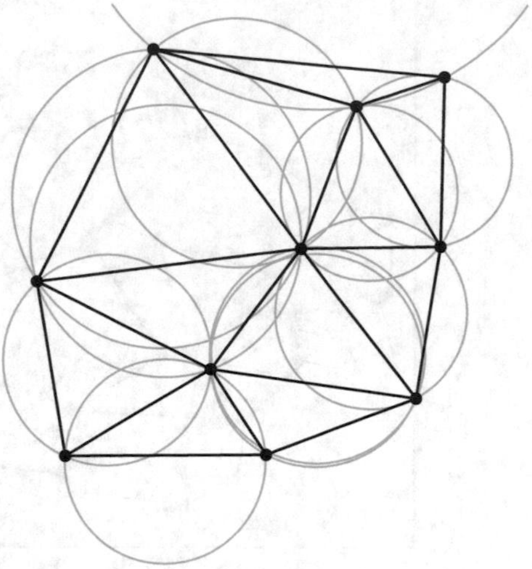

Fig. 16.2 A Delaunay
triangulation in the plane
showing the circumcircle of
the triangles. One can easily
check that they are all empty.
CC BY-SA 3.0 Gjacquenot

Fig. 16.3 The nodes i and j
are Gabriel neighbors: the
disk of diameter $i - j$ does
not contain any other nodes

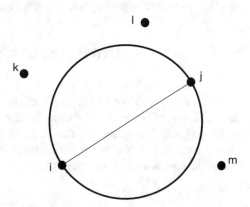

nodes i and j we construct the disk of diameter $i - j$ and if this disk does not contain
any other nodes, we call i and j Gabriel neighbors (see Fig. 16.3).

Having defined this notion of neighborhood, we can then construct the random
geometric graph and keep links between Gabriel neighbors only. An example of the
resulting 'Gabriel graph' is shown in Fig. 16.4). It is less dense than other graphs
such as the random geometric network for example. Also, these Gabriel graphs can
naturally be generalized to higher dimensions, with the empty disks replaced by
empty closed balls.

The Gabriel graph is obviously a subgraph of the Delaunay triangulation as the
connection rule is a particular case of the Delaunay condition. It also contains the
Euclidean minimum spanning tree, the relative neighborhood graph, and the nearest
neighbor graph. In fact, for a given set of points P, we have an inclusion sequence
between these different graphs and which is [4]

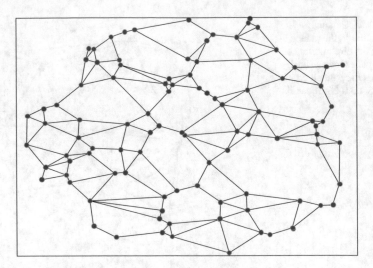

Fig. 16.4 Example of a gabriel graph constructed over 100 points

$$\text{MST(P)} \subseteq \text{RNG(P)} \subseteq \text{GG(P)} \subseteq \text{DT(P)} \tag{16.1}$$

Also, as we will see in the next section, the Gabriel graph is an instance of a beta-skeleton with $\beta = 1$. Like beta-skeletons, and unlike Delaunay triangulations, it is not a geometric spanner: for some point sets, distances within the Gabriel graph can be much larger than the Euclidean distances between points. More precisely, we define the spanning ratio of a graph as

$$S = \max_{i,j} \frac{d(i, j)}{d_E(i, j)} \tag{16.2}$$

where $d(i, j)$ is the length of the shortest path between i and j on the graph and $d_E(i, j)$ is the euclidean distance between them. For a t-spanner, we then have $S = t$. For the Gabriel graph with N nodes, it can be shown [5] that $S \sim \sqrt{N}$.

We briefly mention here a generalization of this graph, the witness Gabriel graph [6]. For this graph denoted by $GG(P, W)$, in addition to the set P of points, we have another set W of 'witnesses'. There is then an edge between two nodes i and j of P if and only if there is no point of W in the disk of diameter $i - j$. When the set W of witnesses coincides with the set P of vertices, we recover the usual Gabriel graph: $GG(P) = GG(P, P)$. This generalization can be applied to many other graphs [7] and allows witnesses to play a negative role on the graph construction, and seems to have a great potential for practical applications.

16.3 Relative Neighborhood Graph

The relative neighbourhood graph (RNG) of a set of N points on the plane was defined by Toussaint in 1980 [8]. The main point in this study was to extract a structure from a set of points that would match human perception. The connection rule is the following: two nodes i and j are connected by an edge, if there does not exist a third point k that is closer to both i and j than they are to each other (see an example in Fig. 16.5).

As we will see in the next section, the RNG is a particular case of beta-skeletons, obtaine for $\beta = 2$. It has been shown for these graphs [5], that the spanning ratio diverges as $\sqrt{\log N}$ for large N.

16.4 β-Skeletons

The β-skeletons generalize the previous graphs and constitute an important family of excluded volume graphs governed by a single parameter $\beta > 0$. These β-skeletons were defined by Kirkpatrick and Radke [4] in order to facilitate the reconstruction of shapes from sample points (and providing a topological skeleton). Their applications span now a wide spectrum of areas going from geographic information systems to wireless networks, image processing, and machine learning. In particular, it has been shown that the value $\beta = 1.7$ correctly reconstructs the entire boundary of any smooth surface with some constraints [9]. Interestingly enough, it has also been shown that

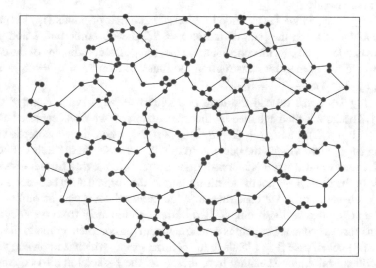

Fig. 16.5 Relative neighborhood graph constructed over 200 points

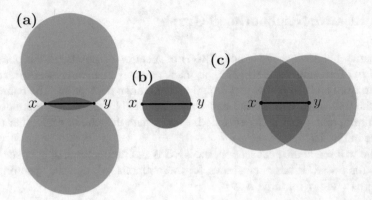

Fig. 16.6 The geometry of the lune-based β-skeleton for **a** $\beta = 1/2$, **b** $\beta = 1$ (which corresponds to the Gabriel graph), and **c** $\beta = 2$ which corresponds to the relative neighbourhood graph. For $\beta < 1$, nodes within the intersection of two disks each of radius $|x - y|/2\beta$ preclude the edges between the disk centers, whereas for $\beta > 1$, we use instead radii of $\beta|x - y|/2$. Thus, whenever two nodes are nearer each other than any other surrounding points, they connect, and otherwise do not. Figure taken from [11]

the value $\beta = 1.2$ allows for a more effective reconstruction of streets maps from the set of points corresponding to the center point of streets [10].

The construction of the graphs introduced above rely on the definition of a certain area 'between' two points. These cases can be generalized with the concept of *lune* also called a 'lens' in this field. A lune is the intersection of two disks of equal radius, and has a midline joining the centres of the disks and two corners on its perpendicular bisector (see Fig. 16.6).

For $\beta \leq 1$, we define the excluded region of each pair of points (i, j) separated by a distance d to be the lune of radius $r = d/2\beta$ with corners at i and j. For these two nodes i and j (with coordinates $(x_{i(j)}, y_{i(j)})$), the coordinates of the centers of the two disks needed for constructing the lune are then $x_1 = x_2 = (x_i + x_j)/2$, $y_{1(2)} = (y_i + y_j)/2 \pm \sqrt{r^2 - (d/2)^2}$.

For $\beta \geq 1$ we use instead the lune of radius $r = \beta d/2$, with i and j on the midline. The lune is then obtained by the intersection of two disks centered at $\vec{u}_1 = \beta/2\vec{u}_i + (1 - \beta/2)\vec{u}_j$ and $\vec{u}_2 = (1-)\beta/2)\vec{u}_i + \beta/2\vec{u}_j$ where $\vec{u}_{i(j)}$ denote the vectors pointing to $i(j)$ and $\vec{u}_{1(2)}$ to the centers. In this $\beta > 1$ case, these circles always go through i and j and when β is increasing the centers are moving further away from i and j. In the limit $\beta \to \infty$, the resulting lune is the vertical strip between i and j.

For each value of β we construct an edge between each pair of points if and only if its excluded region is empty. If β varies continuously from 0 to ∞, the β-skeletons form a sequence of graphs extending from the complete graph to the empty graph. The special case $\beta = 1$ leads to the Gabriel graph, which is known to contain the Euclidean minimum spanning tree; therefore, the β-skeleton also contains the Gabriel graph and the minimum spanning tree whenever $\beta \leq 1$. For $\beta = 2$ we have the relative neighbourhood graph. We note that for $\beta \leq 2$, the graph is necessarily

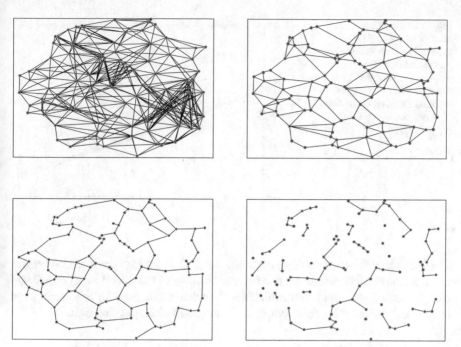

Fig. 16.7 Evolution of the β-skeleton when β is varied (the set of points is the same for all figures). The values of β are: $\beta = 0.7$, $\beta = 1.0$ (GG), $\beta = 2.0$ (RNG), and $\beta = 5$

connected, otherwise, it is typically disconnected. We show in Fig. 16.7 β-skeletons for different values of β and constructed over the same set of points.

For the β-skeletons, we have the following inclusion sequence for $1 \leq \beta \leq \beta' \leq 2$

$$\text{MST(P)} \subseteq \text{RNG(P)} \subseteq G_{\beta'}(P) \subseteq G_{\beta}(P) \subseteq \text{GG(P)} \subseteq \text{DT(P)} \qquad (16.3)$$

For increasing β, the β-skeleton loses its edges and in [12, 13], Adamatzky showed that this decrease can be fitted by a power law of the form $E \sim \beta^{-\alpha}$ where α depends on the properties of the random set of points. For a Poisson distribution we found the result shown in Fig. 16.8 and a power law fit indicates that the number of edges $E(N, \beta)$ varies as $E \sim 1/\sqrt{\beta}$.

Finally, we note that β-skeletons are not $t-$spanners as their spanning ratio diverges with their size N. Indeed, the spanning ratio of the $\beta-$skeletons have been calculated in [5] and the results are

$$\begin{cases} \text{For } \beta \in [0, 1]: \ S \leq N^{\gamma}, \ \gamma = [1 - \log_2(1 + \sqrt{1 - \beta^2})] \\ \text{For } \beta = 2: \ S \sim N \\ \text{For } N \to \infty \text{ and } \beta \in [1, 2]: \ S \sim \sqrt{\log N} \end{cases} \qquad (16.4)$$

Fig. 16.8 Variation of the number of edges $E(N, \beta)$ present in a β-skeleton when β is varied (for $N = 1000$ points). The straight line is a power law fit showing that we have a scaling of the form $E \sim \beta^{-\alpha}$ with $\alpha \approx 0.50$ ($r^2 > 0.99$)

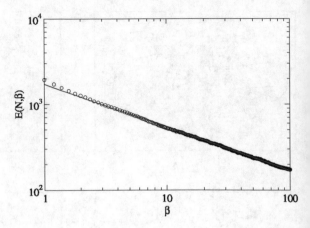

The main idea [5] here is that for any constant β, it is possible to construct a sequence of point sets (which ressembles to a flattened version of the fractal Koch snowflake) with the unit square and over which the β-skeleton can have an arbitrarily large length. This implies that these graphs have an unbounded spanning ratio.

References

1. G. Narasimhan, M. Smid, *Geometric spanner networks* (Cambridge University Press, 2007)
2. G. Xia, *Proceedings of the Twenty-Seventh Annual Symposium on Computational Geometry* (2011), pp. 264–273
3. K.R. Gabriel, R.R. Sokal, Syst. Zool. **18**(3), 259 (1969)
4. D.G. Kirkpatrick, J.D. Radke, *Machine Intelligence and Pattern Recognition*, vol. 2 (Elsevier, 1985), pp. 217–248
5. P. Bose, L. Devroye, W. Evans, D. Kirkpatrick, SIAM J. Discret. Math. **20**(2), 412 (2006)
6. B. Aronov, M. Dulieu, F. Hurtado, Comput. Geom. **46**(7), 894 (2013)
7. M. Dulieu, Witness proximity graphs and other geometric problems. Ph.D. thesis, Polytechnic Institute of New York University (2012)
8. G. Toussaint, Pattern Recognit. **12**, 261 (1980)
9. N. Amenta, M. Bern, D. Eppstein, Graph. Models Image Process. **60**(2), 125 (1998)
10. J. Radke, A. Flodmark, Geogr. Inf. Sci. **5**(1), 15 (1999)
11. A.P. Kartun-Giles, M. Barthelemy, C.P. Dettmann, Phys. Rev. E **100**(3), 032315 (2019)
12. A. Adamatzky, Inf. Sci. **254**, 213 (2014)
13. A. Adamatzky, Comput. Geom. **46**(6), 805 (2013)

Chapter 17
Loops and Branches

A problem that can be faced when studying networks is the abundance of measures. This is particularly true for spatial networks where the combination of spatial and topological metrics contributes to the explosion of possible measures, and it is obviously worse when these networks evolve in time. In order to select the most relevant measures it is then important to have in mind a benchmark graph. We will illustrate this on the case of evolving subway networks which are spatial, time-evolving networks and whose growth can be difficult to understand without an appropriate template. We show that the structure which describes well these systems is made of a core delimited by a loop, and branches 'radiating' from it. This image reduces drastically the number of parameters needed for describing these graphs and allows us to understand and quantify their growth.

In a second part, we focus on the particular structure made of radial branches and a loop, that seems to be quite relevant in many real-world networks. We will first consider betweenness centrality properties of this structure and show that, under certain conditions, the loop can be more central than the spatial barycenter of nodes in agreement with measures on real-world networks. We then consider the effect of congestion on this structure allowing us to discuss the competition between centralized and decentralized transport pathways.

17.1 Reducing the Complexity of a Spatial Network

As noted above, a problem that can be faced when studying networks is the abundance of measures. This is particularly true for spatial networks where multiple combinations of spatial and topological metrics contribute to the explosion of the number of possible measures, a situation that is obviously worse when these networks evolve in time.

© The Author(s), under exclusive license to Springer Nature Switzerland AG 2022
M. Barthelemy, *Spatial Networks*,
https://doi.org/10.1007/978-3-030-94106-2_17

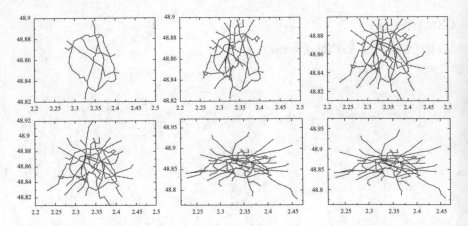

Fig. 17.1 From **a** to **f**: Paris subway for the years 1910, 1930, 1950, 1970, 1990, 2009 (note that the scale is not the same for the different figures)

As discussed thoroughly in Chap. 10, subway networks evolve in time and grow over large areas and during very long times (see for example, the evolution of the subway in Paris over almost a century shown in Fig. 17.1). As discussed in Chap. 10, large subway networks seem to converge to a long time limit shape which exhibits several typical topological and spatial features. Indeed, by inspection, we observe that in most large urban areas, the subway network consists of a set of stations delimited by a 'ring' that constitute the 'core'. From this core, quasi-one dimensional branches grow and reach out to areas of the city further away from the core. More formally, branches are defined as the set of stations which are iteratively built from a terminal station, i.e. a station with degree 1. New neighbors are added to a given branch as long as their degree is 2—continuing the line, or 3—defining a fork. In this latter case, the aggregative process continues *if and only if* at least one of the two possible new paths stemming from the fork is made up of stations of degree 2 or less. Note that the core of a network with no such fork is thus a k-core with $k = 2$ [1]. The general structure can schematically be represented as in Fig. 17.2. Once we have this structure in mind, it is not difficult to characterize it quantitatively and to identify the relevant parameters. In particular, we can characterize this branch and core structure with the parameter $\beta(t)$ defined as

$$\beta(t) = \frac{N_B(t)}{N_B(t) + N_C(t)} \tag{17.1}$$

where $N_B(t)$ and $N_C(t)$ respectively represent the number of stations on branches and the number of stations in the core at time t.

We can then easily characterize the branches by measuring their spatial extension and to determine the average euclidean distance from the geographic barycenter of

Fig. 17.2 Template used for analyzing subway networks. A large 'ring' or 'loop' encircles a core of stations. Branches radiate from the core and reach further areas of the urban system. Figure taken from [2]

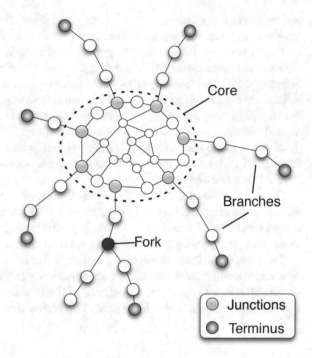

the system to all the nodes of the core and the branches, denoted by $\overline{D}_C(t)$ and $\overline{D}_B(t)$, respectively. The barycenter is computed as the center of mass of all nodes, or in other words, the average location of all the nodes. In particular, the distance $\overline{D}_B(t)$ provides information about the spatial extension of the branches and we construct the ratio $\mu(t)$

$$\mu(t) = \frac{\overline{D}_B(t)}{\overline{D}_C(t)} \tag{17.2}$$

which gives a spatial measure of the amount of extension of the branches.

We also need information about the structure of the core. If we assume that the core is a planar graph, it can be characterized by many parameters [3]. It is important to choose those which are not simply related but ideally represent different aspects of the network (such as those proposed in the form of various indicators, see for example [3–5]). For example, we will characterize the core structure at each time step t, by the following two parameters. The first parameter is simply the average degree of the core which characterizes its 'density'

$$\langle k_{\text{core}} \rangle(t) = \frac{2E_C(t)}{N_C(t)} \tag{17.3}$$

where $N_C(t)$ is the number of core nodes and $E_C(t)$ the number of its edges. We note that the average degree is connected to the standard gamma index $\gamma(t) = E_C(t)/(3N_C(t) - 6)$ where the denominator is the maximum number of links admissible for a planar network [4]. The average degree of the core contains a useful information about it, and there are many other quantities (such as standard indices such as α, etc., see for example [4] and Chap. 4) which can give additional information. In the case of subway networks, we can use another simple quantity which describes in more detail the level of interconnections in the core and which is given by the fraction f_2 of nodes in the core with $k = 2$. In the case of well-interconnected systems, this fraction will tend to be small, while sparse cores with a few interconnections will have a larger fraction of $k = 2$ nodes.

Once we know this fraction f_2 of $k = 2$ nodes in the core which characterizes the level of interconnection and the parameter $\mu(t)$ which characterizes the relative spatial extension of branches, we have key information about the entanglement of topological and geographical features in such 'core/branch' networks.

This simple example shows how a template can help in reducing the complexity of a real-world network and allows to identify a small number of parameters that can characterize its structure. The choice of the template depends obviously on the system under study and should be guided by physical considerations.

17.2   A Loop and Branches Toy Model

As discussed in Chap. 10, the spatial distribution of the BC in random graphs seems to contain a lot of information about their organization. In particular, we observe non-trivial objects made of central links, such as loops or other motifs. It is therefore important to understand the formation of these structures and the conditions for their existence. In particular, it seems that randomness can induce very large perturbation in the spatial distribution of the BC and can dramatically modify it with respect to ordered lattices. We can indeed observe cases where the barycenter is not the most central node, or equivalently, that the BC is not a simple decreasing function of the distance to the barycenter. In order to understand this phenomenon, we discuss here a simple toy model proposed in [6], made of a star network composed of N_b branches, where each branch is composed of n nodes. We then add a loop at distance ℓ from the center (see Fig. 17.3 for a sketch of this graph). We also consider here the more general case where links are weighted: we assume that links on branches have a weight equal to one and the loop segment between two consecutive branches has a weight given by w. The purely topological case then corresponds to the case $w = 1$. We then compute the BC using weighted shortest paths. This generalization allows us to discuss for example the impact of different velocities on a street network. In this case, w can be seen as the time spent on the segment (and the weighted shortest path is then the quickest path). In this section, we will discuss the conditions under which

Fig. 17.3 Representation of the toy model discussed here. The number of branches is here $N_b = 5$, the number of nodes on each branch is $n = 11$ and the loop is located at a distance $\ell = 6$ from the center 0. The node C is at the intersection of a branch and the loop and T is the terminal node of a branch. Figure taken from [6]

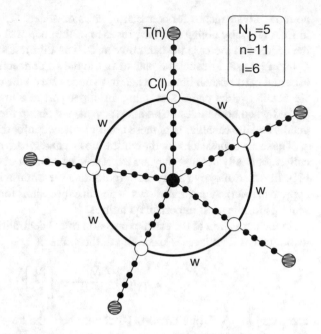

the loop is more central than the 'origin' at the center for this toy model. Intuitively, for very large w, it is always less costly to avoid the loop, while for $w \to 0$, loops are very advantageous. The two main quantities of interest are therefore the centrality at the center denoted by $g_0(\ell, n, w)$ and the centrality, denoted by $g_C(\ell, n, w)$, at the intersection C of the branch and the loop (by symmetry all intersections have the same BC). We then compute the difference $\delta g = g_0 - g_C$ and will detail under which condition it can be negative.

17.2.1   Exact and Approximate Formulas for the BC

The interest of this toy model lies in the fact that we can estimate analytically the BC for the center $g_0(\ell, n, w)$ and for the intersection nodes on the loop $g_C(\ell, n, w)$. Formally we can write these quantities as

$$g_0(\ell, n, w) = g_0(\ell, n, \infty) - (a_1^0 + a_2^0 + a_3^0)$$
$$g_C(\ell, n, w) = g_C(\ell, n, \infty) + (a_1^C + a_2^C + a_3^C) \qquad (17.4)$$

where the a_x^i are positive. We distinguish two parts in these centralities. First, we estimate the BC when there is no loop which is represented by the case where $w \to \infty$. This part is modified by the presence of the loop that under certain conditions can be more advantageous for connecting pairs of nodes. We can understand the signs in Eq. (17.4), by noting that the presence of the loop will decrease the centrality at the center and increase the centrality at C. The different terms a_i^x (where $x = 0$ or C and $i = 1, 2, 3$) count the paths (that avoid 0) connecting two nodes that lie on different parts of their branch: when both nodes are on the upper part of the branches we obtain a_1^x; the paths connecting an upper part to a lower part are described by a_2^x and when both nodes lie on a lower part, we obtain the coefficient a_3^x. For more details and the calculation of these coefficients, we refer the interested reader to [6].

The exact expressions for the centralities g_0 and g_C are difficult to handle analytically, essentially because they are expressed as sums of complicated arguments (see [6]). In order to derive analytical predictions we discuss in the following a simple approximation scheme that allows to obtain expressions for g_0 and g_C and the correct scalings for the most important quantities.

In the derivation of the exact expression of the centralities Eq. (17.4), we have to distinguish different cases according to the value of

$$\chi \equiv \min\left(\frac{N_b - 1}{2}, \left[\frac{2\ell}{w}\right]\right) \tag{17.5}$$

compared to $j - 1$ (the brackets $[\cdot]$ denote here the lowest nearest integer) which denotes the number of loop segments between the first branch and the branch j. This essentially amounts to compare the cost of the path between a node on the lower part (with $0 < s < \ell$) of the first branch B_1 to a node on the lower part ($0 < t < \ell$) of another branch B_j. If $[2\ell/w] > j - 1$ the cost of the path which goes through 0 is larger than going directly via the loop (given by $(j - 1)w$) and therefore produces a negative contribution to g_0. We see that this discussion allows to distinguish for a given value of w 'near' from 'far-away' branches (Fig. 17.4). The nearest branches are then defined by the condition $j - 1 \leq \chi$ and the remote branches by $\chi < j - 1 \leq (N_b - 1)/2$ (for simplicity we assume here that N_b is odd and by symmetry we can discuss only one half for the branches from $j = 2$ to $j = (N_b - 1)/2$). We will then use the following simplification: for the χ near branches, going through the center is always the best choice for $s < \ell$ and $t < \ell$ only and gives

$$g_{near} = \chi \frac{(\ell - 1)^2}{2} \tag{17.6}$$

For the $(N_b - 1)/2 - \chi$ far-away branches, we consider that for all nodes $s, t \in [0, n]$, the paths are going through the center leading to

$$g_{far} = \left(\frac{N_b - 1}{2} - \chi\right) n^2 \tag{17.7}$$

Fig. 17.4 Schematic representation of the approximation used to compute the centrality $g_0(w)$ at the center 0. Figure taken from [6]

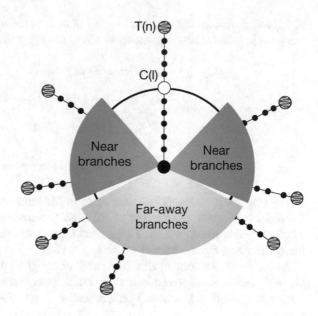

Taking into account the factor 2 for not counting twice the same path, we obtain for $g_0(w) = N_b(g_{near} + g_{far})$ the following expression

$$g_0(w) \approx N_b \left\{ \left(\frac{N_b - 1}{2} - \chi \right) n^2 + \chi \frac{(\ell - 1)^2}{2} \right\} \qquad (17.8)$$

which recovers both exact limits

$$g_0 \simeq \begin{cases} N_b \frac{(N_b-1)}{2} \frac{(\ell-1)^2}{2} & \text{for } w \to 0 \\ n^2 N_b \frac{(N_b-1)}{2} & \text{for } w \to \infty \end{cases} \qquad (17.9)$$

In the following it will also be useful to consider the limit $\ell, n \to \infty$ with $x = \ell/n$ fixed which gives for $g_0(x, \chi) = g_0(\ell, n, w)/n^2$ (up to terms of order $1/n$)

$$g_0(x, \chi) \approx N_b \left\{ \left(\frac{N_b - 1}{2} - \chi \right) + \frac{1}{2} \chi x^2 \right\} \qquad (17.10)$$

where the only dependence on w is now encoded in χ, hence the change of argument for clarity.

We can produce the same type of arguments for the BC on the loop. First the value without the loop is easy to compute and we obtain

$$g_C(\ell, n, w = \infty) = (n - \ell) [\ell + n(N_b - 1)] \qquad (17.11)$$

which simply counts the number $(n - \ell)$ of nodes 'above' C and all the others (C being excluded). Similar arguments as above then give the following result

$$
\begin{aligned}
g_C(\ell, n, \chi) = {} & g_C(\ell, n, w = \infty) \\
& + 2\chi\Big[(n - \ell + 1)(\ell - 1) + \frac{(\ell - 1)(\ell - 2)}{2}\Big] \\
& + \frac{\chi(\chi - 1)}{2}\Big[(n - \ell + 1)^2 \\
& + 2(n - \ell + 1)(\ell - 1) + \frac{(\ell - 1)(\ell - 2)}{2}\Big]
\end{aligned} \tag{17.12}
$$

where χ is given by Eq. (17.5). In particular the term proportional to χ counts all the paths between the lower part of the branch containing C and all the nodes of a branch that is close enough. The second term (proportional to $\chi(\chi - 1)$) counts the paths going from a branch B_j with $j \in [1, \chi - 1]$ to the other branches $j' = 1, 2, \ldots, j - 1$. The sum of all these contributions gives the factor $\chi(\chi - 1)/2$. The counting factor is not trivial here and comes from evaluating all the paths from a node s in a branch j to a node t on a branch j' (j and j' are different from 1) such that

$$
s + t > |\ell - s| + |\ell - t| + w\Delta j \tag{17.13}
$$

The left hand side of this inequality corresponds to the distance from s to t through the center and $w\Delta j$ is the distance on the loop (for the exact expression of the centrality and how to recover this approximate formula, we refer the interested reader to [6]).

Similarly to the case of the BC at 0, it will be convenient for analyzing these expressions to consider the limit $n, \ell \to \infty$ such that $\ell/n = x$. Up to terms of order $1/n$ we then obtain for $g_C(x, \chi) = g_C(\ell, n, w)/n^2$

$$
\begin{aligned}
g_C(x, \chi) = {} & (1 - x)(x + N_b - 1) \\
& + 2\chi x\Big(1 - \frac{x}{2}\Big) \\
& + \frac{\chi(\chi - 1)}{2}\Big(1 - \frac{x^2}{2}\Big)
\end{aligned} \tag{17.14}
$$

We show in Fig. 17.5 the comparison of the exact result with the approximations exposed here. For large values of ℓ the approximation is not excellent and can certainly be improved. However as we will see in the following, these simple approximations allow to understand and to predict the correct scaling for the important quantities ℓ_{opt} and w_c.

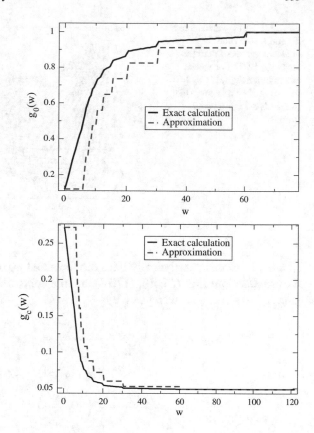

Fig. 17.5 Comparison between the exact result and the approximation for $g_0(w)$ (top) and $g_C(w)$ (bottom) (the BC are here normalized). The parameter values are here $N_b = 21$, $n = 60$ and $\ell = 30$. Figure taken from [6]

17.2.2 Threshold Value of w and Optimal ℓ

The fundamental quantity that we wish to understand is the difference $\delta g(x, \chi) = g_0(x, \chi) - g_C(x, \chi)$ given by equations Eqs. (17.10), (17.14). We plot this quantity versus ℓ for different values of w and the result is shown in Fig. 17.6. For w sufficiently small, δg can be negative and this demonstrates the existence of a threshold value w_c such that at $w = w_c$ the minimum is $\min_\ell \delta g = 0$. For $w < w_c$, the minimum of δg is negative and we can define an optimal value ℓ_{opt} which corresponds to this smallest value of δg and gives the position of loop that maximizes the difference between the BC of the loop and the center.

In order to estimate this optimal value ℓ_{opt}, we note (using the expression Eq. (17.5) for χ) that the difference $\delta g(x, \chi)$ gives

$$\delta g(x, \chi) = \begin{cases} \delta g(x, \frac{2\ell}{w}) & \text{for } \ell \in [0, \frac{(N_b-1)w}{4}] \\ \delta g(x, \frac{N_b-1}{2}) & \text{for } \ell \in [\frac{(N_b-1)w}{4}, 2n] \end{cases} \qquad (17.15)$$

Fig. 17.6 $\delta g(\ell)$ versus ℓ for N_b and n fixed and for different values of w in the range [0, 12.5]. For values less than a threshold ($w_c \approx 4$ here) there is a minimum that is negative. Figure taken from [6]

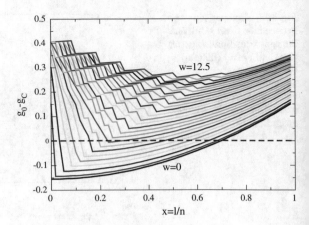

In order to estimate analytically both the threshold w_c and the optimal value ℓ_{opt}, we use equations Eqs. (17.10), (17.14) and the approximate difference $\delta g(x, \chi) = g_0(x, \chi) - g_C(x, \chi)$ is then given by

$$
\delta g(x, \chi) = N_b \left[\frac{N_b - 1}{2} - \chi + \frac{1}{2} \chi x^2 \right]
$$
$$
- (1 - x)(x + N_b - 1) - 2\chi x \left(1 - \frac{x}{2} \right)
$$
$$
- \frac{\chi(\chi - 1)}{2} \left(1 - \frac{x^2}{2} \right) \tag{17.16}
$$

In the domain $\ell < [(N_b - 1)w/4]$ and for large N_b and n (we treat here ℓ as a continuous variable), we have

$$
d\delta g/d\ell < 0 \tag{17.17}
$$

A similar calculation shows that in the domain $(N_b - 1)/4 < \ell < 2n$, the function $\delta g(\ell, n, \chi)$ is increasing with ℓ (at least for N_b large enough). These results thus show that the minimum of δg is actually reached at the intersection of the two curves which occurs for

$$
\ell_{opt} = \frac{(N_b - 1)w}{4} \tag{17.18}
$$

This expression for ℓ_{opt} is actually independent from the exact form of δg as long as it is decreasing for $\ell < \ell_{opt}$ and increasing above ℓ_{opt} which is verified numerically [6]. The theoretical prediction Eq. (17.18) is compared with numerical results in Fig. 17.7, and for N_b large enough (here, typically $N_b > 10$) this prediction is in excellent agreement with data.

Fig. 17.7 Comparison between the theoretical prediction Eq. (17.18) and numerical results for ℓ_{opt} (for $w < w_c$). For large value of N_b the prediction is excellent. We note that ℓ_{opt} exists for $w < w_c$ and w_c decreases with N_b which implies that the range over which we can see a linear behavior is decreasing as $1/N_b$ (here $n = 40$). Figure taken from [6]

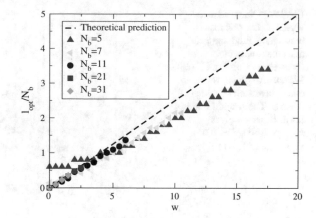

We can understand this value of ℓ_{opt} with the following simple argument. If ℓ is small most paths connecting nodes from different branches will go through 0 and we expect $\delta g > 0$. When ℓ is increasing more paths will go through the loop and will increase the value of g_C. However, when ℓ is too large, paths connecting the (large) fraction of nodes located on the lower branches will go through 0 again. In order to get a sufficient condition on ℓ_{opt}, we consider the path between the node C on the branch B_1 and the corresponding node C' on the furthest branch $(N_b - 1)/2$. The optimal value for ℓ_{opt} is then such that the cost of the path from C to C' through 0 and which is 2ℓ is equal to the cost on the loop which is given by $w(N_b - 1)/2$. This immediately gives the result $\ell_{opt} \approx w(N_b - 1)/4$.

The threshold quantity w_c is obtained by imposing that the minimum of $\delta g(\ell = \ell_{opt})$ is equal to zero. Using the approximate form Eq. (17.16), it can be shown that the minimum is obtained for $\ell = \ell_{opt}$ and for $\chi = (N_b - 1)/2$. We thus have to consider the quantity $\delta g(\ell_{opt}, n, \chi = (N_b - 1)/2)$ which for large N_b is behaving as

$$\delta g(\ell_{opt}) \approx \frac{N_b^2}{8} \left[\frac{5}{2} \left(\frac{w N_b}{4n} \right)^2 - 1 \right] \tag{17.19}$$

(details of this calculation are given in [6]) and we therefore obtain

$$w_c \approx \kappa \frac{n}{N_b} \tag{17.20}$$

where $\kappa = 4\sqrt{\frac{2}{5}}$ in this approximation. We can understand the scaling for w_c with the simple following argument. Indeed, a necessary condition on w is that ℓ_{opt} must be less than n. This gives the condition

$$w < \tilde{w}_c = 4 \frac{n}{N_b} \tag{17.21}$$

Fig. 17.8 Value of $w_c N_b$ versus n. The collapse is reasonably good and is in agreement with our theoretical result Eq. (17.20). We observe plateaus that are due to the discrete values of ℓ and n. The straight line is a linear fit which gives $\kappa_{emp} \approx 0.66$ ($r^2 = 0.96$). Figure taken from [6]

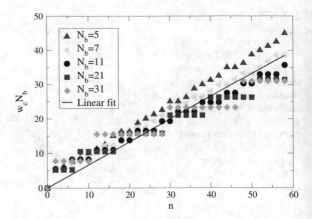

This threshold $\tilde{w}_c$ is a priori larger than the exact value, as we imposed here a necessary condition, but allows to understand in a simple way the scaling of w_c with n and N_b. The scaling for w_c is tested by plotting $w_c N_b$ versus n (Fig. 17.8) and which should be linear. We indeed observe a reasonable agreement with the linear behavior predicted by this analysis. The differences are probably due to the small values of N_b used for the numerical calculations. The linear fit however gives a prefactor $\kappa_{emp} \approx 0.66$ which is far from the value obtained within the simple approximation scheme presented here. The important fact is that this approximation is able to predict the correct scaling and it could maybe be possible to find more refined approximations in order to get a better estimate for the prefactor κ.

Finally, we note that when $w_c > 1$, the case $w = 1$ displays then a negative minimum and we can observe a very central loop. This case is particularly interesting as it corresponds to the 'topological' case for which the distance is the minimum number of hops. This will then happen when there are few branches, or if the branches are large enough.

17.2.2.1   Discussion: Disorder and Centrality

The main purpose of the toy model presented here is to shed light on the appearance of non-trivial patterns made of very central nodes (or links) in real-world planar graphs. In particular, this model shows that a loop at a certain distance from the center can be more central than the physical center itself. The condition for the existence of such a phenomenon is that the weight on the loop has to be smaller than a threshold value w_c. This threshold depends on the size and number of radial branches, highlighting their crucial role in planar graphs. In particular, this result allows us to understand the appearance of very central loop even in the topological case where the shortest topological distance is used for computing the BC: if the extension of the network is large compared to the number of radial branches, w_c can be larger than one $w_c > 1$ and central loops for $w = 1$ can be observed. In ordered systems—such as lattices—

the effective number of branches is too large leading to a very small w_c and therefore prohibits the appearance of central loops in the 'topological' case ($w = 1$). In real-world planar graphs where randomness is present, the absence of some links can lead to a small number of 'effective' radial branches which in the framework of this toy model implies a large value of w_c and therefore a large probability to observe central loops.

17.3 Analyzing the Impact of Congestion Cost

In many real-world cases the pure hub-and-spoke structure is not present and we observe a ring structure around a complicated core or an effective hub (see for example Fig. 17.9). An interesting discussion on centralization versus decentralization from the perspective of the minimum average shortest path and of the effect of congestion can be found in [7, 8] and we will reproduce here the main arguments and results of these papers.

The idea is then to study the competition between the centralized organization with paths going through a single central hub and decentralized paths going along a ring and avoiding the central hub in the presence of congestion. A simple model of hub-and-spoke structure together with a ring was proposed in [9]. In this model, N nodes are on a circle and there is a hub located at the center of the circle (see Fig. 17.10), and radial links—the spokes—are present with probability p.

In a first part we will consider this model without congestion and reproduce the results obtained in [9] for the distribution of the shortest path and its average. In a second part, following [7], we will include congestion cost and compute these quantities, allowing for a discussion about the competition between centralized and decentralized transport pathways.

17.3.1 An Exactly Solvable Hub-and-Spoke Model

The model described in Fig. 17.10 was originally proposed by Dorogovtsev and Mendes [9] who focused on the shortest path ℓ and computed both its average $\ell(p, n)$ and its distribution $P(\ell)$. At that time the motivation was to find a simple model for small-world networks that would display a crossover from a large world behavior (with an average shortest path scaling as $\bar{\ell} \sim N$) to a small-world behavior where $\bar{\ell} \sim \log N$ (or smaller). However, this model turns out to be also a good toy for understanding loops and spokes in spatial networks, and in this respect we find it interesting to report these results here.

We start with the simpler case where the loop is oriented as shown in Fig. 17.10a. The central point is connected with undirected links of weight $1/2$ added with probability p. This amounts to connect random pairs of nodes by undirected links of length 1. Obviously $\bar{\ell}(p = 0) \sim N$ while $\bar{\ell}(p = 1) = 1$ showing that we have for

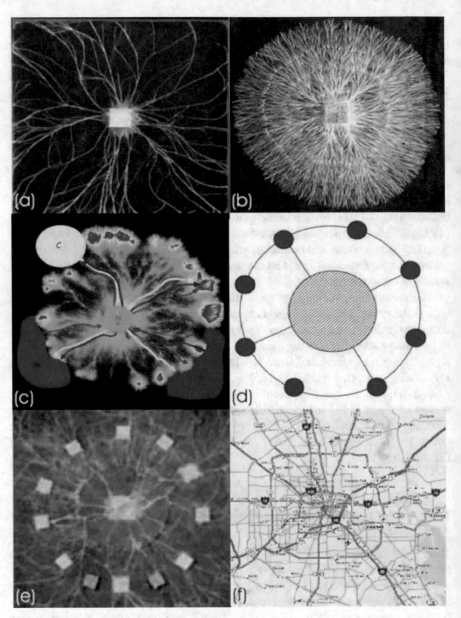

Fig. 17.9 Examples of hub-and-spoke structures with rings. **a–c**, **e**: Typical fungi networks, in **c** a schematic representation of the nutrient flow is shown. **d** The model studied in [7–9] with spokes radiating from a hub. **f** Road network in Houston showing an inner hub with a complicated structure. Figure taken from [8]

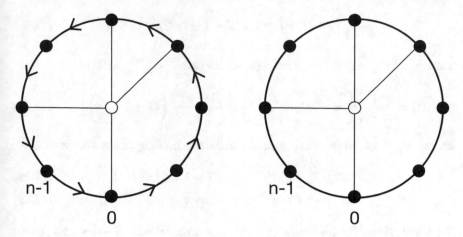

Fig. 17.10 Models proposed in [9] and studied in [7, 8] with congestion. A central site is connected to a site on a ring with probability p. In **a** all the links on the ring are directed and in **b** these links are not directed

this simple model a crossover (or a transition) between a large and a small-world behavior when p is varied. The basic quantity considered in [9] is $P(\ell, k)$ which is the probability that the shortest path between two nodes separated by a distance k counted on the ring, is equal to ℓ when we take the spokes into account. By definition, we have $P(\ell > k, k) = 0$ and the following probability conservation condition

$$\sum_{\ell=1}^{k} P(\ell, k) = 1 \tag{17.22}$$

It is easy to see that $P(1, 1) = 1$, $P(1, 2) = p^2$, $P(2, 2) = 1 - p^2$, etc., and more generally that

$$P(\ell < k, k) = \ell p^2 (1 - p)^{\ell - 1} \tag{17.23}$$

$$P(\ell = k, k) = 1 - p^2 \sum_{i=0}^{k-1} i(1 - p)^{i-1} \tag{17.24}$$

which simply state that we need two spokes connecting the two nodes to the hubs and no spokes connecting the nodes on the loop in between. The distribution is simply given by

$$P(\ell) = \frac{1}{N - 1} \sum_{k=\ell}^{N-1} P(\ell, k) \tag{17.25}$$

In the case of directed links, we obtain for the shortest path distribution [9]

$$P(\ell) = \frac{1}{N-1}\left[1 + (\ell-1)p + \ell(N-1-\ell p^2)p^2\right](1-p)^{\ell-1} \tag{17.26}$$

The general expression for the average shortest path $\bar{\ell} = \sum_{\ell=1}^{N} \ell P(\ell)$ is

$$\langle\ell\rangle = \frac{1}{N-1}\left[\frac{2-p}{p}N - \frac{3}{p^2} + \frac{2}{p} + \frac{(1-p)^N}{p}\left(N - 2 + \frac{3}{p}\right)\right] \tag{17.27}$$

and we can check on these expressions that the two limiting case are correct

$$\text{for } p \to 0: \quad P(\ell) \to 1/N-1 \Rightarrow \bar{\ell} = N/2 \tag{17.28}$$

$$\text{for } p \to 1: \quad P(\ell) \to \delta_{\ell,1} \Rightarrow \bar{\ell} = 1 \tag{17.29}$$

We can also consider the continuous limit of this model with $N \to \infty$ and $p \to 0$ such that $\rho = pN$ and $z = \ell/N$ are fixed. In this limit the continuous distribution is

$$P_{dir}(z,\rho) = \lim_{N\to\infty, p\to 0} NP(\ell) = \left[1 + \rho z + \rho^2 z(1-z)\right]e^{-\rho z} \tag{17.30}$$

and the average shortest path is

$$\bar{z} = \frac{\bar{\ell}}{N} = \frac{1}{\rho^2}\left[2\rho - 3 + (\rho+3)e^{-\rho}\right] \tag{17.31}$$

These analytical results are qualitatively similar to what is obtained for the Watts-Strogatz model [10] justifying in this way the interest for this toy model. Various limits can be considered and discussed, and we refer the interested reader to the original paper [9]. This case is however a little bit artificial as links on the loop are directed. In the undirected case, we have more paths going from one site to the other and the enumeration is a little bit more tedious. It can nonetheless be solved and the result for $P(\ell, k)$ is now [9]

$$P(1,1) = 1 \tag{17.32}$$

$$P(\ell = 1, k) = p^2 \tag{17.33}$$

$$P(2 \le \ell < k, k) = p^2(1-p)^{2\ell-4}(2-p)(2\ell - 2 - \ell p) \tag{17.34}$$

$$P(\ell = k, k), 1) = 1 - \sum_{\ell=1}^{k-1} P(\ell, k) \tag{17.35}$$

The result for the shortest path distribution now reads [9]

$$P(\ell = 1) = \frac{2}{N-1}\left(1 + \frac{N-3}{2}p^2\right) \tag{17.36}$$

$$P(\ell \geq 2) = \frac{1}{N-1}[a_0 + a_1 p + a_2 p^2 + a_3 p^3$$
$$+ a_4 p^4](1-p)^{2\ell-4} \tag{17.37}$$

where

$$\begin{cases} a_0 &= 2 \\ a_1 &= 4(\ell - 2) \\ a_2 &= 2(\ell - 1)(2N - 4\ell - 3) \\ a_3 &= -2(2\ell - 1)(N - 2\ell - 1) \\ a_4 &= \ell(N - 2\ell - 1) \end{cases} \tag{17.38}$$

We can also consider for this undirected model the continous limit and the distribution becomes $P_{un}(z, \rho)$ which can be shown to be

$$P_{un}(z, \rho) = 2P_d(2z, \rho) \tag{17.39}$$

which demonstrates that in fact there is only a difference of a scaling factor equal to 2 between both directed and undirected models.

For both these models a continuous limit can be defined by taking the limit $N \to \infty$ and $p \to 0$ with $\rho = pN$ and $z \equiv \ell/N$ fixed. The shortest path distribution then converges to (in the undirected model)

$$NP(\ell) \to 2[1 + 2\rho z + 2\rho^2(1 - 2z)]e^{-2\rho z} \tag{17.40}$$

17.3.2 Congestion and Centralized Organization

The interesting observation made in [7] is that if we now add a cost c each time a path goes through the central hub, we expect some sort of transition between a decentralized regime where it is less costly to stay on the peripheral ring to a centralized regime where the cost is not enough to divert paths from the central hub. The cost could in general depend on how busy the center is and could therefore grow with the number of connections to the hub. In the case of a constant cost c (and in the directed case), we can estimate the shortest path distribution (N is here the number of nodes on the ring)

$$P(\ell, \ell \le c) = \frac{1}{N-1} \tag{17.41}$$

$$P(\ell < N, \ell > c) = (\ell - c)p^2(1-p)^{\ell-c-1} \tag{17.42}$$

$$P(\ell = N, \ell > c) = 1 - \sum_{i=c+1}^{\ell-c-1} P(\ell < N, \ell > c) \tag{17.43}$$

which leads to

$$P(\ell) = \begin{cases} \frac{1}{N-1} & \text{for } \ell \le c \\ \frac{1}{N-1}[1 + b_1 p + b_2 p^2](1-p)^{\ell-c-1} & \text{for } \ell > c \end{cases} \tag{17.44}$$

where $b_1 = \ell - c - 1$ and $b_2 = (N - 1 - \ell)(\ell - c)$. For paths of length $\ell \le c$, there is no point to go through the central hub. In the opposite case, when $\ell > c$, we recover a distribution similar to the $c = 0$ case in [9]. The average shortest path is now

$$\bar{\ell} = \frac{(1-p)^{N-c}[3 + (N-2-c)p]}{p^2(N-1)}$$
$$+ \frac{p[2 - 2c + 2N - (c-1)(c-N)p] - 3}{p^2(N-1)} + \frac{c(c-1)}{2(N-1)} \tag{17.45}$$

In the continuous limit ($p \to 0$, $N \to \infty$ and $z \equiv \ell/N$ and $\rho = pN$ fixed), the average shortest path is a function of ρ, c, and N.

In the case of costs increasing linearly with ρ, the average shortest path displays a minimum when ρ is varied (N and c being fixed). Indeed for $\rho \to 0$, there are no spokes and $\bar{\ell}$ scales as N. In the opposite case, ρ large, the cost is also large and it is less costly to go along the ring. In [7], the authors use a simple approximation and found (with $c \equiv k\rho$) that the optimal value of ρ is

$$\rho^* \approx \sqrt{\frac{N}{k}} \tag{17.46}$$

a result that is confirmed numerically (see Fig. 17.11). This result can actually be rewritten as

$$pc(\rho) \sim 1 \tag{17.47}$$

(with $pN \sim \ell$) which means that the optimal situation is obtained when the average cost of a radial trip through the central hub is of order one: when c is too large, this trip is too costly and when p is too small, the existence of this path is too unlikely. The same argument applied to nonlinear cost $c \sim k\rho^2$ gives the scaling

$$\rho^* \sim (N/k)^{1/3} \tag{17.48}$$

Fig. 17.11 Minimal shortest path length $\bar{\ell}_{min}$ (i.e. minimum value of $\bar{\ell}$) as obtained from Eq. (17.45). **a** Optimal number of connections $\rho \equiv pn$ as a function of the cost-per-connection k to the hub. Results are shown for $n = 1000$ and $n = 10000$. **b** Optimal number of connections ρ as a function of the network size. Results are shown for $k = 2$ and $k = 4$. Figure taken from [7]

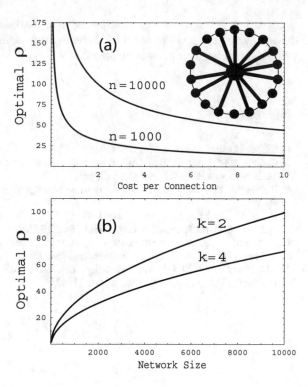

For this optimal value, the minimum average shortest path is then of the order the cost

$$\bar{\ell}_{min} \sim c(\rho^*) \tag{17.49}$$

In the linear case $c = k\rho$, one obtains

$$\bar{\ell}_{min} \sim \sqrt{kN} \tag{17.50}$$

and in the nonlinear case $c \sim k\rho^2$, one obtains

$$\bar{\ell}_{min} \sim (kN^2)^{1/3} \tag{17.51}$$

These expressions and arguments apply essentially both to the directed and non-directed model.

This study [7] was generalized in [8] to the case of a more complicated cost function such as $c(\rho) = C\rho + B\rho^2 + A\rho^3$ where the authors observe different behaviors and a phase transition according to the values of the coefficients A, B, and C.

These studies on a simple toy model show how congestion could have an important impact on the a priori optimal hub-and-spoke structure and favorizes the transport along a ring. From a more general perspective it would be interesting to observe the emergence of rings—as observed in real-world examples—in general structures without imposing their presence a priori.

References

1. S.B. Seidman, Soc. Netw. **5**(3), 269 (1983)
2. C. Roth, S.M. Kang, M. Batty, M. Barthelemy, J. R. Soc. Interface **9**(75), 2540–2550 (2012)
3. M. Barthelemy, Phys. Rep. **499**(1), 1 (2011)
4. P. Haggett, R.J. Chorley, *Network Analysis in Gography*, vol. 67 (Edward Arnold London, 1969)
5. F. Xie, D. Levinson, Geogr. Anal. **39**(3), 336 (2007)
6. B. Lion, M. Barthelemy, Phys. Rev. E **95**(4), 042310 (2017)
7. D. Ashton, T. Jarrett, N. Johnson, Phys. Rev. Lett. **94**, 058701 (2005)
8. T. Jarrett, D. Ashton, M. Fricker, N. Johnson, Phys. Rev. E **74**, 026116 (2006)
9. S.N. Dorogovtsev, J.F.F. Mendes, Europhys. Lett. **50**, 1 (2000)
10. D.J. Watts, D.H. Strogatz, Nature **393**, 440 (1998)

Chapter 18
Optimal Networks

Variational approaches have been largely disregarded in complex network studies although they frequently provide an alternative and possibly more meaningful point of view. This important class of network models is obtained by looking for graphs that optimize a given quantity, functional of the graph. The simplest case is for example the minimum spanning tree (MST) that minimizes the total length for a given set of points. We will also discuss here t-spanner network which connect in some 'good' way a set of points.

Most existing spatial networks in the real-world do not seem to result from a global optimization, but rather from the progressive addition of nodes and segments resulting from a local optimization (see next chapters). By modeling (spatial) networks as resulting from a global optimization, one overlooks the usually limited time horizon of planners and the self-organization underlying their formation. The interest of these optimal networks lies then rather in the fact that they constitute interesting benchmarks to compare actual networks with. The comparison with the MST for example indicates how far we are from the minimum cost possible, and is therefore a very important example and was largely studied. In particular, there is an extensive mathematical litterature on this tree and we will try to present here the most important results. We will also discuss other optimal trees that generalize the MST to the case where a more complex quantity than the total length is minimized. In this chapter, we will also discuss geometric t-spanners that guarantee a spanning ratio less than t.

18.1 Optimization, Complexity, and Efficiency

18.1.1 Complexity

Although one of the main pillars in complex systems studies is the emergence of a collective behavior without relying on any central planning, it is usually a matter of time scale compared with the typical time horizon of planners. On a short time scale

© The Author(s), under exclusive license to Springer Nature Switzerland AG 2022   347
M. Barthelemy, *Spatial Networks*,
https://doi.org/10.1007/978-3-030-94106-2_18

it is reasonable to assume that planning operations play an important role and that the system under consideration evolves through an optimization process. On a larger time scale however, most systems result from the addition of these successive layers and even if each of these layers is the result of an optimization process, it is very likely that the long time result is not an optimum.

Optimization is however of great importance in many practical engineering problems and both the problem of optimal networks [1] and of optimal traffic on a network [2, 3] have a long tradition in mathematics and physics. It is well known, for example that the laws that describe the flow of currents in a resistor network [4] can be derived by minimizing the energy dissipated by the network [5]. Optimal spatial networks are also relevant in the study of mammalians circulatory system [6], food webs [7], general transportation networks [8], metabolic rates [9], river networks [10], and gas pipelines or train tracks [11].

We note here that there is another broad class of optimal networks where spatial constraints are absent. For example, it has been shown that optimization of both the average shortest path and the total length can lead to small-world networks [12], and more generally, degree correlations [13] or scale-free features [14] can emerge from an optimization process. Cancho and Sole [15] showed that the minimization of the average shortest path and the link density leads to a variety of networks including exponential-like graphs and scale-free networks. Guimera et al. [16] studied networks with minimal search cost and found two classes of networks: star-like and homogeneous networks. Finally, Colizza et al. [17] studied networks with the shortest route and the smallest congestion and showed that this interplay could lead to a variety of networks when the number of links per node is changed.

Optimal networks therefore appear in many different branches such as mathematics, physics and also in engineering. This subject is in fact so broad that it would deserve a whole review to explore its various aspects. In this chapter, we thus made the choice to restrict ourselves to the most recent and relevant statistical studies involving optimal networks and space.

18.1.2   Efficiency of Transport Network

An important discusion that appeared in the context of biological networks is related to optimization in transport networks [8]. The problem is to understand the relation between the metabolic rate (measured by the number of nodes) and the typical size of the system. The focus is on distribution networks which indeed corresponds to many biological systems of interest such as cardiovascular and respiratory networks, plant vascular systems, or other natural distribution systems such as drainage networks of river basins.

Very generally, for a system of linear size L in dimension d, the number of nodes N which is proportional to the mass M is scaling as the volume $V \sim L^d$ (assuming a constant density)

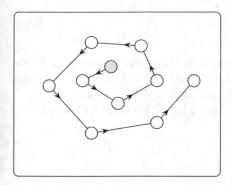 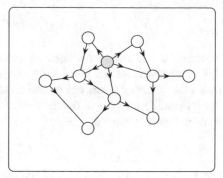

Fig. 18.1 Different spanning networks from the source (shown in grey). (Left) 1d topology leading to the scaling $C \sim L^{2d}$. (Right) 'Standard' topology with lattice-like behavior $\ell \sim L$ leading to the efficient behavior $C \sim L^{d+1}$. Figure inspired from [8]

$$N \sim L^d \tag{18.1}$$

This implies that any geometrical feature of the system will scale as a power of $N^{1/d}$. For example the surface will scale as $N^{(d-1)/d}$. For three-dimensional objects we should thus observe exponents related to $1/3$. In light of empirical data that suggest power multiple of $1/4$ and not $1/3$, this simple argument has been rediscussed in [8] and we follow here this important paper.

The problem for many distribution systems is to understand how nutrients are delivered from a central source to the L^d sites of the system. The total amount of nutrient is then proportional to L^d and the total volume C of the transportation network for a given organism depends on its structure and the authors of [8] discuss bounds on this quantity C. In particular the most efficient class of networks is such that C is as small as possible. The theorem in [8] identifies the following bounds on C

$$C_{min} \sim L^{d+1} \leq C \leq C_{max} \sim L^{2d} \tag{18.2}$$

We can understand this result by first noting that we can rewrite C as

$$C \sim N\ell \tag{18.3}$$

where ℓ is the average distance of the nodes to the source. The less efficient value C_{max} is obtained for a 1d topology such as a spiral starting at the center and connecting all nodes, and which display the one-dimensional behavior $\ell \sim N$ (see Fig. 18.1(left)). In contrast, for most spanning networks $\ell \propto L$—for example in 2d, the average shortest path is scaling $\ell \sim N^{1/2}$—which then leads to $C_{min} \sim L^{d+1}$. For efficient networks, $C \sim L^{d+1}$ and since the mass is proportional to C and the metabolic rate B proportional to N we have

$$\left.\begin{array}{l} B \sim N \sim L^d \\ M \sim C \sim L^{d+1} \end{array}\right\} \Rightarrow B \sim M^{\frac{d}{d+1}} \qquad (18.4)$$

which for $d = 3$ gives a scaling with exponent $3/4$. This result thus accounts for the $1/4$ power observed for many living organisms, leading some authors [9] to speak about the 'fourth dimension' of life as scalings involve $d + 1 = 4$ and not $d = 3$, but technically the root of this behavior is the scaling of the volume C for efficient transportation networks.

18.2 Minimum Spanning Tree

The minimum spanning tree (MST) can be defined for any graph with edge weights: it is the tree subgraph which connects all nodes at a minimal total weight (also called cost). In other words, the MST computed over N nodes is the tree T (with $E = N - 1$ edges) which minimizes the total weight W of the graph

$$W(T) = \sum_{e \in T} w(e) \qquad (18.5)$$

where $w(e)$ is the weight on edge e. If the edge weights are all different the MST is unique (which is not the case if weights are integers for example).

The MST has therefore a clear meaning and is potentially important for many applications that involve connecting many points at a minimal cost. The practical importance of this object and its (relative) simplicity explains the large number of studies in mathematics and in computer science, and we can present here only a small selection of all these studies.

There are several algorithms for finding the MST and the greedy version for weighted undirected graph is given by Prim's algorithm [18] that runs in $\mathcal{O}(N \log N)$. It is a greedy algorithm based on the local optimal choice hoping that it will lead to the global optimum. It can however be demonstrated that Prim's algorithm finds correctly the minimum spanning tree for a weighted undirected graph. This algorithm constructs the tree by adding successively the cheapest link connecting the existing tree at that time and the unconnected vertices. The main idea for showing that Prim's algorithm produces the MST is to assume that at a certain 'time' t a link $u - v$ that does not belong to the MST is added to the existing MST tree $T(t)$ at time t. We assume that the node u belongs to $T(t)$ and v is outside it. The MST over all nodes is a tree that connects all of them and there is therefore a path between u and v. This path necessarily contains a link e that crosses the frontier of $T(t)$ and the weight should be $w(e) < w(u - v)$ contradicting the initial assumption that $u - v$ is chosen.

In the following we will focus on some important examples. We will start with the MST on a complete graph, followed by the 'Euclidean' MST constructed on a set of points in $d-$dimensional space and where the weight is given by the length of links.

18.2.1 *Minimum Spanning Tree on a Complete Graph*

In this case, we consider that all $N(N-1)/2$ links exist and are characterized by given weights. Frieze [19] showed that for this problem with weights that are independent random variables distributed according to the same cumulative function F with $F'(0) > 0$, the total weight of the MST is given in the large N limit by

$$W_{tot}(N \gg 1) \sim \frac{\zeta(3)}{F'(0)} \tag{18.6}$$

where ζ is the Riemann zeta function and $\zeta(3) \approx 1.2$. Additionally, Steele [20] proved the convergence in probability and Janson proved the existence of a central limit theorem governing the fluctuations of $W_{tot}(N)$ (see also below).

We can understand this—surprising—result Eq. (18.6) with the following rough argument.[1] We denote by X_1 the minimum of n independent random variables identically distributed. The cumulative distribution of the minimum $F_{(1)}(y) = P(X_1 \le y)$ of n identically distributed random variables is given by

$$F_{(1)}(y) = 1 - P(X_1 > y)P(X_2 > y) \ldots P(X_n > y) = 1 - P(X > y)^n \tag{18.7}$$

where $P(X > y) = 1 - F(y)$ is the cumulative distribution of the X variables. The probability density of the minimum is then given by

$$P(X_1 = x) = n\rho(x)(1 - F(x))^{n-1} \tag{18.8}$$

where $\rho(x) = F'(x)$ is the distribution of the random variables X. The average minimum is then given by

$$\overline{X_1} = \int dx \, x n\rho(x) \left[1 - F(x)\right]^{n-1} \tag{18.9}$$

where the bar $^-$ denotes the average. In the limit $n \to \infty$, assuming that the mimimal bound of X is zero and that $F'(0) = \rho(0)$ exists, we then write

$$\overline{X_1} \simeq \int_0^\infty dx \, x n\rho(x) e^{-nxF'(0)} \tag{18.10}$$

where we expanded around $x \approx 0$: $F(x) \approx xF'(0)$. This integral can be estimated and leads to

---

[1] I thank J.-M. Luck for useful discussions on this point.

$$\overline{X_1} \simeq n\rho(0) \int_0^\infty \mathrm{d}x\, x \mathrm{e}^{-nx\rho(0)} \tag{18.11}$$

$$\simeq \frac{1}{nF'(0)} \tag{18.12}$$

This result recovers in particular the case of the uniform distribution that can be computed exactly and which is

$$\overline{X_1} \sim \frac{1}{n+1} \tag{18.13}$$

This discussion can actually be extended to the kth smallest variable whose distribution $f_k(x)$ is given by

$$f_k(x) = \frac{n!}{(n-k)!(k-1)!}\rho(x)F(x)^{k-1}(1-F(x))^{n-k} \tag{18.14}$$

The average value of the kth variable is then given by

$$\overline{X_k} = \frac{n!}{(k-1)!(n-k)!} \int \mathrm{d}x\, x\rho(x)F(x)^{k-1}[1-F(x)]^{n-k} \tag{18.15}$$

where we recall that $F(x) = \mathrm{Prob}(X \le x)$. We assume that $n \gg k$ and expand F to the first order around 0: $F(x) \approx xF'(0)$. We then obtain

$$\overline{X_k} \approx \frac{n!}{(k-1)!(n-k)!}\rho(0)^k \int \mathrm{d}x\, x^k \mathrm{e}^{-nxF'(0)} \tag{18.16}$$

Changing variable ($u = nF'(0)x$) and taking the limit $n \to \infty$ leads to

$$\overline{X_k} \approx \frac{n!}{(k-1)!(n-k)!n^{k+1}} \frac{1}{F'(0)} \int_0^\infty \mathrm{d}u\, u^k \mathrm{e}^{-u} \tag{18.17}$$

We recognize the last integral as the Gamma function $\Gamma(k+1) = k!$ and we finally obtain in the large n limit (allowing to use $n! \sim n^n \mathrm{e}^{-n}$)

$$\overline{X_k} \approx \frac{k}{n} \frac{1}{F'(0)} \tag{18.18}$$

signalling a universal behavior when $\rho(0)$ is non-zero. If we now assume that the MST is constructed with the set of the N smallest values among all the $n = N(N-1)/2$ possible links we obtain

$$W_{tot}(N) = \sum_{e \in MST} d(e) \approx \frac{1}{F'(0)} \sum_{i=1}^{N} \frac{2i}{N(N-1)}$$

$$\approx \frac{1}{F'(0)} \frac{2}{N(N-1)} \frac{1}{2} N(N+1)$$

$$\approx \frac{1}{F'(0)} \tag{18.19}$$

for large N. This rough argument does not allow to recover the numerical value $\zeta(3)$ (which is however not too far from 1) but helps understanding the (absence of) scaling of $W_{tot}(N)$ with N.

18.2.2 Properties of the Euclidean Minimum Spanning Tree

The euclidean minimum spanning tree (EMST) is defined as the minimum spanning tree (MST) constructed on a set of points in a $d-$dimensional space and where the weight of a link is its length. This network provides a lower bound for the cost needed to connect these nodes and can serve as a null model for many practical applications. As an illustration of this network we show in Fig. 18.2, the EMST constructed on the set of stations present in the Paris subway in 2009. In this subway case the total length is directly connected to the cost of the network and the MST represents the most economical network. The MST has however some drawbacks such as a large average shortest path and a large vulnerability to failure, and the real subway network has obviously many redundant links. It is however interesting to understand the interplay between costs and efficiency by comparing the actual network with the MST.

The MST is a subgraph of the Delaunay graph (see Chap. 16) and we reproduce here briefly the argument. We assume that a link between nodes i and j of the MST is not in the Delaunay graph. This means that in the circle of diameter $i - j$ there is at least another point k and the points $i - j - k$ form a triangle where $i - j$ is the longest edge. It is clear that in any cycle the longest edge cannot belong to the MST: removing this edge will keep the connectivity properties and is always the most beneficial move in terms of the total weight. This implies that the link $i - j$ cannot be in the MST. We therefore proved here that the MST is included in the Delaunay graph. In fact, it is possible to show that there is a nested sequence of graphs (see for example [21] and references therein)

$$\text{MST} \subseteq \text{Relative neighborhood} \subseteq \text{Gabriel} \subset \text{Delaunay} \tag{18.20}$$

This EMST has been the subject of numerous mathematical studies and we will discuss in the following the most salient properties of this object: the total length, the longest link and the existence of central limit theorems.

Fig. 18.2 Top: Paris subway
network (2009). Bottom:
Corresponding euclidean
minimum spanning tree (the
scale is not exactly the same
for both figures)

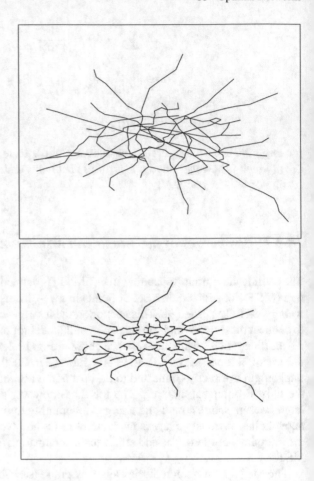

18.2.2.1   The Total Length

If the tree contains N nodes that are uniformly ditributed over a $d-$dimensional
space of volume $V = 1$, the typical distance between two nodes is of the order

$$\ell_1 \sim \frac{1}{N^{1/d}} \tag{18.21}$$

The total length of the MST is then

$$L_{tot}(N) = \sum_{e \in MST} \ell(e) \sim N\ell_1 \sim N^{1-1/d} \tag{18.22}$$

where $\ell(e)$ is the length of linkn e. For $d = 2$ we then obtain $L_{tot}(N) \sim \sqrt{N}$. This
argument can be made rigorous and Beardwood, Halton, Hammersley [22] demon-

strated the following result for the length

$$L_{tot}(N) \sim_{N\to\infty} C_d N^{1-1/d} \int_{\mathbb{R}^d} \rho(x)^{1-1/d} dx \qquad (18.23)$$

where $\rho(x)$ is the density of points and C_d a constant that depends on the dimension only. In the Poisson case (with density $\rho = 1/\Omega$ where Ω is the volume containing the N points), we define the MST constant $\beta(d)$ in the limit $N \to \infty$

$$\lim_{N\to\infty} \frac{L_{tot}(N)}{N^{1-1/d}} = \beta(d)\Omega^{1/d} \qquad (18.24)$$

where Ω is the volume containing the N points. More generally, Kesten and Lee [23] considered nodes randomly distributed in $[0, 1]^d$ and discussed the weighted minimum spanning tree such that the total weighted length

$$L_{tot}(N, \alpha) = \min \sum_{e \in T_\alpha} \ell(e)^\alpha \qquad (18.25)$$

is mimimum for edges in the tree T_α (the parameter $\alpha \in \mathbb{R}$). Using a argument similar to the one above, this length scales as $N^{1-\alpha/d}$ and we define the generalized MST constant by

$$\lim_{N\to\infty} \frac{L_{tot}(N, \alpha)}{N^{1-\alpha/d}} = \beta(d, \alpha)\Omega^{\alpha/d} \qquad (18.26)$$

These constants $\beta(d, \alpha)$ are independent from the volume or the shape of the domain and the disorder configuration, but are not exactly known. Early numerical simulations [24] showed that

$$\beta(2, 1) \approx 0.656 \pm 0.002 \qquad (18.27)$$
$$\beta(3, 1) \approx 0.668 \pm 0.002. \qquad (18.28)$$

and more recent ones [25] found

$$\beta(2, 1) \approx 0.6331 \pm 0.0013 \qquad (18.29)$$
$$\beta(3, 1) \approx 0.6232 \pm 0.0017 \qquad (18.30)$$

Avram and Bertsimas [26] discussed the MST constant and showed that $\beta(d, 1)$ can be expressed as a series where the different terms involve intersections of spheres and are extremely difficult to compute. By expanding to the third term Avram and Bertsimas were able to show that

$$\beta(2, 1) \geq 0.600822 \qquad (18.31)$$

and more generally that the following bounds hold (and can be used for large d expansions)

$$\frac{\Gamma(1/d)}{dV_d^{1/d}} \leq \beta(d, 1) \leq \frac{2^{1/d}\Gamma(1/d)}{dV_d^{1/d}} \tag{18.32}$$

where $V_d = \pi^{d/2}/\Gamma(d/2+1)$ is the volume of the unit radius ball in dimension d. These bounds imply that in the large d limit

$$\beta(d, \alpha) \sim \left(\frac{d}{2\pi e}\right)^{\alpha/2} \tag{18.33}$$

in agreement with the result shown by Bertsimas and Van Ryzin [27].

18.2.2.2   The Longest Link

The longest link in the MST is an interesting quantity for many applications. Indeed if the nodes represent infected individuals or wireless devices, if the longest link is above the range of disease spread or the radio range of the devices, the MST will be disconnected in two or more parts. In contrast if it is below we can be sure that the MST will connect all the nodes in the plane.

Penrose [28] demonstrated that the length M_N of the longest link in the minimum spanning tree behaves as

$$M_N \sim \sqrt{\frac{\log N}{\pi N}} \tag{18.34}$$

and more precisely that

$$\mathrm{Prob}(\pi N M_n^2 - \log N \leq x) = \mathrm{e}^{-\mathrm{e}^{-x}} \tag{18.35}$$

He also showed that the length of the longest link for the random geometric graph is actually the same as for the MST.

We write the following simple argument in order to understand this scaling. We first rewrite the expression for the total length as

$$\sqrt{N}L_N = \sum_{e=1}^{N} X_e \tag{18.36}$$

where $X_e = \sqrt{N}\ell(e)$ and we assume that these random variables X_e are uncorrelated Gaussian variables of average $\overline{X} = 1$ and of variance $\sigma^2 = \overline{X^2} - \overline{X}^2$. The probability of the maximum is then

$$P(X_{max} = x) \simeq e^{-N \int_0^x e^{-(x-1)^2/2\sigma^2}} \tag{18.37}$$

$$\simeq e^{-Ne^{-x^2/2\sigma^2}} \tag{18.38}$$

The typical value of the maximum is then $x_{max} \sim \sqrt{\log N}$ and the typical longest link in the MST is then

$$\ell_{max} \sim \sqrt{\frac{\log N}{N}} \tag{18.39}$$

18.2.2.3 Central Limit Theorems and Geometrical Probabilities

The minimum spanning tree and other problems such as the traveling salesman, the matching problem, etc. (see for example [29, 30] for a statistical physics introduction to these problems) are all combinatorial optimization problems. These different graphs are all obtained by minimizing a quantity such as the total length $L_{tot}(N)$ constructed over a set of N points that are usually distributed in $[0, L]^d$ (the volume is then $\Omega \sim L^d$) according to a Poisson process where d is the dimension of the embedding space. The law of large numbers [22] discussed above holds in fact for many of these combinatorial problems and reads as

$$L_{tot}(N) \to_{n \to \infty} \beta(d)\Omega^{1/d} N^{1-1/d} \tag{18.40}$$

where $\beta(d)$ is a constant that depends on the dimension only and on the graph considered. This result was generalized to a class of combinatorial problems called subadditive euclidean functionals [31]. Examples are the minimum matching, the Steiner tree, etc. We note that in general the function $\beta(d)$ is not known (except for the MST, see above) and its determination in general is an important open problem in this field.

As discussed above, Kesten and Lee [23] considered the weighted minimum spanning tree such that the total weighted length

$$L_{tot}(N) \equiv L(X_1, \ldots, X_N; \alpha) = \min \sum_{e \in T_\alpha} \ell(e)^\alpha \tag{18.41}$$

is mimimum for edges in the tree T_α. In this case, they could go beyond the law of large numbers and showed that there is a central limit theorem which reads

$$\frac{L_{tot}(N) - \overline{L_{tot}(N)}}{N^{(d-2\alpha)/2d}} \to \mathcal{N}(0, \sigma^2) \tag{18.42}$$

where the $\bar{}$ denotes the average and where $\sigma^2(\alpha, d) > 0$. We note here that in the 'usual' case of the $d = 2$ embedding and $\alpha = 1$ the first correction to the dominant term if of order $\mathcal{O}(1)$.

This type of central limit theorem was actually extended to other graphs and we cite here the case of the radial spanning tree where each point is connected to its nearest neighbor that is closer to the origin [32].

18.2.2.4   A Dynamical Version of the Euclidean Minimum Spanning Tree

In Prim's algorithm, the MST is constructed by successive addition of links with the minimal weight. In the case of the euclidean MST, the nodes are in the $d-$dimensional space and we can grow the EMST by adding nodes to the existing tree such that their distance to the set of nodes already connected is minimal. When we start from nodes in $2-$dimensional space, we then connect successively the nodes that are the closest to the existing network (see 18.3). If in contrast we select at random a node that is not yet connected and choose to creaste the shortest edge from this node to the existing network, we obtain a dynamical version of the EMST. This dynamical euclidean minimum spanning tree (dEMST) can contain longer links and can be significantly different from the usual EMST. In particular, the total length displays the same behavior as the EMST but with a different constant (here for $d = 2$)

$$L_{tot, dMST}(N) \sim \beta_{dEMST}(2)\sqrt{\Omega}\sqrt{N} \tag{18.43}$$

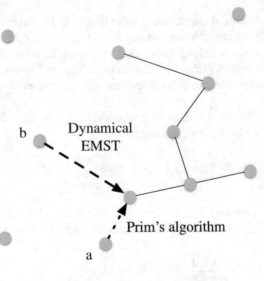

Fig. 18.3 Growth of network on a set of nodes (shown in grey). In Prim's case, the closest node a is connected to the existing network, leading to the euclidean minimum spanning tree. In contrast, if we choose the non-connected node at random (in this figure, node b) and connect it to the nearest connected node, we obtain a dynamical version of the EMST different from its usual version

with $\beta_{dEMST} \approx 1.0$ (the quantity Ω is the volume of the embedding domain). The fact that $\beta_{dEMST} > \beta_{EMST}$ (as seen above $\beta_{EMST} \approx 0.6$ for $d = 2$) comes from the existence of longer links. More precisely, we observe numerically [33] that the length $\ell(t)$ of a link at time t displays the following behavior

$$\ell(t < t^*) \sim 1/\sqrt{t} \tag{18.44}$$

$$\ell(t > t^*) \sim 1/\sqrt{N} \tag{18.45}$$

where N is the final number of nodes and where the crossover size is given by $t^*(N) \sim \alpha N$ where $\alpha \approx 1/2$. The total length of the dEMST is then given by

$$L_{tot,\,dEMST} = \sum_{t < t^*(N)} \frac{1}{\sqrt{t}} + L_{tot,\,EMST}(N - t^*(N)) \tag{18.46}$$

which leads to

$$L_{tot,\,dEMST} = \sum_{t < t^*(N)} \frac{1}{\sqrt{t}} + L_{tot,\,EMST}(N - t^*(N)) \tag{18.47}$$

$$\Rightarrow \beta_{dEMST} = 2\sqrt{\alpha} + \beta_{EMST} \tag{18.48}$$

This result indeed shows that $\beta_{dEMST} > \beta_{EMST}$ and this simple approximation leads to $\beta_{dEMST} \approx \sqrt{2} + \beta_{EMSR} \approx 2.0$ which is larger than the numerical result.

18.3   Geometric t-Spanners

18.3.1   Definition

The t-spanner network idea finds its origin in a simple network design question: given a set of N points, how can we construct a 'good' network connecting them? Obviously, the solution depends on what we call 'good'. If we look for the network with the smallest total weight (where the weight of a link is given by its length for example), the result is the minimum spanning tree (MST) discussed above. There are many other choices such as minimizing the diameter of the graph, for example, and the t-spanner comes from minimizing the detour index (or spanning ratio, or stretch factor, see Chap. 4). The detour index is given by the average ratio of the route's length on the network and the euclidean distance between two nodes. In other words, we would like to find a network that doesn't involve large detours to go from one point to another. This is obviously important for transportation networks for example.

The problem is thus to find networks whose detour index is bounded [34] and these networks are called spanners. More precisely, a t-spanner path with $t \geq 1$ between two nodes i and j is a path such that its length on the network—denoted by

$d_G(i, j)$—is less than the euclidean distance denoted by $d_E(i, j)$

$$d_G(i, j) \leq t d_E E(i, j) \tag{18.49}$$

The stretch factor of the entire spanner is the maximum stretch factor over all pairs of points within it and the graph is then called a t-spanner. The parameter t is then called the stretch factor or dilation factor of the graph.

Instead of trying to find a t-spanner constructed over a set of points, we can also start from a given graph and ask if it is a t-spanner. More generally, a t-spanner of a graph G is a spanning subgraph H which contains vertices all connected to each other by a t-spanner path. In general, determining if a given graph is a t-spanner is a difficult problem and determing the value of t is even more difficult. Few exact results are known and we refer the interested reader to the review book [34] (and for a survey of open problems, see [35]). In particular, Delaunay triangulation are t-spanners [36], and the current best bounds for its stretch factor are [37, 38]

$$1.5932 < t < 1.998 \tag{18.50}$$

Other graphs such as β-skeletons are not t-spanners and their spanning ratio diverges with the number of nodes (see Chap. 16).

18.3.2   The Theta Graph

It is in general very difficult to find a t-spanner on a given set of points, and a 'good' solution (maybe not optimal) is given by Theta graphs [34]. The main idea for constructing this graph is that if for most vertices there are many edges pointing to many directions, it will be easy to find a t-spanner path with t not too large. More formally, the Theta graph or Θ-graph belongs to the family of cone-based spanners. The basic method of construction involves partitioning the space around each vertex into a set of cones, which themselves partition the remaining vertices of the graph and to have at most one edge per cone. For the Θ-graph, one defines a fixed ray contained within each cone (conventionally the bisector of the cone) and selects the nearest neighbor with respect to orthogonal projections to that ray. For the so-called Yao graphs [39], we select the nearest vertex according to the metric space of the graph (see Fig. 18.4 for an example of a Yao graph).

More precisely, if $\kappa \geq 2$ denotes an integer, and $\theta = 2\pi/\kappa$ the undirected Θ-graph $\Theta(S, \kappa)$ constructed over a set S of points is defined as follows

1. The vertices of $\Theta(S, \kappa)$ are the points of S
2. For each point i of S, we construct κ cones and their bisector lines $B_a, a = 1, \ldots, \kappa$. For each cone a, we select the node whose orthogonal projection on B_a is the closest to i.

For Yao graphs for $\kappa \geq 9$ (with $\theta = 2\pi/\kappa$), the stretch factor is bounded by [34]

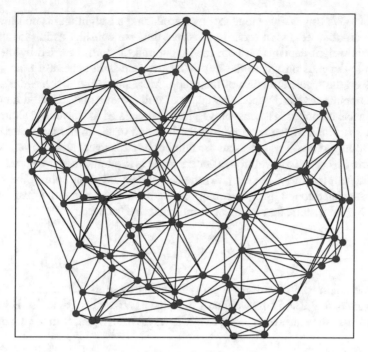

Fig. 18.4 Example of a Yao graph constructed for $\kappa = 6$ and over a set of 100 points uniformly distributed in the disk of radius 1

$$t \leq \frac{1}{\cos \theta - \sin \theta} \tag{18.51}$$

while for Θ-graphs, we obtain the following results. First, the Θ-graph is a sparse spanner and the number of edges is of order N (it is at most κN). Then, the spanning ratio for $\kappa \geq 7$ is at most [40]

$$t \leq \frac{1}{1 - 2\sin(\pi/\kappa)} \tag{18.52}$$

Various results exist for κ going from 2 to 7 and we refer the interested reader to [41–44]. In particular, it can be shown [44] that Θ_6 is a 2-spanner.

18.4 Optimal Trees: Generalization

The MST discussed in the previous sections is defined as the tree which minimizes the total cost given by a sum over all edges belonging to this tree. The weight of a link is here a simple local function and in general does not depend on the other links. We can easily extend this by introducing a weight that is non-local and depends on other edges. In order to illustrate this, we consider as in [45] the case where this weight depends on both the length and the traffic carried by the links. The traffic depends on where the link is located and is in general a function that depends on the whole structure of the graph. More precisely, we consider a set of N nodes randomly distributed in a square of unitary area and we build a network that connects all points. The cost or weight w_{ij} associated to 'traveling' along a link $i - j$ is a function of the length d_{ij} of the link and of the traffic t_{ij} it carries. The t_{ij}, in a transport analogy, represent the number of passengers and is assumed to be symmetric $t_{ij} = t_{ji}$. To travel from a generic node i_0 to another generic node i_p along a specific path $\{i_0, i_1, i_2, \ldots, i_{p-1}, i_p\}$ the cost to be paid is the sum of the weights $w_{k,k+1}$ associated to the links that compose the path, and when more than one path is available we assume that the most economical one is chosen

$$E_{i_0,i_p} = \min_{p \in \mathscr{P}(i_0,i_p)} \sum_{e \in p} w_e \qquad (18.53)$$

where the minimisation is over all paths p belonging in the set of paths $\mathscr{P}(i_0, i_p)$ going from i_0 to i_p (w_e is the weight of the edge e). The global quantity E we wish to minimize is then the average cost to pay to travel from a generic node to any other node.

$$E_0(\{t_{ij}\}) = \frac{2}{N(N-1)} \sum_{i<j} E_{i,j} \qquad (18.54)$$

Our purpose here is therefore to find the traffic $\{t_{ij}^*\}$ carried by the links and which minimises Eq. (18.54), with the only constraints that all $t_{ij} \geq 0$ and that the total traffic $T = \sum_{i<j} t_{i,j}$ is fixed. We assume here that the weight of a link e is given by the ratio of its length to traffic

$$w_e = \frac{d_e}{t_e} \qquad (18.55)$$

Although this choice is not the most general, it naturally verifies the expectation that the weight increases with d_{ij} and decreases with t_{ij}. This last condition can be easily understood in the case of transportation networks and means that it is more economic to travel on links with a large traffic, reducing the effective distance of the connection.

In order to find the optimal network, we use a Monte-Carlo algorithm where the elementary move consists in transferring a random fraction of the traffic carried by a link to another one. The minimum-cost path between two points is recalculated at any step with Djikstra's algorithm [1]. Starting with the topology of the complete graph, the Monte-Carlo simulation converges to the optimal solution characterized by a majority of links that carry no traffic and allows the emergence of a non-trivial topology. The optimal network results then as the compromise of two opposing forces: the need for short routes and the traffic concentration on as few paths as possible. The interplay between topology and traffic naturally induces the observed correlations between degree, distance and traffic itself (see for example [46]).

Most importantly, simulations show that the optimal network is a tree and a simple argument supports this finding: Consider for example an isocele triangle ABC with $d(A, C) = d(B, C) = d$ and $d(A, B) = d'$, optimization leads to the values $t_{AC} = t_{BC} \approx T/2$ and $t_{AB} \approx 0$ when $d \gg d'$. The minimum energy is thus $E \approx d/t_{AC} + d/t_{BC} \approx 4d/T$. When we remove the link BC (and therefore killing the loop), the traffic on AC becomes approximately twice $t_{AC} \approx t_{BC} \approx T$ but the minimum energy is $E' \approx 2d/t_{AC} \approx 2d/T$ which is lower than E. This simple argument shows that optimization reduces the number of links joining nodes in the same regions and increases the traffic on the remaining links. Loops between nodes in the same neighborhood become then redundant.

The optimal network being a tree enormously simplifies the computaton of the energy. Since only a single path exists between any two nodes in a tree, the energy (18.54) can be rewritten as

$$E_0 = \sum_{e \in \mathscr{T}} b_e \frac{d_e}{t_e} \tag{18.56}$$

where b_e is the edge-betweenness centrality [47] (see Chap. 5) and counts the number of times that e belongs to the shortest path between two nodes. The optimal traffic (with the same constraints as above) is given by $t_e = T\sqrt{b_e d_e}/\sum_e \sqrt{b_e d_e}$. The 'optimal traffic tree' (OTT) can then be obtained by minimizing

$$E = \sum_{e \in \mathscr{T}} \sqrt{b_e d_e} \tag{18.57}$$

The minimal configuration can now be searched by rewiring links. Replacing link $i - j$ with $i - j'$ modifies only the centralities along the path between j and j'. This implies that the calculation has a complexity of order $\mathscr{O}(N)$ and allows computation over very large network. We expect to obtain something very different from the classical euclidean minimum spanning tree discussed in the previous section, since this energy Eq. (18.57) involves a combination of metric (the distance) and topological (the betweenness) quantities. The expression Eq. (18.57) suggests an interesting generalization given by the optimization of

$$E_{\mu\nu} = \sum_{e \in \mathscr{T}} b_e^{\mu} d_e^{\nu} \tag{18.58}$$

where μ and ν control the relative importance of distance against topology as measured by centrality. Figure 18.5 shows examples of spanning trees obtained for different values of (μ, ν).

For $(\mu, \nu) = (0, 1)$ we recover the euclidean minimum spanning tree (Fig. 18.5a) which can also be obtained by minimizing the total weight $\sum_e w_e$ and gives a traffic

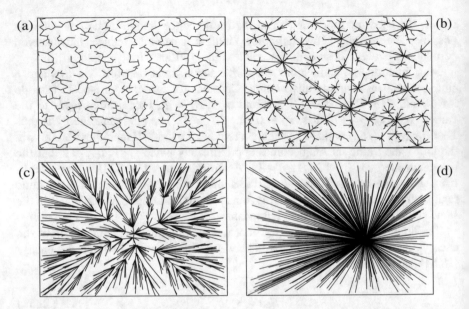

Fig. 18.5 Different spanning trees obtained for different values of (μ, ν) in Eq. (18.58) obtained for the same set of $N = 1000$ nodes. **a** Minimum spanning tree obtained for $(\mu, \nu) = (0, 1)$. In this case the total distance is minimized. **b** Optimal traffic tree obtained for $(\mu, \nu) = (1/2, 1/2)$. In this case we have an interplay between centralization and minimum distance resulting in local hubs. **c** Minimum euclidean distance tree obtained for $(\mu, \nu) = (1, 1)$. In this case centrality dominates over distance and a 'star' structure emerges with a few dominant hubs. **d** Optimal betweenneess centrality tree obtained for $(\mu, \nu) = (1, 0)$. In this case we obtain the shortest path tree which has one star hub (for the sake of clarity, we omitted some links in this last figure). Figure taken from [45]

$t_e = T\sqrt{d_e}/\sum \sqrt{d_e}$. For $(\mu, \nu) = (1/2, 1/2)$ we obtain the OTT (Fig. 18.5b) which displays an interesting interplay between distance and shortest path minimization (see below). For $(\mu, \nu) = (1, 1)$, the energy is proportional to the average shortest weighted path (with weights equal to euclidean distance (Fig. 18.5c). When $(\mu, \nu) = (1, 0)$, the energy (18.58) is proportional to the average betweenness centrality and therefore to the average shortest path $\sum_e b_e \propto \ell$. The tree $(1, 0)$ shown in (Fig. 18.5d) is thus the shortest path tree (SPT) with an arbitrary "star-like" hub (a small non zero value of ν would select as the star the closest node to the gravity center). The minimization of Eq. (18.58) thus provides a natural interpolation between the MST and the SPT, a problem which was addressed in previous studies [48]. The degree distribution for all cases considered above (with the possible exception $(\mu, \nu) = (1, 1)$) is not broad, possibly as a consequence of spatial constraints. A complete inspection of the plane (μ, ν) would be very interesting.

It has been shown that trees can be classified in 'universality classes' [49, 50] according to the distribution of the sizes of the two parts in which a tree can be divided by removing a link (or the distribution of sub-basins areas, in the language of river network). We define A_i and A_j as the sizes of the two parts in which a generic tree is

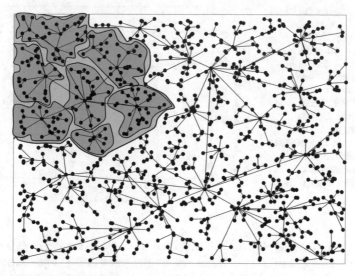

Fig. 18.6 Hierarchical organization emerging for the optimal traffic tree $(\mu, \nu) = (1/2, 1/2)$ (for $N = 1000$ nodes). Longer links lead to regional hubs which in turn connect to smaller hubs distributing traffic in smaller regions. Figure taken from [45]

divided by removing the link $i - j$. The betweenness b_{ij} of link $i - j$ can be written as $g_{ij} = \frac{1}{2}[A_i(N - A_i) + A_j(N - A_j)]$, and the distributions of $A's$ and $b's$ can be easily derived one from the other. It is therefore not surprising that the exponent $\delta = 4/3$ of the Minimum Spanning Tree [49] also characterizes the distribution $P(g) \sim g^{-\delta}$ (we note however that a random spanning tree $(\mu, \nu) = (0, 0)$ displays $\delta = 3/2$ to be compared with $\delta = 11/8$ [50] when only 'short' connections are present). In contrast, for the OTT we obtain an exponent $\delta \simeq 2$, a value also obtained for trees grown with preferential attachment mechanism [51]). The OTT thus tends to have a more uniform centrality with respect to the MST [52], with important consequences on the vulnerability of the network since there is no clearly designated 'Achille's heel' for the OTT.

The spatial properties of the OTT are also remarkable. The OTT displays (Fig. 18.6) a hierarchical spatial organization where long links connect regional hubs, that, in turn are connected to sub-regional hubs, etc. This hierarchical structure can be probed by measuring the average euclidean distance between nodes belonging to the cluster obtained by deleting recursively the longest link. For the OTT (Fig. 18.7), we observe a decrease of the region size, demonstrating that longer links connect smaller regions, a feature absent in non-hierarchical networks such as the MST, the shortest path tree or the random tree (Fig. 18.7).

It is interesting to note that the emergence of complex structures in traffic organization could also be explained by an optimization principle. In particular, strong correlations between distance and traffic arise naturally as a consequence of optimizing the average weighted shortest path. In the optimal network, long-range links carry

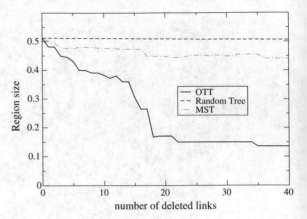

Fig. 18.7 Average euclidean size of the largest cluster remaining after deleting links ranked according to their length (in decreasing order) obtained for one typical configuration of size $N = 1000$ of different networks. The decrease observed for the OTT is consistent with a hierarchical spatial organization as it is visually evident from Fig. 18.6. Figure taken from [45]

large traffic and connect regional hubs dispatching traffic on a smaller scale ensuring an efficient global distribution. These results suggest that the organization of the traffic on complex networks and more generally architecture of weighted networks could in part result from an evolutionary process.

18.5 Beyond Optimal Trees: Noise and Loops

In most examples studied in literature (including the ones presented above), optimal networks are trees. However in many natural networks such as veins in leaves or insect wings, one observes many loops. Two studies which appeared simultaneously [53, 54] proposed possible reasons for the existence of a high density of loops in real optimal networks. In particular, it seems that the existence of fluctuations is crucial in the formation of loops. In the context of the evolution of leaves, the resilience to damage also naturally induces a high density of loops (see Fig. 18.8 for an example of flow re-routing after an injury).

In these studies, the model is defined on an electrical network with conductances C_e on each link and the total dissipated power

$$P = \frac{1}{2} \sum_k \sum_{j \in \Gamma(k)} C_{kj} (V_k - V_j)^2 \qquad (18.59)$$

is minimized under the cost condition

$$\frac{1}{2} \sum_k \sum_{j \in \Gamma(k)} C_{jk}^{\gamma} = 1 \qquad (18.60)$$

Fig. 18.8 Re-routing of the flow around an injury (indicated by a dark spot) in a lemon leaf. Figure taken from [54]

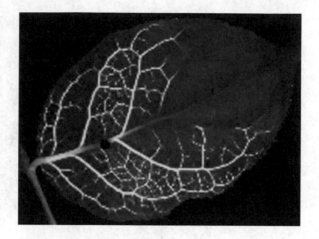

where in this equation it is assumed that the cost of a conductance C_{kj} is given by C_{kj}^{γ} where γ is a real number. The quantity V_i is the potential at node i.

Following [54], we introduce two variants of this model. The first one which represents the resilience to damage is defined as follows. We cut a link e and compute for this system the total dissipated power denoted by P^e. The resilience to this damage can then be rephrased as the minimization of the functional

$$R = \sum_{e \in E} P^e \qquad (18.61)$$

Note that if breaking e disconnects the network, there is one link with infinite resistance in the system and the dissipated power P^e is infinite. The finiteness of R implies the existence of loops in the optimal network. In another model, [54] Katifori et al. introduces time fluctuating load by considering a system with one source at the stem of the leaf and one single moving sink at position a. For this system, one can compute the total dissipated power P^a and the resilience to fluctuations can be rephrased as the minimization of the functional

$$F = \sum_{a} P^a \qquad (18.62)$$

In these models, one observes the formation of loops (see Fig. 18.9), reminiscent of the ones seen in real leaves. These studies shed a some light in the formation and the evolution of real-world networks and opened interesting directions of research. In particular, it would be very interesting to understand more quantitatively the condition of appearance of loops for example. Following this direction, the authors of [55]

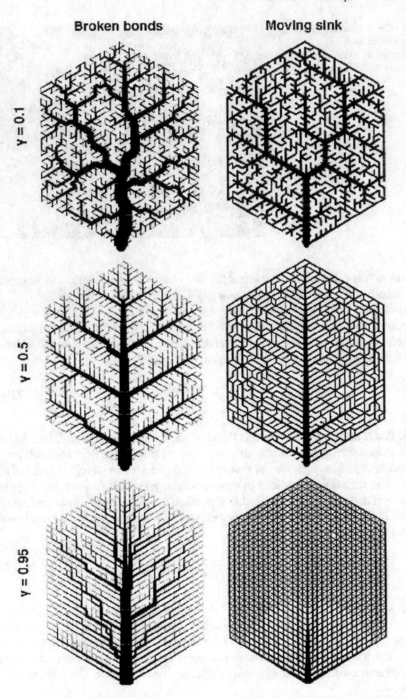

Broken bonds

Moving sink

Fig. 18.9 Optimal networks for the resilience to damage (left) and to a fluctuating load (right panels). Figure taken from [54]

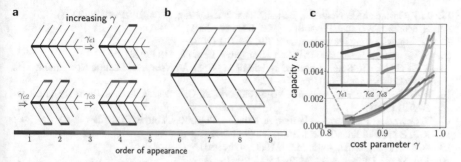

Fig. 18.10 a, b When the cost of new edges decreases (which corresponds here to the increase of the parameter γ), new loops appear according to the color code shown here: the darker the edge colour, the earlier the edge appears. For the i-th loop that appears, we denote its critical cost parameter γ_{ci}. **c** The transition to loop formation is discontinuous for all loops at all levels (all loops appear with a non-zero capacity as shown in the inset). Figure taken from [55]

looked into more details how this transition from a tree to a loopy network arises. The emergence of loops results from a trade-off governing their formation: adding redundant edges to supply networks is costly, yet beneficial for resilience. We can then expect that loops form when costs for new edges are small or inputs random. The authors of [55] showed that loops emerge discontinuously when decreasing the costs for new edges for both an edge-damage model and a fluctuating sink model. In the Fig. 18.10, they analyze this transition in the case of the fluctuating sink model for a larger network formed over a triangular grid (Fig. 18.10). The order in which new loops appear when the cost for new edges decreases is shown with different colors. All the new loops emerge discontinuously with a non-zero capacity (Fig. 18.10c). The ith loop appears at a critical cost parameter γ_{ci} where the loop starts to become beneficial for the dissipation-optimised network. The loop formation is discontinuous for all loops: all loops appear with a non-zero capacity as shown in the inset of Fig. 18.10 (more details in [55] and references therein).

References

1. D. Jungnickel, *Algorithm and Computation in Mathematics*, vol. 5 (Springer, Heidelberg, 1999)
2. J. Wardrop, Proc. Inst. Civ. Eng. **1**, 325 (1952)
3. R. Ahuja, T. Magnanti, J. Orlin, *Network Flows* (Prentice Hall, New Jersey, 1993)
4. G. Kirchoff, Ann. Phys. und Chemie **72**, 497 (1847)
5. P. Doyle, J. Snell, *The Mathematical Association of America* (USA, 1984), pp. 83–149
6. T. McMahon, J. Bonner, *On Size and Life* (Scientific American Library, New York, 1983)
7. D. Garlaschelli, G. Caldarelli, L. Pietronero, Nature **423**, 165 (2003)
8. J.R. Banavar, A. Maritan, A. Rinaldo, Nature **399**(6732), 130 (1999)
9. G. West, J. Brown, B. Enquist, Science **276**, 122 (1997)
10. A. Maritan, F. Colaiori, A. Flammini, M. Cieplak, J. Banavar, Science **272**, 984 (1996)

11. M.T. Gastner, M.E. Newman, J. Stat. Mech: Theory Exp. **2006**(01), P01015 (2006)
12. N. Mathias, V. Gopal, Phys. Rev. E **63**, 021117 (2001)
13. J. Berg, M. Lassig, Phys. Rev. Lett. **89**, 228701 (2002)
14. S. Valverde, R.F. Cancho, R.V. Solé, Europhys. Lett. **60**, 512 (2002)
15. R.F. i Cancho, R. Solé, *Statistical Mechanics of Complex Networks*, Lecture Notes in Physics, vol. 625 (Springer, 2003), pp. 114–125
16. R. Guimerà, A. Diaz-Guilera, F. Vega-Redondo, A. Cabrales, A. Arenas, Phys. Rev. Lett. **89**, 248701 (2002)
17. V. Colizza, J. Banavar, A. Maritan, A. Rinaldo, Phys. Rev. Lett. **92**, 198701 (2004)
18. R.C. Prim, Bell Labs Tech. J. **36**(6), 1389 (1957)
19. A.M. Frieze, Discret. Appl. Math. **10**(1), 47 (1985)
20. J.M. Steele, Discret. Appl. Math. **18**(1), 99 (1987)
21. D.J. Aldous, J. Shun, Stat. Sci. **25**(3), 275 (2010)
22. J. Beardwood, J.H. Halton, J.M. Hammersley, *Mathematical Proceedings of the Cambridge Philosophical Society*, vol. 55 (Cambridge University Press, 1959), pp. 299–327
23. H. Kesten, S. Lee, *The Annals of Applied Probability* (1996) pp. 495–527
24. F. Roberts, Biometrika **55**(1), 255 (1968)
25. M. Cortina-Borja, T. Robinson, Stat. Probab. Lett. **47**(2), 125 (2000)
26. F. Avram, D. Bertsimas, *The Annals of Applied Probability* (1992), pp. 113–130
27. D.J. Bertsimas, G. Van Ryzin, Oper. Res. Lett. **9**(4), 223 (1990)
28. M. Penrose, Ann. Appl. Probab. **7**, 340 (1997)
29. M. Mézard, G. Parisi, M.A. Virasoro, *Spin Glass Theory and Beyond* (World Scientific Publishing Co., Inc, Pergamon Press, 1990)
30. M. Mezard, A. Montanari, *Information, Physics, and Computation* (Oxford University Press, 2009)
31. J.M. Steele, *The Annals of Probability* (1981), pp. 365–376
32. M. Schulte, C. Thäle, Random Struct. Algorithms (2016)
33. A. Bourges, Internship report under the supervision of M. Barthelemy (2017). Unpublished
34. G. Narasimhan, M. Smid, *Geometric Spanner Networks* (Cambridge University Press, 2007)
35. P. Bose, M. Smid, Comput. Geom. **46**(7), 818 (2013)
36. D.P. Dobkin, S.J. Friedman, K.J. Supowit, Discret. Comput. Geom. **5**(4), 399 (1990)
37. G. Xia, L. Zhang, in *CCCG 2011: Proceedings of the 23rd Canadian Conference on Computational Geometry* (2011)
38. G. Xia, SIAM J. Comput. **42**(4), 1620 (2013)
39. A.C.C. Yao, SIAM J. Comput. **11**(4), 721 (1982)
40. J. Ruppert, R. Seidel, in *Proceedings of the 3rd Canadian Conference on Computational Geometry (CCCG 1991)* (1991), pp. 207–210
41. O. Aichholzer, S.W. Bae, L. Barba, P. Bose, M. Korman, A. Van Renssen, P. Taslakian, S. Verdonschot, Comput. Geom. **47**(9), 910 (2014)
42. L. Barba, P. Bose, J.L. De Carufel, A. Van Renssen, S. Verdonschot, in *Workshop on Algorithms and Data Structures* (Springer, 2013), pp. 109–120
43. P. Bose, P. Morin, A. Van Renssen, S. Verdonschot, Comput. Geom. **48**(2), 108 (2015)
44. N. Bonichon, C. Gavoille, N. Hanusse, D. Ilcinkas, in *International Workshop on Graph-Theoretic Concepts in Computer Science* (Springer, 2010), pp. 266–278
45. M. Barthelemy, A. Flammini, J. Stat. Mech. p. L07002 (2006)
46. M. Barthelemy, Phys. Rep. **499**(1), 1 (2011)
47. L.C. Freeman, Sociometry (1977), pp. 35–41
48. S. Khuller, B. Raghavachari, N. Young, Algorithmica **14**, 305 (1995)
49. H. Takayasu, M. Takayasu, A. Provata, G. Huber, J. Stat. Phys. **65**, 725 (1991)
50. S.S. Manna, D. Dhar, S. Majumdar, Phys. Rev. A **46**, R4471 (1992)
51. R. Albert, A.L. Barabasi, Rev. Mod. Phys. **74**, 47 (2002)
52. Z. Wu, L. Braunstein, S. Havlin, H. Stanley, Phys. Rev. Lett. **96**, 148702 (2006)

53. F. Corson, Phys. Rev. Lett. **104**, 048703 (2010)
54. E. Katifori, Phys. Rev. Lett. **104**, 048704 (2010)
55. F. Kaiser, H. Ronellenfitsch, D. Witthaut, Nat. Commun. **11**(1), 1 (2020)

Chapter 19
Optimal Transportation Networks and Network Design

Modern location science [1] concerns problems of facility locations and is highly interconnected with other fields such as mathematics, geography or computer science. The main question is essentially how to find the optimal locations for a set of facilities in order to serve a set of demand. This problem is of course very general and finds its roots in mathematics with the Fermat problem but also in economics with the Von Thunen model. The central mathematical problem in this field is the p-median one which consists in finding the location of p facilities such that the average distance between demand nodes and facilities is minimum [1]. This field evolved pushed by practical applications and now discusses a large variety of situations, ranging from health care, telecommunications or logistics. More generally, location problems can be categorized according to the nature of the location space being continuous, discrete or a network.

We will focus in this chapter on the part that concerns the optimal location of a network. The typical example is the subway: what is the best location of subway stations and more generally what is the best shape of this transportation network (same questions apply for other systems such as bus lines or railways). The difficulty of the network design comes from the large number of degrees of freedom which makes it difficult to explore efficiently the space of possibilities. This is even reinforced in the case of spatial networks, where we have additional variables corresponding to the position of nodes.

There has been extensive study of optimal networks over a given set of nodes (such as the minimum spanning tree [2], or other optimal trees [3]). Some such problems allow extra chosen nodes, for example the Steiner tree problem [4], or geometric location problems in which n given demand points are to be matched with p chosen supply points [5]. At another extreme is the much-studied Monge-Kantorovich mass transportation problem [6], involving matching points from one distribution with points from another distribution. Here we will focus on a fundamentally different problem where the density of origins and destinations are given and we have to find the optimal network connecting them together. More precisely, the general problem

© The Author(s), under exclusive license to Springer Nature Switzerland AG 2022
M. Barthelemy, *Spatial Networks*,
https://doi.org/10.1007/978-3-030-94106-2_19

that we address in this chapter is the following one: Given a set of points distributed in the plane according to some density ρ, what is the graph of given length L that minimizes the total (or average) time between pairs of these points? It is the same problem considered in [7] and which belongs to the more general class of locational optimization problems [8].

A network is intrinsically one-dimensional, in the sense of being a collection of (maybe curved) lines embedded in the plane. In a sense we are studying a coupling between a given distribution over points in the continuum and a network of our choice constrained only by length and connectedness. Some simpler problems of this type have been addressed previously. For instance, the problem of the quickest access between an area and a given point was discussed in [9, 10], and the impact of the shape of the city and a single subway line was discussed in [11]. Algorithmic aspects of network design questions similar to those have been studied within computational geometry (e.g. see for example the Chap. 9 of [8]) and 'location science' (e.g. [12] and references therein), but the specific question—optimal network topologies as a function of population distribution and network length—has apparently not really been explicitly addressed and is largely an open problem.

Although real-world networks are probably not optimal and result from the super-imposition of many different factors, understanding theoretically optimal networks could give some information about the actual structure observed in many cases. For example, it could help us to understand the seemingly universal structure displayed by very large subway networks [13]. In addition, the structure of real-world networks certainly depend on a large number of factors, but we will try here to get some insights with simpler models. Even for simple models, optimizing over all possible topologies is difficult, and here we will discuss various simple shapes only.

We will begin this chapter with empirical results about subway networks which will give us the opportunity to illustrate in a concrete way the problem of optimal networks. We will then discuss briefly the classical problem in this field of the hub-and-spoke structure. We will also consider one-dimensional problems such as the optimal spacing between consecutives stations and the geometry of spatial service cycles which display a transition from a circle to a more complex structure when the length of the cycle increases. We will then introduce more complex problems with the optimal location of a radial system of lines or a system of parallel lines. We will end this chapter with a discussion on the more general problem of the optimal network for a particular distribution of points (such as uniform or gaussian) and for some specific—but simple—objective functions.

19.1 Empirical Motivation: The Structure of Subway Networks

Transportation networks evolve in time and their structure has been studied in many contexts from street networks to railways and subways [13–16]. The specific case of

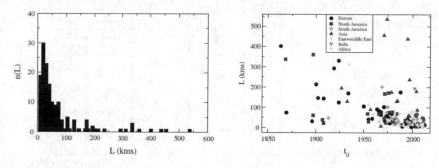

Fig. 19.1 (Left) Histogram of the total length of subway networks. (Right) Total length of the network versus its first construction date. We grouped the networks according to broad regions. Figures taken from [7]

subways is particularly interesting (for network analysis of subways, see for example [13, 17–21]). In most very large cities, a subway system has been built and later enlarged [13], with current lengths L varying from a few kilometers to a few hundred kilometers. We observe that the length of subway networks is distributed over a broad range (see Fig. 19.1(left)). Figure 19.1(right) also shows the total length versus the first construction date for most subway networks worldwide (the data is from various sources, see [21] and references therein): the oldest networks are mostly European and the largest and more recent ones can be found in Asia.

Concerning the geometry of these networks, as L increases we observe more complex shapes and an increase in the number of lines (see Fig. 19.2 and also [22]). Usually for small subways (L of order a couple of 10 kms) we observe a single line or a simple tree (e.g. a single line in the case of Baltimore, Haifa, Helsinki, Hiroshima, Miami, Mumbai, Xiamen, ...; or many radial lines such as in Atlanta, Bangalore, Incheon, Kyoto, Philadelphia, Rome, Sendai, Warsaw, Boston, Budapest, Buenos Aires, Chicago, Daegu, Kiev, Los Angeles, Sapporo, Tehran, Vancouver, Washington DC). For larger L (of order 100 kms), we typically observe the appearance of a loop line, either in the form of a single ring (e.g. Glasgow) or multiple lines with connection stations (Athens, Budapest, Lisbon, Munich, Prague, São Paulo, St. Petersburg, Cairo, Chennai, Lille, Marseille, Montreal, Nuremberg, Qingdao, Toronto). For larger networks (L over 200 kms) we observe in general some more complex topological structures (Berlin, Chongqing, Delhi, Guangzhou, Hong Kong, Mexico City, Milan, Nanjing, New York, Osaka, Paris, Shenzhen, Taipei). For the largest networks, convergence to a structure with a well-connected central core and branches reaching out to suburbs has been observed [13].

In this chapter, we investigate the optimal structure of transportation networks, as a function of the length L, for several related notions of *optimal* involving minimizing travel time. Real-world subway networks have developed under many other factors, of course, rather than resulting from the optimization of some simple quantity, but optimal structures provide interesting benchmarks for comparison with real-world networks.

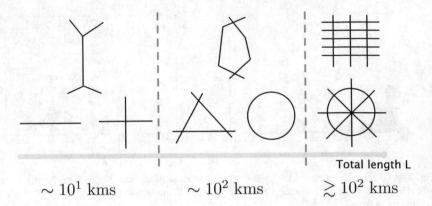

Fig. 19.2 Typical observed shapes when the length L increases. For small L we observe a line or a simple tree. For larger L we observe the appearance of a loop and for much larger L more complex shapes including a lattice like network or a superimposition of a ring and radial lines. Figure taken from [7]

In general we expect an evolution of the shape of optimal networks when its length L increases, with the possible existence of sharp transitions between different shapes. Although there is no proof in general, transitions in simple cases such as branches and a loop are observed: starting from a small value of L the branches first grow smoothly, then suddenly for a value L_c we observe the appearance of a loop. Although the design of a transit system is a very complex problem involving many different players, multiple objectives, etc., simple models of an optimal transit system in a city provide support to empirical results about the structure of subway systems in the world.

19.2   Hub-and-Spoke Structure

An important example in practical applications of an optimization process is the hub-and-spoke structure (see for [23] and references therein). In this organization, direct connections are replaced with fewer connections to hubs which form a network at a larger scale. The hub-and-spoke structure reduces the network costs, centralizes the handling and sorting, and allows carriers to take advantage of scale economies through consolidation of flows. Such networks have widespread application in transportation and in particular became pregnant in airline transportation where a carrier has to minimize its costs even if by doing so the average traveling time for user is not minimized (see the Fig. 19.3 for an illustration of this difference of designs).

One of the first cases where the hub-and-spoke system was observed is in the US with Delta Air Lines which had its hub in Atlanta (GA). FedEx also adopted this system in the 1970s and after the airline deregulation in 1978 most of the airline carriers adopted it. This system is based on the construction of regional hubs and

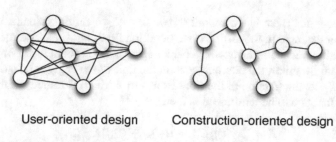

Fig. 19.3 Schematic comparison between a transportation network which optimizes the user travel time and distance (left) and a network which minimizes the construction costs (right)

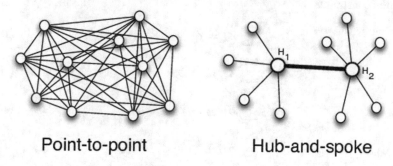

Fig. 19.4 Schematic comparison between a point-to-point transit model which optimizes the user travel time and distance (left) and a hub-and-spoke system (with two hubs H_1 and H_2) which minimizes the total operating costs by bundling together the traffic between the two hubs

to the creation of major routes between these hubs (see Fig. 19.4). One can easily compare the point-to-point (P2P) transit model and the hub-and-spoke model for N nodes. In the hub-and-spoke model, there are of the order N routes to connect all nodes, while in the P2P the number of routes is of the order N^2. In transportation systems such as airlines, this leads to fuller planes, larger benefits, and expensive operations such as sorting and accounting packages are done at hubs only. Obviously, there are also drawbacks. The centrality of hubs is very large and overloading at a hub can have unexpected consequences throughout the whole network. In other words, the hub is a bottleneck in the network and in agreement with general results on networks, hubs also constitute the vulnerable points in the network: a failure at a hub can trigger large delays which can propagate in the whole network.

An important problem in transportation research is to locate the hub(s) in a system of N nodes such that a given cost function is minimum. Analytical research on this *hub location problem* began with the study [24] and the interested reader can find a short review on this subject in [25]. In order to illustrate this approach we will detail here the simplest case, known as the *linearized multiple assignement model* and which can be formulated as follows. We assume that in a system of N nodes, there is a known origin-destination matrix T_{ij} (i.e. the flow between i and j). We denote by C_{ij} the transportation cost per individual to go from i to j and the importance of hubs lies in the fact that there is a *discount factor* $0 \leq \alpha \leq 1$ on links connecting them together. For example in the airline case, bundling flows allows to use for example larger aircrafts and reduce the passenger-mile costs. This discount factor implies that if k and m are two hubs, the interhub cost on the link $k - m$ is given by αC_{km}. If we denote by $X_{ij,km}$ the fraction of the flow from i to j which is routed via the hubs k and m, the function to be minimized is then

$$E = \sum_{ijkm} T_{ij}[C_{ik} + \alpha C_{km} + C_{mj}]X_{ij,km} \qquad (19.1)$$

under the constraints

$$\sum_{k} Z_k = p \qquad (19.2)$$

$$\sum_{km} X_{ij,km} = 1 \qquad (19.3)$$

$$\sum_{m} X_{ij,km} - Z_k \leq 0 \qquad (19.4)$$

$$\sum_{k} X_{ij,km} - Z_m \leq 0 \qquad (19.5)$$

where $Z_k = 1$ if k is a hub and zero otherwise. We look for a solution of this problem with a fixed number p of hubs, and the two last constraints ensure that the flow will not be routed via k and m unless both k and m are hubs. This type of minimization will lead to solutions schematically shown in Fig. 19.4(right) where the reduction of the number of links is made possible by the establishment of hubs and the bundling of flows between the hubs. For these optimal networks, the total network cost is minimum but obviously individual travel times are larger (as compared with a point-to-point network).

We just gave here the flavor of this type of approaches and there is a huge amount of studies on this problem that we cannot discuss here because of lack of space. Basically, there is a more practical research direction which amounts to add more characteristics of real-world networks [25] such as transfer delays for example [26]. We note here that many optimization problems are still unsolved and a statistical approach on the large network properties of this hub-and-spoke minimization problem could be an interesting direction for future research.

19.3 One-Dimensional Problems

19.3.1 A Single Open Line

We discuss here the problem of the location of nodes on a single line. This can be applied on transport problems such as the optimal location of bus or subway stations for a given distribution of demand nodes (or equivalently for a given population density). This is in some way the simplest network problem considering the simplest geometry of the line. The important aspect here is therefore the discrete nature of nodes but not the geometry of the network. This is in contrast with more complex problems that we will consider in the next sections where we forget about the discrete nature of stations but focus on the geometry of the network.

For the sake of simplicity, we consider the problem in the context language of a bus line. We assume that there is a bus route of length L in a certain region of the 2d plane and on which N bus stops need to be located. We index these bus stops from 1 to N. The bus stop 0 denotes the station where everybody wants to go (typically the center, or a train or subway station). We denote by ℓ_i the route distance from station i to $i + 1$. The terminal station is N. We represent the location of a point on the route by the abscissa s and the location of the bus stop i is $s_i = \sum_{j=1}^{i} \ell_j$. The commuters are distributed over this region according to a density $\phi(x)$ (Fig. 19.5).

We make the following assumptions:

1. The commuters want to reach the station 0 as quickly as possible
2. The commuters walk towards the bus stop at velocity v_w and the bus velocity is v_b. The ratio of velocities is denoted by $\eta = v_w/v_b < 1$.

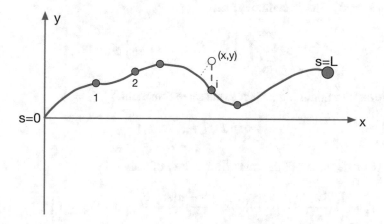

Fig. 19.5 A bus route goes from 0 to the terminal station at a route distance $s = L$. From a point located at (x, y) an individual can walk directly to the nearest bus stop or walk to the nearest point on the line and then follows the line to the nearest stop

3. At each stop, the bus stops during a time b (and we will neglect waiting time problems related to the frequency of buses v).
4. An important assumption is how the commuter reaches the nearest bus station: (i) either the commuter walks directly to the nearest station or (ii) the commuters walks to the nearest point on the bus route and then follows the bus route to the nearest bus stop.

We choose as an objective function to be minimized the total time T (summed other all points distributed in the plane) to reach the central station located at 0. We denote by V_i the catchment area of station i. If we choose assumption (i) and denote by x_i the location of station i this total time can be written as

$$T = \sum_{i=0}^{N} \int_{V_i} \left[\frac{||x - x_i||}{v_w} + \frac{s_i}{v_b} + ib \right] \phi(x) \mathrm{d}x \tag{19.6}$$

together with the constraint

$$\sum_{i=0}^{N} \ell_i = L \tag{19.7}$$

This problem is however too difficult and following [8], we will instead consider the assumption (ii) for which the commuters walks to the nearest point on the bus route (the 'access point') and then walks along the bus line to the nearest bus stop. In this case, the choice of the bus stop does not depend on the distance between the commuter's home and the bus route. We denote $\Phi(s)$ the density of commuters whose access point is at point s on the bus route (we thus construct some mapping $\Phi(x) \to \Phi(s)$). The total time is then

$$T = \sum_{i=0}^{N} \int_{V_i} \left[\frac{|s - s_i|}{v_w} + \frac{s_i}{v_b} + ib \right] \Phi(s) \mathrm{d}s \tag{19.8}$$

that we have to minimize subject to the same constraint $\sum_i \ell_i = L$

$$\min_{\{s_i\}} T(s_1, s_2, \ldots, s_N) \tag{19.9}$$

The precise definition of the catchment area V_i of station i is

$$V_i = \left\{ s | \frac{|s - s_i|}{v_w} + \frac{s_i}{v_b} + ib \leq \frac{|s - s_j|}{v_w} + \frac{s_j}{v_b} + jb, \ \forall j \neq i \right\} \tag{19.10}$$

This is by definition the weighted Voronoi cell generated along the bus line [8].

19.3.1.1 Straight Line Case: Optimal Interstation Spacing

In the straight line case, the catchment areas (i.e. the Voronoi cells) are simple and are simply defined by the bisectors of points x_i and x_{i+1} (see Fig. 19.6).

For N stations, we denote by $e_j = x_{j+1} - x_j$ the interstation distance and the total equivalent walking distance $\ell_N = T v_w$ is then given by (up to a constant that represents the vertical movements along the y-axis)

$$
\ell_N = \int_0^{m_0} dx \rho(x) |x|
$$

$$
+ \sum_{k=0}^{N-1} \int_{m_k}^{m_{k+1}} dx \rho(x) \left[(k+1)\delta + \eta \sum_{j=0}^{k} e_j + |x - x_{k+1}| \right] \tag{19.11}
$$

with the total length constraint $\sum_{j=0}^{N-1} e_j = L$ and where $\delta = b v_w$ corresponds to the equivalent distance walked during the bus stop time. The density of users is assumed to be uniform and given by $\rho(x) = 1/\Lambda$ where $\Lambda > L$ is the linear size of the system.

After a simple calculation, we obtain

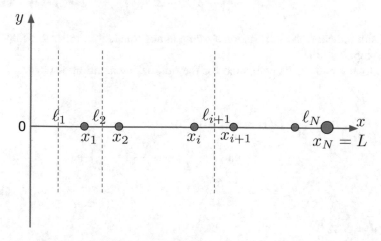

Fig. 19.6 One-dimensional bus line. The dashed lines (at locations $m_0, m_1, \ldots, m_N$) represent the frontier between two adjacent Voronoi cells

$$\ell_N = \frac{1}{2\Lambda} \left(\frac{1}{2} \sum_0^{N-1} e_k^2 + (\Lambda - L)^2 \right)$$

$$+ \frac{\delta}{\Lambda} \left[\frac{1}{2} \sum_{k=0}^{N-1} (2k+1)e_k + N(\Lambda - L) \right]$$

$$+ \frac{\eta}{\Lambda} \left[\frac{1}{2} \left(\sum_{j=0}^{N-1} e_j \right)^2 + (\Lambda - L) \sum_{j=0}^{N-1} e_j \right] \qquad (19.12)$$

We have to optimize this quantity subject to the constraint and introducing the Lagrange multiplier λ we will consider

$$\mathscr{L}_N = \ell_N - \lambda \left(\sum_{j=0}^{N-1} e_j - L \right) \qquad (19.13)$$

and write the equations $\partial \mathscr{L}_N / \partial e_i = 0$. We then obtain for $i = 0, \ldots, N-1$

$$e_i^* = \frac{L}{N} + \delta(N - 2i - 1) \qquad (19.14)$$

This result indicates the interstation spacing is not constant in order to compensate for the stopping time b.

The total length value computed for the optimal interstation distances e_i^* reads then as

$$\frac{1}{\Lambda} \ell(\{e_i^*\}) = \frac{1}{2}(1 - u)^2 + \frac{u^2}{4N}$$

$$+ \eta u \left(1 - \frac{u}{2} \right) + \frac{\delta N}{\Lambda} \left(1 - \frac{u}{2} \right) +$$

$$+ \frac{\delta^2 N}{12\Lambda^2}(N^2 - 1) \qquad (19.15)$$

where $u = L/\Lambda$.

For $\delta = 0$, it is easy to see that (treating N as a continuous variable)

$$\frac{\partial \ell_N}{\partial N} = -\frac{u^2}{4N^2} \le 0 \qquad (19.16)$$

In this case the total length decreases with the number of stations and the optimal number is then equal to its maximal value which is infinite here.

The optimal length of the line is obtained with the condition $\partial \ell / \partial u = 0$ which gives

$$u^* = \frac{1 - \eta}{1 - \eta + \frac{1}{2N}} \approx 1 \tag{19.17}$$

If $\delta > 0$, we can then show that at dominant order in N

$$\frac{\partial \ell_N}{\partial N} \approx -\frac{3\delta^2 N^2}{12\Lambda^2} \leq 0 \tag{19.18}$$

$$\frac{\partial \ell_N}{\partial u} \approx -\frac{\delta N}{2\Lambda} \leq 0 \tag{19.19}$$

These results imply then that there is a finite optimal number of stations given by

$$N_{opt} = N_{\max} = \sqrt{\frac{L}{\delta}} \tag{19.20}$$

$$u^* = 1 \tag{19.21}$$

indicating a non-trivial scaling with the size of the line.

19.3.2 Transition for a Cyclic Service Line

This is the problem of the locational optimization of a service route described for example in [8] (p. 564). In particular, results about an optimal bus route connecting two points show that the optimal cycle for a bus is not a circle but display oscillations. This problem is different from the one discussed in the previous section: here stations or stops are not given (the user just need to reach any point of the network) and the geometry of the network is not fixed.

This optimization problem for the cycle can be written under the following form. We assume that the curve is described by the polar equation $r = r(\theta)$ and the length L of the curve is then

$$L = \int_0^{2\pi} d\theta \sqrt{r^2 + \dot{r}^2} \tag{19.22}$$

(where $\dot{r} = dr/d\theta$). We assume that the points are distributed in the disk of radius $R = 1$ according to a distribution $\rho(r)$ independent from θ. The total time is then given by

$$T = \int r \, dr \, d\theta \, |r - r(\theta)| \rho(r) \tag{19.23}$$

We can compute this last quantity for either uniform density ($\rho = const.$) or a decreasing function (for example ($\rho(r) \propto e^{-r/r_0}$)) and we will focus here on the uniform case (for a disk of radius R) and obtain

$$T = \rho \int d\theta \left[\frac{1}{3} r(\theta)^3 - \frac{1}{2} r(\theta) R^2 + \frac{1}{3} R^3 \right] \tag{19.24}$$

We are interested in the existence of a transition from a circle to another curve with some oscillations. We then assume that the curve can be described by

$$r = a + bf(\theta) \tag{19.25}$$

where $b \ll a$ and $f(0) = f(2\pi) = 1$ (and integrals such as the average $\int_0^{2\pi} d\theta f$ is equal to zero). We write the optimization in a perturbation expansion in b (the lowest non trivial order is $\mathcal{O}(b^2)$). The average time to reach the line is given by

$$T = \frac{2\pi}{3} - \pi a + \frac{2\pi}{3} a^3 + ab^2 \int f^2 d\theta \tag{19.26}$$

and the total length of this cycle is given by

$$L = 2\pi a + \frac{b^2}{2a} \int \dot{f}^2 d\theta \tag{19.27}$$

We have to optimize the quantity

$$\mathcal{L} = T + \lambda \left(2\pi a + \frac{b^2}{2a} \int \dot{f}^2 d\theta - L \right) \tag{19.28}$$

where λ is a Lagrange multiplier. The optimization equations read

$$\frac{\delta \mathcal{L}}{\delta f} - \frac{\partial}{\partial \theta} \frac{\delta \mathcal{L}}{\delta \dot{f}} = 0 \tag{19.29}$$

$$\frac{\partial \mathcal{L}}{\partial a} = 0 \tag{19.30}$$

$$\frac{\partial \mathcal{L}}{\partial b} = 0 \tag{19.31}$$

The first equation gives

$$\ddot{f} = \frac{2\rho a^2}{\lambda} f \tag{19.32}$$

while the second and third ones give

$$0 = -\pi + 2\pi a^2 + b^2 \int f^2 + \lambda \left(2\pi - \frac{b^2}{2a^2} \int f^2 \right) \quad (19.33)$$

$$0 = b \left(2a^2 \int f^2 + \lambda \int \dot{f}^2 \right) \quad (19.34)$$

The trivial solution is $b = 0$, leading to $a = L/2\pi$: this is the circular solution. When $b \neq 0$, Eq. (19.34) implies that $\lambda < 0$ and Eq. (19.32) implies that f is a periodic solution, combination of a cosine and sine. The condition $f(0) = f(2\pi)$ implies that

$$\frac{2\rho a^2}{|\lambda|} = n^2 \quad (19.35)$$

where n is a positive integer.

We thus see here the possibility to a transition from a circle to a solution with oscillations. We can make a simplified calculation with $f(\theta) = \cos(n\theta)$ and compute explicitly all integrals. We get the constraints

$$\frac{2a^2}{|\lambda|} = n^2 \quad (19.36)$$

$$L = 2\pi a + ab^2 \frac{\pi}{|\lambda|} \quad (19.37)$$

$$0 = -1 + 2a^2 + 2b^2 - 2|\lambda| \quad (19.38)$$

Solving this set of equations for a gives

$$a = \frac{1}{2(6/n^2 - 1)} \left(\frac{2L}{\pi n^2} \pm \sqrt{\Delta} \right) \quad (19.39)$$

where the discriminant Δ is

$$\Delta = \left(\frac{2L}{\pi n^2} \right)^2 - 2 \left(\frac{6}{n^2} - 1 \right) \quad (19.40)$$

For $n = 1$, we thus obtain $L \geq L_c$ with $L_c = \pi \sqrt{5/2} \approx 4.967...$. We test these results with numerical simulations on the 'Limaçon' curves of the form $r = a + b \cos(n\theta)$ (where n is an integer). In this case we do observe a transition for $L = L_c \approx 3$ in agreement with the simple analytical argument presented here. In the Fig. 19.7, we show the corresponding curves for the different values of L and we indeed observe a transition to oscillating curves.

Fig. 19.7 Numerical
calculation of the optimal
'Limaçon' curves. We
observe a transition for
$L = L_c \approx 3$ from a circular
solution to oscillating ones,
in agreement with the
analytical discussion
presented in the text

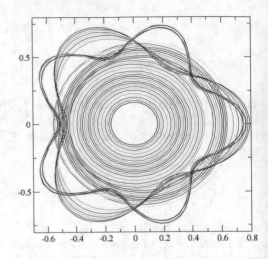

19.4   Multiple Transit Lines

In this section, we will consider some simple geometries that can be worked out
mostly analytically. In these cases, we focus on a given geometry that depends on
a small number of parameters such as the number of lines, their location, etc. We
will consider for example parallel transit lines and then radial lines. In these cases
we have the additional parameter of headways that can be different from one line to
another. For the parallel lines, we will follow the continuous approach used in [27]
while in radial case, we will follow the discrete approach discussed in [28]. In this
way, we can present two different methods that can be helpful for other problems.

19.4.1   Parallel Transit Lines

We consider here an area where individuals can access (by foot for example) a feeder
line which is connected to a transit line on the x-axis (Fig. 19.8). It could however
be a linear central business district or shopping area, etc. and we won't consider
here synchronization problems between the feeder lines and the destination. The
passenger can then walk to the feeder line and then ride to the transit line (such a
path is shown in Fig. 19.8).

This problem has been addressed by different methods and we follow here a
continuous limit approach [27] which is interesting and might allow for further
generalization (the discrete method discussed in [29] will be discussed in the next
section for the radial case).

The population density will induce a demand which can be described by the
number $\rho(x, t)$ of requests for service per unit time and per unit distance in the
vicinity of (x, t). All costs will be expressed in units of passenger waiting time and

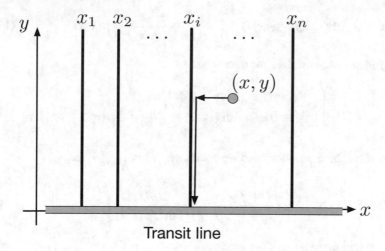

Fig. 19.8 Geometry considered in [27]. A transit line is connected to n parallel feeder lines at locations $x_1, x_2, \ldots, x_n$ that can be accessed by foot for example

are: (i) the cost λ of operating one vehicle for one round trip, and (ii) the cost γ per unit distance incurred by one passenger to access the feeder line. We will also introduce the density $u(x)$ of feeder lines (we assume here that feeder lines do not change in time) and $g(x, t)$ the rate of dispatching vehicles in the vicinity of x at time t. In other words $1/u$ is the typical distance between lines and $1/g(x, t)$ the headway (for lines near x and at time t). The calculation below assumes that the demand (or equivalently the population density) varies slowly on the scales $\sim 1/u$ and $1/g$. More formally this can be written as $[\rho(x + 1/u(x, t), t) - \rho(x, t)]/\rho(x, t) \ll 1$. In particular, this implies that the headways between adjacent lines are similar. It follows that the passengers will walk to the nearest line. From this we see that the average access cost between two lines in the vicinity of x will be

$$\frac{2\gamma}{1/u} \int_0^{1/2u} y \, dy = \frac{\gamma}{4u}$$

where y is the distance to the nearest line. In addition, the waiting time will be on average $1/2g(x, t)$. Finally, we take into account the operating cost $\lambda u(x)g(x, t)$ (which can be seen as $1/(1/u)/(1/g)$ where the first term gives the number of lines and the second the number of vehicle par time cycle). The objective function Z can then be written as the sum of operating cost for the feeder vehicles, and access and waiting costs for all the passengers

$$Z = \int dx dt \, J(x, t) \tag{19.41}$$

where

$$J(x.t) = \lambda u(x)g(x,t) + \left[\frac{\gamma}{4u(x)} + \frac{1}{2g(x,t)}\right]\rho(x,t) \tag{19.42}$$

Using functional derivatives we then obtain

$$\begin{cases} \frac{\delta Z}{\delta g(x,t)} = 0 = \lambda u(x) - \frac{\rho(x,t)}{2g^2(x,t)} \\ \frac{\delta Z}{\delta u(x)} = 0 = \lambda \int dt\, g(x,t) - \frac{\gamma}{4u^2(x)} \int dt\rho(x,t) \end{cases} \tag{19.43}$$

We can solve these equations and we obtain $g(x,t) = \sqrt{\frac{\rho(x,t)}{2\lambda u(x)}}$ and

$$u(x) = \frac{1}{2}\left(\frac{\gamma^2}{\lambda}\right)^{1/3}\left[\frac{\int dt\rho(x,t)}{\int dt\sqrt{\rho(x,t)}}\right]^{2/3} \tag{19.44}$$

and we obtain the inverse of the headways

$$g(x,t) = (1/\lambda\gamma)^{1/3}\left[\frac{\int dt\sqrt{\rho(x,t)}}{\int dt\rho(x,t)}\right]^{1/2}\rho^{1/2}(x,t) \tag{19.45}$$

We thus recover here a result obtained for a single line [30]: the optimal flow rate of vehicles vary as the square root of the arrival rate of passengers.

19.4.2 Radial Lines

We discuss here the radial geometry proposed in [28]. The system is composed of n radial transit lines (Fig. 19.9). The transit line i has angle θ_i and has an headway h_i (the headway is the average time interval between two successive buses). The population density is described by a function $P(r, \theta)$ and the variables are then here h_i and θ_i for $i = 1, \ldots, n$.

The total cost Z is as usual in these problems composed of a user cost C_u and operating costs C_o: $Z = C_u + C_o$. We define the access cost per unit time γ_a (the access velocity is v_a), the waiting time cost γ_w, the riding cost per unit time γ_r (the riding velocity is v_o) and the operating cost per unit time γ_o. We assume that all users want to go to the center O. For a user located at (r, θ) in the sector between lines $i - 1$ and i, there are two choices: either the less costly path is along line i or line $i - 1$. The total cost for choosing line i and for line $i - 1$ read

$$C_{i-1}(r, \theta) = (\theta - \theta_{i-1})\frac{r\gamma_a}{v_a} + c\gamma_w h_{i-1} + \frac{\gamma_r r}{v_o} \tag{19.46}$$

$$C_i(r, \theta) = (\theta_i - \theta)\frac{r\gamma_a}{v_a} + c\gamma_w h_i + \frac{\gamma_r r}{v_o} \tag{19.47}$$

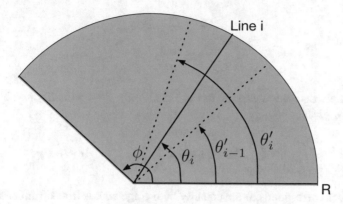

Fig. 19.9 Geometry considered in [28]. Transit lines are radial and span a sector of angle ϕ and radius R

where the average waiting time is ch_i with $c \leq 1$ (and ch_{i-1} for line $i - 1$). The boundary defined by the angle θ_i' (see Fig. 19.9) between the catchment area for $i - 1$ and i is then defined by $C_{i-1} = C_i$ which gives

$$\theta_i' = \frac{1}{2}(\theta_{i-1} + \theta_i) + \frac{K}{2r}(h_i - h_{i-1}) \tag{19.48}$$

This expression means that the boundary is at the bisectrix (first term of the r.h.s.) corrected by a term that depends on the headway difference $\Delta h = h_i - h_{i+1}$. For example, if $\Delta h < 0$, the boundary is then closer to θ_i as the waiting time will be shorter due to a smaller headway.

We also see on Eq. (19.48) that the boundary depends on r and is not a simple straight line. However, for an individual close to the center (small $r < R'$) will find it faster to walk directly to the center. We will thus consider as in [28] that all the following discussion is valid for $r > R'$.

The total user cost is then given by

$$C_u = \sum_{i=1}^{n} \int_{R'}^{R} r\,dr \int_{\theta_{i-1}'}^{\theta_i'} d\theta\, P(r, \theta) \left[\frac{\gamma_a r}{v_a} |\theta - \theta_i| + c\gamma_w h_i + \frac{\gamma_r r}{v_o} \right] \tag{19.49}$$

and the operating cost per line is given by the sum over all lines of the operating cost γ_o per vehicle times the number of vehicles T/h_i

$$C_o = \sum_{i=1}^{n} \gamma_o \frac{T}{h_i} \tag{19.50}$$

where T is the cycle time (typically a day). We recall that the variables are here the headways $\{h_i\}$ and positions of lines $\{\theta_i\}$. The optimality conditions then read

$$\frac{\partial C_u}{\partial \theta_i} = \frac{\partial C_o}{\partial \theta_i} = 0 \tag{19.51}$$

$$\frac{\partial C_u}{\partial h_i} = \frac{\partial C_o}{\partial h_i} = 0 \tag{19.52}$$

After a simple derivation and the use of Eq. (19.48), the optimality condition with respect to θ_i gives

$$\int_{\theta'_{i-1}}^{\theta_i} d\theta \int_{R'}^{R} r^2 dr \, P(r, \theta) = \int_{\theta_i}^{\theta'_i} d\theta \int_{R'}^{R} r^2 dr \, P(r, \theta) \tag{19.53}$$

which in fact corresponds to the equality of average access times from each side of the boundary defined by $\theta = \theta_i$. The derivative with respect to h_i gives

$$\frac{\partial Z}{\partial h_i} = c \gamma_w \int_{\theta'_{i-1}}^{\theta'_i} \int_{R'}^{R} P(r, \theta) r dr - \gamma_o \frac{T}{h_i^2} \tag{19.54}$$

leading to the equation

$$c \gamma_w \int_{R'}^{R} \int_{\theta'_{i-1}}^{\theta'_i} r dr d\theta \, P(r, \theta) = \gamma_o \frac{T}{h_i^2} \tag{19.55}$$

These Eqs. (19.53), (19.55) cannot be solved in the general case and we now consider the much simpler isotropic case $P(r, \theta) = P(r)$ where all calculations can be done. We first define the following quantities

$$Q_1 = \int_{R'}^{R} r dr \, P(r) \tag{19.56}$$

$$Q_2 = \int_{R'}^{R} r^2 dr \, P(r) \tag{19.57}$$

and Eqs. (19.53), (19.55) read

$$\theta_i - \theta'_{i-1} = \theta'_i - \theta_i \tag{19.58}$$

$$c \gamma_w (\theta'_i - \theta'_{i-1}) = \frac{\gamma_o}{h_i^2} \tag{19.59}$$

The boundary conditions are $\theta'_0 = 0$ and $\theta'_n = \phi$ and due to the isotropy we have naturally $h_i = h$. We then obtain for the positions of the lines and the boundaries

$$\theta_i = \frac{2i-1}{2n}\phi \tag{19.60}$$

$$\theta'_i = \frac{i}{n}\phi \tag{19.61}$$

which correspond to an equal spacing between lines. The headway is given by

$$h = \sqrt{\frac{\gamma_o Tn}{c\gamma_w Q_1 \phi}} \tag{19.62}$$

The headway is then varying as the square root of the number of lines, a non-trivial result that can be recovered by noting that normal operation of the system can be achieved when the total operating cost for all vehicles on all lines is of order the waiting cost

$$n\gamma_o \frac{T}{h} \sim \gamma_w h \tag{19.63}$$

leading to $h \sim \sqrt{n}$.

Using these results, the total cost reads

$$Z = \frac{\gamma_a Q_2 \phi^2}{4v_a}\frac{1}{n} + 2(c\gamma_w\gamma_o\phi Q_1 T)^{1/2}\sqrt{n} + \gamma_r Q_2\phi/v_o \tag{19.64}$$

This is a function of the number of lines and its maximum will be obtained for $\partial Z/\partial n = 0$ which gives the optimal number of lines

$$n_o = \left[\frac{\gamma_a Q_2 \phi^{3/2}}{4v_a\sqrt{c\gamma_w\gamma_o Q_1 T}}\right]^{2/3} \tag{19.65}$$

$$\sim \left[\frac{\gamma_a^2}{\gamma_w\gamma_o}\right]^{1/3} \tag{19.66}$$

The optimum number of lines depends thus on the ratio of the access cost and the waiting and operating costs. It is indeed natural to expect a larger number of lines for a larger access cost or a smaller waiting or operating cost. Note however that both the combination of these costs and the exponent $1/3$ are not trivial.

This result is however modified if we impose a constraint of the number of vehicles in the fleet. Indeed, if we write this constraint as

$$\sum_{i=1}^{n}\frac{T}{h_i} = N \tag{19.67}$$

where N is the fleet size, we have to minimize

$$C_u + C_o + \lambda \left(\sum_i^n T/h_i - N \right) \tag{19.68}$$

where λ is a Lagrange multiplier. The minimization equations leads to a constant headway $h_i = h \sim \sqrt{n}$ and by optimizing $Z(n)$ with respect to n, the optimal number of lines in this case reads

$$n_0 \propto \left(\frac{\gamma_a}{\gamma_w} \frac{N\phi}{T} \right)^{1/2} \tag{19.69}$$

We then see on this result that when the fleet size is constrained, the optimal number of lines depends on the ratio of access and waiting cost (and with another non-trivial exponent $1/2$) and not on the operating cost γ_o, as expected.

19.5   The Optimal Subway Problem

We follow here [7] and discuss the following general problem: we consider a distribution of points in the plane and ask for the network G of given length L that is 'optimal' (see below). We will mostly consider the uniform distribution and also the isotropic population density Gaussian density

$$\rho(x, y) = (2\pi)^{-1} \exp(-r^2/2), \ \ r^2 = x^2 + y^2 \tag{19.70}$$

Individuals distributed according this density want to reach as quickly as possible other points. In the subway picture, these individuals can move anywhere in the plane at speed 1 and on the subway network at speed $S > 1$ (this quantity can therefore be seen as the ratio $S = v_s/v_w$ between walking and subway velocities). In this simplified setting, we assume that each point of the network can be seen as a 'station' where the subway is accessible (an approximation done in [27, 28] as well, see previous sections).

The general problem is as follows: we consider the shortest time $\tau(x_1, x_2)$ to go from any point x_1 to any other one x_2. For a given network G, the average journey time is then

$$\overline{\tau}(G) = \int \int \tau(x_1, x_2) \rho(x_1) \rho(x_2) dx_1 dx_2 \tag{19.71}$$

The problem is then to find the graph G of given length L that minimizes this quantity $\overline{\tau}(G)$. In order to address this general problem, we will consider simpler objective functions essentially based on average user time. This will allow us to get some general results about the shape of G. For more complex, objective functions, we will test simple shapes made of radial branches and loops. More precisely, we will consider the following three variants:

1. First, we consider the problem of minimizing the average (over starting points from the given distribution) distance to the network, that is to the closest point in the network. In this case, we don't compute the average time between all pairs of points but just consider the access time to the network from each point (this is then almost similar to the $S = \infty$ case).
2. The second variant that we consider is the problem of minimizing the journey time from all the points to a single destination, which we may take to be the origin O.
3. In the third variant, we engage the general issue of routes between arbitrary points which typically (but not always) involve entering and exiting the subway network, but now require these entrances and exits to be the closest positions to the starting and ending points, rather than the time-minimizing positions.

19.5.1 A First Simplification: Optimal Placement

We seek here the network of given length L that minimizes the average distance from a point to the network. Note that this problem bear some similarities with the notion of space-filling curves [31], although the latter are fractal curves whereas our networks cannot be fractal as their length is finite.

In this case, we can derive a few rigorous results [7]. Indeed, for a small line segment, the area within a small distance ε from the line is 2ε per unit length which for a network of length L, gives a total area within that distance at most $2\varepsilon \times L$. This area is reduced by curved lines and intersections, which implies that the optimal network is a single curve, or in graph theory terms, a tree. If the density ρ is not isotropic, we might have a more complex optimal network but we won't consider this case here.

In the large L limit, the optimal network density (edge length per unit area) near a point z should be of the form $L\phi(\rho(z))$ for some increasing function ϕ. The total length constraint then reads

$$\int \phi(\rho(z))\, dz = 1 \tag{19.72}$$

The average distance from a typical point near x to the network should then scale as $c_0/(L\phi(\rho(z)))$ for some constant c_0. The overall distance to the network is then

$$\overline{d}(L) = \frac{c_0}{L} \int \frac{1}{\phi(\rho(z))} \rho(z) dz. \tag{19.73}$$

We thus have to minimize over ϕ the following function

$$\mathscr{L} = \frac{c_0}{L} \int \frac{1}{\phi(\rho(z))} \rho(z) dz - \lambda \left(\int \phi(\rho(z))\, dz - 1 \right) \tag{19.74}$$

where λ is a Lagrange multiplier. This leads to $\phi(\rho) \propto \rho^{1/2}$, and the minimal average distance reads

$$\overline{d}(L)_{opt} = \frac{c_0}{L} \left(\int \rho^{1/2}(z) \, dz \right)^2. \tag{19.75}$$

The constant c_0 can be re-interpreted as the minimum average distance to network in the context of networks on the infinite plane with network density $= 1$. Intersections being not optimal, the optimal network in this infinite context consists of parallel lines spaced one unit apart for which we can easily calculate $c_0 = 1/4$. We thus finally obtain the following result for $L \to \infty$ [7]

$$\overline{d}(L)_{opt} \sim \frac{1}{4L} \left(\int \rho^{1/2}(z) \, dz \right)^2 \tag{19.76}$$

In the Gaussian case, the integral in (19.76) equals $\sqrt{8\pi}$ and so $\overline{d}(L)_{opt} \sim \frac{2\pi}{L}$.

This argument suggests that a network is asymptotically optimal if the local pattern around a position z is made of parallel lines with spacing proportional to $\rho^{-1/2}(z)$, and orientations that can depend arbitrarily on z. The typical image for this would be something like a fingerprint. For isotropic densities, we can thus expect such a network to be a spiral. Indeed, if we consider a spiral of length L starting at point $(a, 0)$ and with rings at radius r separated by $b \exp(r^2/4)$, the numerical optimization over (a, b) displays a slow convergence to the asymptotic behavior (see Fig. 19.10). For comparison, the optimal star shape is $\overline{d}(L)_{opt} \sim 2^{1/2}\pi^{3/2}/L$, and we see that the value at which such spiral networks out-perform star networks in the Gaussian model is around $L = 110$.

This asymptotic mathematical results is interesting but says nothing about optimal shapes for smaller and more realistic values of L. When L grows we however expect the following scenario: for very small L, the optimal network is a line segment centered at the origin, and for larger L we expect a smooth transition to a slightly curved path, such as 'C-shape'. For much larger L, we know that we obtain the spiral but it is unclear at this point how. It is however not difficult to test numerically simple shapes (for the standard Gaussian distribution) and we consider (Fig. 19.11):

- The line segment $[-L/2, L/2]$.
- The 'cross' (two length $L/2$ lines crossing at the origin).
- The 'hashtag' or '2 × 2 grid' (Fig. 19.11a).
- The 'ring' (circle centered on the origin).
- The 'C-shape' (off-centered partial circle, with arc length 2θ removed, see Fig. 19.11b).
- The 'S-shape': two arcs of circle of radius R and of angle 2θ, connected by a straight line of length $2R$ (see Fig. 19.11c).
- The 'star' with n_b branches of length r^* (so $r^* n_b = L$, see Fig. 19.11d).
- The (Archimedean) 'spiral', $r = a\theta + b$.

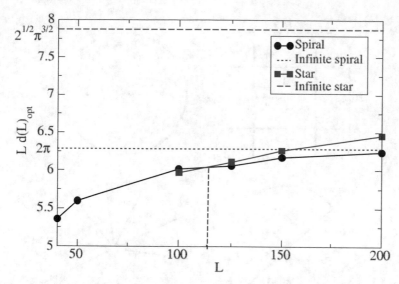

Fig. 19.10 Rescaled average distance $L\bar{d}(L)_{opt}$ versus L for the spiral and the star network. For the spiral, we observe a slow convergence towards the theoretical limit 2π for gaussian disorder. The convergence is even slower for the star network which converges to $2^{1/2}\pi^{3/2}$. For $L \approx 110$ (dashed vertical line) the spiral outperforms the star network. Figure from [7]

We know that the optimal shape is a tree which implies that shapes such that the 2×2 grid or ring can never be overall optimal, but it is nonetheless interesting to include them for the sake of comparison. Also, in some cases, we can compute analytically the average distance. For example, for any isotropic distribution, the optimal ring has radius equal to the median of the radial component of the underlying distribution, which gives in the Gaussian case $\sqrt{2\log 2} \approx 1.18$.

For these different shapes, we compute numerically for each size L the optimal value of their parameters and we show the results in Fig. 19.12. From these numerical results, we thus observe that for small $L \lesssim 3.2$ the optimal shape is the C-shape. We note that for this shape the optimal θ decreases with L: for $L \approx 6$ we have $s = 0$ and for $L \approx 8.0$ the optimal $\theta = 0$. When $\theta = 0$ and $s = 0$ we then recover the ring result. For $3.2 \lesssim L \lesssim 5.8$, the cross (star network with 4 branches) is optimal and for $5.8 \lesssim L \lesssim 8.7$, the S-shape is optimal. We note that for this shape, as L increases, the optimal angle θ increases from 0 to $\pi/2$, but with a jump at $L \approx 5$. For $L \gtrsim 8.7$ the star network with n_b branches is the optimal shape and the number of branches is roughly increasing with L: $n_b \sim pL + q$ with $p \approx 0.4$. The simple Archimedean spiral was slightly less efficient than the star network over the range of L considered above: For $L = 20$, the average time is $\bar{d}_{opt} \approx 0.215$ for the star network, while for the spiral we have $\bar{d}_{opt} \approx 0.233$. These numerical results suggest that there are 4 sharp transitions (see Fig. 19.13): C-shape to cross near $L = 3.2$, cross to S-shape near $L = 5.8$, S-shape to star near $L = 8.7$ and finally a transition from the star network to the spiral near $L = 110$. We note that the star only slightly out-performs

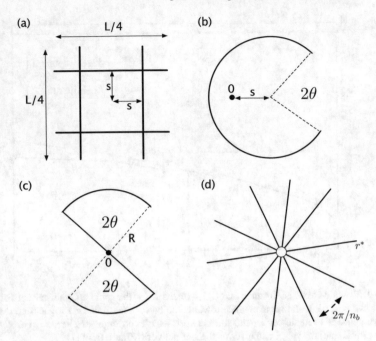

Fig. 19.11 Main shapes considered for the optimization (in addition to the line and the spiral). **a** The 'hashtag' of 2×2 grid with parameter s. **b** The 'C-shape' with paramerers s and θ. **c** The 'S-shape' with parameter θ. **d** The star network with n_b branches of size r^*. Figure from [7]

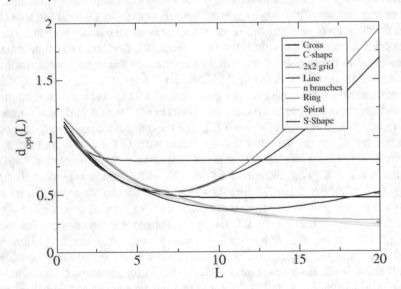

Fig. 19.12 Average minimum distance to the network versus L for various shapes. Figure from [7]

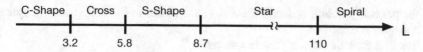

Fig. 19.13 Representation of the different optimal shapes found in [7] versus the total length L of the network

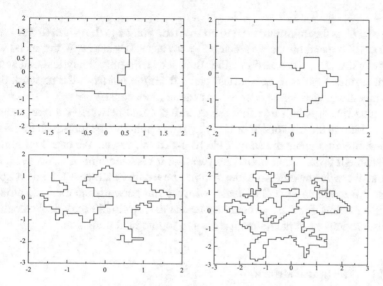

Fig. 19.14 Best configurations obtained with simulated annealing for different values of $L = 5, 10, 20, 50$. The simulations are obtained with a lattice polymer and using the pivot algorithm [32]. Figure from [7]

these curves, so it is possible that in fact there is a smooth evolution of curves as optimal networks.

Finally, we mention that there are alternative numerical approaches for computing the optimal network which do not necessitate to put 'by hand' specific shapes. This is the case of simulated annealing for example. Some results for optimal curves using this method are shown in Fig. 19.14, which are roughly consistent with the qualitative summary discussed above.

19.5.2 The Minimum Average Distance to the Center

The second variant considers the average time to reach the center. In the case of street networks, this was studied by Bejan [9, 10] who found an optimal tree. This problem was also considered in the context of a single bus line [8] (and references therein). Even if this problem seems too simplified as we are in general interested in reaching many points in the system, we will see that it however captures the essence of the general problem and might constitute a useful toy model where analytical calculations are feasible.

The objective function is then in this case

$$\bar{\tau} = \int \tau(z, 0)\rho(z)dz \tag{19.77}$$

where $\tau(z, 0)$ is the minimum time to go from the point z to the origin 0. The optimal network will depend on the parameter $S = v_s/v_w > 1$, where v_s is the speed within the network, and v_w the speed outside. In order to simplify analytical calculations, we will assume here that the shortest paths from the points to the network can be made only along circular ($r =$ const.) or radial ($\theta =$ const.) lines.

We note that it is still true that the overall optimal network is a tree. Indeed, if there is a loop in the network, there is at least one point on the loop such that starting in either direction takes the same time to get to the center. We can then remove a small segment around that point and reassign it elsewhere in order to get a better network. We will however consider simple shapes that include a loop as it appears in most real-world cases. We will start with the star network and then add a ring. As we will see below, for these structures we can develop simple analytical calculations and observe important phenomena such as a topological transition.

19.5.2.1   The Star Network

We start with the star network with n_b branches of lengths r^*, outward from the origin, and evenly spaced with angle $2\pi/n_b$ spacing (see Fig. 19.11d). The total length of this graph is $L = n_b r^*$ and n_b is a free parameter to be optimized over (for given L). Thanks to the isotropy, the average time $\bar{\tau}$ to the center is

$$\frac{1}{2n_b}\bar{\tau} = \int_0^{\pi/n_b} d\theta \left[\int_0^{r^*} dr r\rho(r) \left(\frac{\theta r}{v_w} + \frac{r}{v_s} \right) + \int_{r^*}^R dr r\rho(r) \left(\frac{r - r^*}{v_w} + \frac{\theta r^*}{v_w} + \frac{r^*}{v_s} \right) \right] \tag{19.78}$$

The first integral in the r.hs. corresponds to points located at $r < r^*$: there is some time needed to reach the branch (at walking speed v_w), and then there is some time needed to reach the center (at subway speed v_s). The second integral corresponds to points located at larger distances $r > r^*$. In this case, there is some time to reach the terminal r^* and then to go from r^* to the center.

We begin with the uniform density of points in a disk of radius R: $\rho(r) = \rho_0 = 1/(\pi R^2)$. If the network is absent, the average time to reach the center is $\tau_0 = 2R/(3v_w)$. We denote by $\eta = 1/S = v_w/v_s$, and we assume that taking the subway is always better than walking directly to the center, which is the condition that $\eta \leq 1 - \pi/n_b$. We introduce the reduced variables $u^* = r^*/R$ (the branch length relative to city radius) and $u_0 = L/R$ (network length relative to city radius) and $\chi = \pi/n_b$. Evaluating the integrals in (19.78), the average time $\bar{\tau}$ to reach the center via subway satisfies

$$\frac{\overline{\tau}}{\tau_0} = \frac{u^{*3}}{2}\left[-\frac{\chi}{2} - \eta + 1\right] - \frac{3}{2}u^*\left[-\frac{\chi}{2} - \eta + 1\right] + 1.$$

For given L we want to optimize over the free parameter r^*, that is over u^*. From $L = n_b r^*$ we obtain $\chi = \pi u^*/u_0$ and then the average time as a function of u^* reads

$$\frac{\overline{\tau}}{\tau_0} = \frac{1}{2}(u^{*3} - 3u^*)\left[-\eta + 1 - \pi\frac{u^*}{u_0}\right] + 1. \qquad (19.79)$$

Minimizing this quantity over u^* leads to a polynomial of degree 3 that can easily be solved numerically. For large $u_0 = L/R$, it is optimal to use roughly u_0 branches of length almost R, and more precisely for $u_0 \gg 1$ we obtain

$$\begin{cases} u^* = 1 - \frac{\pi}{3(1-\eta)u_0} + \mathcal{O}\left(\frac{1}{u_0^2}\right) \\ n_b = u_0 + \mathcal{O}(1). \end{cases} \qquad (19.80)$$

For smaller length (here for $0 \le u_0 \le 10$), the length of branches is increasing faster than their number: when L is increasing in this range, we first observe a radial growth and then an increase of the number of branches.

These results can be extended to the exponential density case where $\rho(r) = \rho_0 \exp(-r/r_0)$ on the infinite plane. We observe (numerically) in this case that for large resources, the number of branches scales as $n_b \sim a u_0$ with $a \approx 0.1$ and the solution u^* seems to converge slowly to some value that depends on η.

19.5.2.2 Loop and Branches

We now add to the star network (with n_b branches of length r^*) a loop of radius ℓ (see Fig. 19.15). This case is essentially motivated by subway networks that seem to display this type of structure when they are large enough (see [13] and Chap. 10). We have 3 parameters describing this structure (n_b, ℓ and r^*, where $\ell \le r^*$) and its total length is given by

$$L = 2\pi\ell + n_b r^*. \qquad (19.81)$$

This simple network enables to understand the relative contributions of the loop and branches for reaching the center in the optimal network. It is however analytically difficult but can easily be studied numerically.

We consider here the uniform distribution on a disc of radius r_0 and $\eta = 1/8$ which corresponds to the reasonable values $v_w \approx 5$ km/h and $v_s \approx 40$ km/h. We first fix the number n_b of branches and optimize the network over r^* and ℓ. In Fig. 19.16,

Fig. 19.15 Schematic of the
star network combined with
a loop. We have now three
parameters: the number of
branches n_b, the length r^* of
the branches, and the radius
ℓ of the loop. Figure from [7]

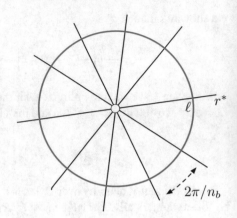

we show the results for the optimal value of r^* and ℓ versus $u_0 = L/r_0$ normalized
by n_b. We observe that when resources are growing from 0, we have only a radial
network ($\ell = 0$). At $u_0/n_b = 1$ we have a 'transition' point where a loop appears.
This means that until $u_0 = n_b$ all the available resource is converted into the radial
structure. When the radial structure is at its maximum ($r^* = r_0$) we observe the
appearance of a loop whose diameter is then increasing with u_0. Even if both the
number of branches and the size of the loop undergo a discontinuous transition, the
average minimum time displays a smooth behavior. Also, if we increase further the
total length L, it will result in a larger number of branches n_b.

In the uniform density case, the domain is finite and at fixed value of n_b, there
is therefore a maximum value of $L_{max} = n_b R + 2\pi R$. For larger values of L, the
optimal network will increase its number of branches n_b. It is different in the Gaussian
disorder case: the domain is infinite and there is no obstacle to have a fixed value of n_b
with size r^* growing indefinitely with L. We could thus expect some differences with
the uniform density case but numerical calculations [7] show that we still observe the
different regimes separated by an abrupt transition: the first regime where the size
of branches grows with L and the second regime where there is a ring whose size
grows very slowly with L (in contrast with the uniform density case, the transition
takes place for a value u_0/n_b that fluctuates in the range [2.5, 3.0]).

19.5.3   Average Minimum Time Between All Pairs of Points

We now consider the 'general' setting of routes between arbitrary points z_1 and z_2.
The route can either go straight (speed v_w) from z_1 to z_2 without using the network,
or follow the trip $z_1 \to A_1 \to A_2 \to z_2$, where A_1 and A_2 are 'stations' (any points
on the network) and where the travel from A_1 to A_2 is within the network at speed

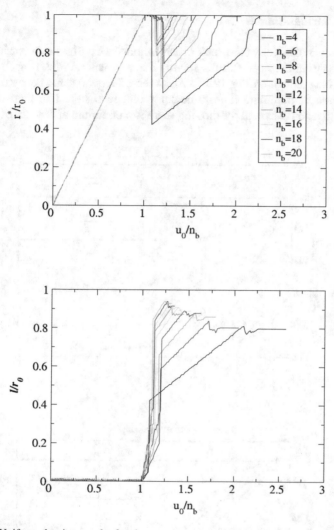

Fig. 19.16 Uniform density: results for the star+loop network for different values of n_b. (Top) Normalized length of branches. (Bottom) Normalized radius of the ring. We normalized u_0 by n_b in order to get the same 'transition' point at $u_0/n_b = 1$. These results are obtained for $\eta = 1/8$. Figure from [7]

v_s, and the other journey segments are straight at speed v_w. We take here each A_i as the closest station to z_i. Unlike previous cases, the overall optimal network is not necessarily a tree.

19.5.3.1   Various Shapes

We will study numerically this problem and as above we will consider various shapes (for the Gaussian density). Results for the line segment and the ring for different values of S are shown on Fig. 19.17. We observe here the same behavior as in the previous case of minimum time to center problem: for the line there is a quick saturation to a constant, and for the ring there is a minimum at $L \approx 2\pi\sigma$.

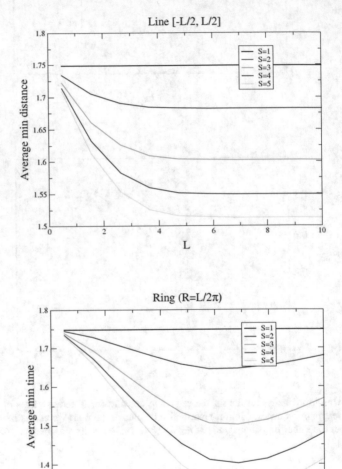

Fig. 19.17 Average journey time in the Gaussian case for different values of S and for (Top) a line and (Bottom) a ring. Figure from [7]

For the star network with n_b branches, we observe that for $0 < L \lesssim 1.15$ the line is optimal. For $1.15 \lesssim L \lesssim 6.6$, the cross $n_b = 4$ is the optimal choice, while for $L \gtrsim 6.6$, the solution $n_b = 6$ is better. Very likely we will have (as in the previous case of the average time to the center) an optimal network with $n_b \propto L$.

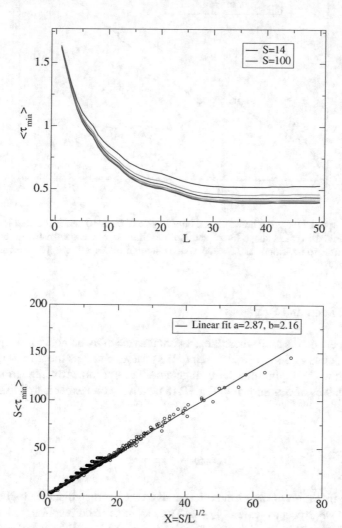

Fig. 19.18 **a** Average time (Gaussian density) for n_b branches plus a loop for values of S from 14 to 100. **b** Rescaled average time $S\overline{\tau}_{min}$ versus the rescaled variable $S/\sqrt{L}$. Figure from [7]

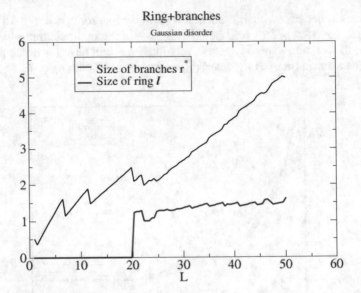

Fig. 19.19 Optimal length of branches and optimal size of the loop versus the total length L (for gaussian disorder): typical shape for r^* and ℓ. As in the case of the quickest path to the origin, we observe a sharp transition for $L \approx 20\sigma$ where a loop of radius $\ell \approx \sigma$ appears (here $\eta = 1/2$). Figure from [7]

19.5.3.2   Loop and Branches

We focus here on the case where the network is made of n_b branches of length r^* and a loop of radius ℓ. So $L = n_b r^* + 2\pi \ell$. We take n_b and ℓ as the 2 free parameters over which we will minimize the average time. The optimized average time $\overline{\tau}_{min}$ for different values of S is shown in Fig. 19.18a. Naively one expects that this quantity behaves as

$$\overline{\tau}_{min} = \frac{a}{\sqrt{L}} + \frac{b}{S} \qquad (19.82)$$

where the first term of the r.h.s. corresponds to the average distance to the network and which we expect to scale as $1/\sqrt{L}$. The second term corresponds to the shortest path distance within the network. In principle a and b could depend on L. If we assume this form to be correct then $S\overline{\tau}_{min}$ versus $X = S/\sqrt{L}$ should be a straight line. We tested this assumption on the data from Fig. 19.18a and the result is shown in Fig. 19.18b. This good collapse except for deviations observed for large values of $S/\sqrt{L}$ supports the assumption Eq. (19.82).

We also observe for this model a transition (Fig. 19.19): r^* grows almost steadily with $\ell = 0$, until a transition value at $L_c \approx 20$ where the ring appears. The size of the ring stays then stable with $\ell \approx \sigma$ (the Gaussian s.d.).

Naively, we can say that the ring appears when the branches can have a length $r^* \approx \sigma$ and a loop of size $\ell \approx \sigma$ which gives the condition $L_c \approx n_b \sigma + 2\pi\sigma$. At the transition, we observe that we have $n_b \approx 8$ which then gives $L_c \approx 14$, not too far from the value $L_c = 20$ observed here. The quantity L_c is independent from S which is expected, as it is essentially controlled by the topology of the network. We note that this transition was already observed in the previous case 'minimum distance to the center'. As L increases the optimal number of branches grows roughly from 2 to 10. The picture that emerges here is consistent with the empirical study [13]: we observe a ring around the 'core' of the city and then branches radiating from it. More generally, these results suggest that the distance to center problem is a reasonably good proxy for the more general problem.

References

1. G. Laporte, S. Nickel, F. Saldanha da Gama (Eds.), *Location Science* (Springer, 2015)
2. R.L. Graham, P. Hell, Ann. Hist. Comput. **7**(1), 43 (1985)
3. M. Barthelemy, A. Flammini, J. Stat. Mech. L07002 (2006)
4. F.K. Hwang, D.S. Richards, P. Winter, The steiner tree proble, vol. 53 in Annals of Discrete Mathematics (1992)
5. N. Megiddo, K.J. Supowit, SIAM J. Comput. **13**(1), 182 (1984)
6. S.T. Rachev, L. Rüschendorf, *Mass Transportation Problems: Theory*, vol. 1 (Springer Science & Business Media, 1998)
7. D. Aldous, M. Barthelemy, Phys. Rev. E **99**(5), 052303 (2019)
8. A. Okabe, B. Boots, K. Sugihara, S.N. Chiu, *Spatial Tessellations: Concepts and Applications of Voronoi Diagrams*, vol. 501 (Wiley, 2009)
9. A. Bejan, J. Adv. Transp. **30**(2), 85 (1996)
10. A. Bejan, G. Ledezma, Phys. A **255**, 211 (1998)
11. M. Mc Gettrick, Public transport **12**(1), 233 (2020)
12. G. Laporte, J.A. Mesa, *Location Science* (Springer, 2019), pp. 687–703
13. C. Roth, S.M. Kang, M. Batty, M. Barthelemy, J. R. Soc. Interface **9**(75), 2540–2550 (2012)
14. F. Xie, D.M. Levinson, *Evolving Transportation Networks*, vol. 1 (Springer, 2011)
15. M. Barthelemy, *Morphogenesis of Spatial Networks* (Springer, 2018)
16. A. Bottinelli, M. Gherardi, M. Barthelemy, J. R. Soc. Interface **16**(154), 20190101 (2019)
17. V. Latora, M. Marchiori, Phys. A **314**(1), 109 (2002)
18. K. Lee, W.S. Jung, J.S. Park, M. Choi, Phys. A **387**(24), 6231 (2008)
19. S. Derrible, C. Kennedy, Phys. A **389**(17), 3678 (2010)
20. S. Derrible, PLoS ONE **7**(7), e40575 (2012)
21. R. Louf, C. Roth, M. Barthelemy, PLoS ONE **9**(7), e102007 (2014)
22. Rapid transit. https://en.wikipedia.org/wiki/Rapid_transit. Accessed 30 Aug 2019
23. M.E. O'Kelly, J. Transp. Geogr. **6**(3), 171 (1998)
24. M. O'Kelly, J. Oper. Res. **32**, 393 (1987)
25. D. Bryan, M. O'Kelly, J. Reg. Sci. **39**, 275 (1999)
26. M. O'Kelly, Netw. Spat. Econ. **10**, 173 (2010)
27. V. Hurdle, Transp. Sci. **7**(4), 340 (1973)
28. B.F. Byrne, Transp. Res. **9**(2–3), 97 (1975)
29. B.F. Byrne, V.R. Vuchic, Traffic flow and transportation (1972)
30. G.F. Newell, Transp. Sci. **5**(1), 91 (1971)
31. H. Sagan, *Space-Filling Curves* (Springer Science & Business Media, 2012)
32. N. Madras, A.D. Sokal, J. Stat. Phys. **50**(1), 109 (1988)

Chapter 20
Greedy Models

In the Chaps. 18 and 19, we discussed models of networks defined by the optimization of a single quantity that depends on the global structure of the network. In contrast, we consider here the growth of networks where nodes are added one by one, located at random and connected to the network in an optimal way. In general, if we denote by i the new node, we will connect it to the node j such that a quantity of the form

$$Z(i, j) = \text{Cost}(i, j) - \text{Benefit}(i, j) \tag{20.1}$$

is minimum. This quantity is the balance between the cost of constructing the link $i - j$ and the benefit that it will create. The optimization is therefore not global— the resulting network does not necessarily optimize some quantity—but is local. In this respect these models can be qualified as 'greedy' as they rely on a local optimal choice, but with no guarantee that the system as a whole will reach a global optimum.

This type of models was proposed by computer scientists [1] for describing the Internet growth and which predicts correctly a scale-free degree distribution as observed empirically (see for example [2] and references therein). In this model, the functional minimized at each node addition reads

$$Z(i, j) = g(j) + \lambda c(i, j) \tag{20.2}$$

and if we allow only one link per new node the resulting network is a tree (if the initial conditions are tree-like). The quantity $g(j)$ is in general a measure of the 'centrality' of the node j such as the average number of hops to other nodes, or to a given central node, etc. and that we wish to be small. The quantity $c(i, j)$ is a cost function, in general proportional to the euclidean distance $d(i, j)$. The quantity λ controls the relative importance of centrality versus distance. For $\lambda \gg 1$, only the cost (distance) is minimized and each new node will connect to the nearest node in the growing cluster; the resulting network will be akin to a dynamical version of the minimum spanning tree (see Chap. 18). For $\lambda \simeq 0$, cost has no importance and the

© The Author(s), under exclusive license to Springer Nature Switzerland AG 2022
M. Barthelemy, *Spatial Networks*,
https://doi.org/10.1007/978-3-030-94106-2_20

new node will connect to the most central node, producing in general some sort of star graph (denoted by the complete bipartite graph $K_{n,1}$). For N nodes distributed uniformly in the plane, the distance between two points is typically of order $1/\sqrt{N}$ and what distinguishes large from small values of λ is $\mathcal{O}(\sqrt{N})$. Fabrikant et al. [1] showed in addition that if λ has some intermediate values (λ growing slower than $\sqrt{N}$ but is larger than a certain constant), we can obtain a network with a power law degree distribution whose exponent γ depends on λ. For instance if $\lambda \sim N^{1/3}$ then $\gamma = 1/6$ (see [1] for other results on this model). More generally, we observe a large variety of networks according to the choice of the functions $g(j)$ and $c(i,j)$ and in this chapter we will explore some interesting possibilities. In the first part, we will discuss a model where the cost is proportional to the distance and the centrality depends on the detour. We will then consider the general class of cost-benefit models where each node maximizes the difference between a benefit and a cost. This cost-benefit framework allows to discuss the structure of networks and also, as we will see in a last part of this chapter, how the network relates to the substrates where it grows. In particular, we will present the case of subways and railways where we relate socio-economical factors to the properties of these networks and the dynamics that takes place on them.

20.1 A Model for Distribution Networks

Many networks, including transportation and distribution networks evolved in time and increase their service area. Clearly, in these situations the resulting networks are growing and cannot result from a global optimization but instead, local optimization could be a reasonable mechanism for explaining the organization of these structures. In the example of a transportation network such as the train system, the nodes represent the train stations and the edges the rail segments between adjacent stations. In many of these systems, there is also a *root node* which acts as a source of the distribution system or in the case of the railway as the 'central station'. During the evolution of the network at least two factors could be considered. First, the total length of the system which represents the cost of the infrastructure should not be too large. Space has another important role here: the transportation system should also allow to connect two nodes in the network through a shortest path whose length is not too far from the 'as crow flies' distance. This efficiency is for example measured by the route factor—or detour index—which for two nodes i and j of the network reads (see also Chap. 4)

$$\eta(i,j) = \frac{d_R(i,j)}{d_E(i,j)} \tag{20.3}$$

where $d_E(i,j)$ is the euclidean distance from i to j and $d_R(i,j)$ is the distance between these nodes but computed on the network. For a system with a root node 0, one can then compute the route factor as the average over all nodes except 0

$$q = \frac{1}{N} \sum_{i \neq 0} \frac{d_R(i, 0)}{d_E(i, 0)} \tag{20.4}$$

and which measures how efficiently the network connects all nodes to the root node 0.

Following these two requirements, Gastner and Newman [3] proposed a model of a growing network where vertices are initially randomly distributed in the two-dimensional plane and where one vertex is designated as the root node 0. A network is then grown from its root by adding an edge between an unconnected node i to a vertex j which belongs to the network. The edge is chosen according to a local minimization process such that the quantity

$$E_{ij} = d_E(i, j) + \alpha \frac{d_E(i, j) + d_R(j, 0)}{d_E(i, 0)} \tag{20.5}$$

is minimum and where $\alpha > 0$ is here a parameter controlling the importance of the route factor. For $\alpha = 0$, the algorithm adds always a link to the closest vertex and the resulting network is similar to the MST and has a poor route factor [4]. When α increases, nodes that allow the straightest alignement of the nodes $(i, j, 0)$ are preferred. In this case, the route factor decreases and the average edge length $\bar{l}$ (not to be confused with the average shortest path ℓ) increases (see Fig. 20.1). In this figure, we see that the route factor q decreases sharply when α increases from zero, while the average edge length—which is a measure of the building cost of the network—increases slowly. This suggests that it is possible to grow networks with a small cost but with a good efficiency.

Fig. 20.1 Route factor q and average edge length $\bar{l}$ versus α in Eq. (20.5). These results are obtained for $N = 10^4$ vertices and in the inset the network is obtained for $\alpha = 12$. Figure taken from [3]

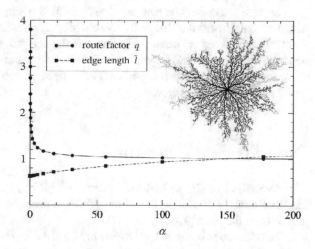

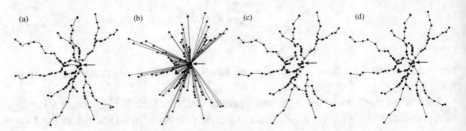

Fig. 20.2 a Commuter rail network in the Boston area. The arrow marks the assumed root of the network. **b** Star graph. **c** Minimum spanning tree. **d** The model of Eq. (20.6) applied to the same set of stations. Figure taken from [3]

Gastner and Newman [3] also studied a simpler version of this model where the local minimization acts on the quantity

$$E'_{ij} = d_E(i, j) + \beta d_R(j, 0) \tag{20.6}$$

which is similar to the model for the Internet proposed by Fabrikant, Koutsoupias, Papadimitriou [1] discussed briefly above. This model produces networks similar to the one described by Eq. (20.5) and self-organizes to networks with small q which is not imposed here. This model can be applied to the set of stations of the Boston rail network and produces a network in good correspondence with the real one (see Fig. 20.2). Also, the small value of q is confirmed in different empirical examples such as sewer systems, gas pipelines, and the Boston subway [3] where the ratio ℓ/ℓ_{MST} is in the range [1.12, 1.63] while the route factor is less than 1.6 (and compared to the MST is improved by a factor in the range [1.4, 1.8]).

The networks obtained here are trees which is a simplification for many of the real-world networks which usually contain loops. In addition, there is also usually an interaction between the density of points and the network and this co-evolution is not taken into account here. However, this simple greedy model with local optimization seems to capture important ingredients and could probably serve as a good starting point for further improvements.

20.2   Cost-Benefit Analysis

We discuss here a simple cost-benefit analysis (CBA) framework for the formation and evolution of spatial networks and which is described by the minimization of a quantity of the form given by Eq. (20.1).

The connection between local constraints and the large-scale structure is in general not elucidated and the cost-benefit framework allows to explore the effect of various parameters and mechanisms. In particular, this model depends essentially on one single scale and produces a family of networks which range from the star-graph to the minimum spanning tree and which are characterised by a continuously varying exponent [5, 6]. In addition, there is a strong path dependency in spatial networks and the properties of a network at a certain time can be explained by the particular historical path leading to it. It thus seems reasonable to model spatial networks in an iterative way. In addition, cost-benefit analysis provides a systematic method to evaluate the economical soundness of a project and has been officially used to assess transport investments since 1960 [7].

20.2.1   Theoretical Formulation

We consider here the simple case where all the nodes are distributed uniformly in the plane (for a rail network, the nodes would correspond to cities). The edges are added sequentially to the graph—as a result of a cost-benefit analysis—until all the nodes are connected [5] and for the sake of simplicity, we limit ourselves to the growth of trees which allows to focus on the emergence of large-scale structures due to the cost-benefit ingredient alone. At each time step we thus build the link connecting a new node i to a node j which belongs to the network, such that the following quantity is maximum

$$Z_{ij} = B_{ij} - C_{ij} \tag{20.7}$$

The quantity B_{ij} is the expected benefit associated with the construction of the edge between node i and node j and C_{ij} is the cost associated with such a construction. Equation (20.7) defines the general framework of this model and we now discuss specific forms of Z_{ij}. In the case of transportation networks, the cost will essentially correspond to maintenance cost and will typically be proportional to the euclidean distance d_{ij} between i and j. We thus write

$$C_{ij} = \kappa d_{ij} \tag{20.8}$$

where κ represents the cost of a line per unit of length and per unit of time.

Benefits are more difficult to assess. For rail networks, a simple yet reasonable assumption is to write the benefits in terms of distance and expected traffic T_{ij} between cities i and j

$$B_{ij} = \varepsilon T_{ij} d_{ij} \tag{20.9}$$

where ε represents the benefits per passenger and per unit of length. We have to estimate the expected traffic between two cities and we will follow the common and simple assumption of the so-called gravity law (see [8] and references therein)

$$T_{ij} = K \frac{P_i \, P_j}{d_{ij}^a} \tag{20.10}$$

where $P_{i(j)}$ is the population of city $i(j)$, and K is the rate associated with the process. We will choose here a value of the exponent $a > 1$ ($a < 1$ would correspond to an unrealistic situation where the benefits associated with passenger traffic would increase with the distance). This parameter a determines the range at which a given city attracts traffic, regardless of the density of cities. The accuracy and relevance of this gravity law is still controversial and improvements have been proposed [9, 10], but it has the advantage of being simple and to capture the essence of the traffic phenomenon: the decrease of the traffic with distance and the increase with population. Within these assumptions, the cost-benefit budget (up to a $K\varepsilon$ factor) now reads

$$Z_{ij} = K\frac{P_i P_j}{d_{ij}^{a-1}} - \beta d_{ij} \qquad (20.11)$$

where $\beta = \frac{\kappa}{K\varepsilon}$ represents the relative importance of the cost with regards to the benefits.

In the case of railways, node are cities and the population is distributed according to a power-law with exponent $1 + \mu$ with $\mu \approx 1.0$. The model thus depends essentially on the two parameters a, and β (see [6] for details), and their exact values are not important (within a certain range) and the obtained graphs have similar properties.

20.2.2   Crossover Between the Star Graph and the MST

We denote by $\overline{P}$ the average population and by $\ell_1 \sim 1/\sqrt{\rho}$ the typical inter-city distance ($\rho = N/L^2$ denotes the city density, and L is the typical size of the whole system). The two terms of Eq. (20.11) are thus of the same order for $\beta = \beta^*$ defined as

$$\beta^* \propto \overline{P}^2 \rho^{a/2} \qquad (20.12)$$

Another way of interpreting β^* which makes it more practical to estimate from empirical data, is to say that it is of the order of the average traffic per unit time

$$\beta^* = < T > \qquad (20.13)$$

From Eq. (20.12) we can guess the existence of two different regimes depending on the value of β:

- $\beta \ll \beta^*$ the cost term is negligible compared to the benefits term. Each connected city has its own influence zone depending on its population and the new cities will tend to connect to the most influent city. In the case where $a \approx 1$, every city connects to the most populated cities and we obtain a star graph constituted of one single hub connected to all other cities.

$\beta/\beta^* = 0.0$ $\beta/\beta^* = 1.0$ $\beta/\beta^* = 1000.0$

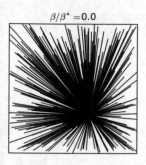 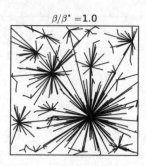 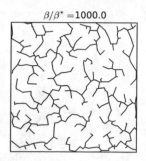

Fig. 20.3 Graphs obtained with the cost-benefit algorithm for the same set of cities (nodes) for three different values of β^* ($a = 1.1$, $\mu = 1.1$, 400 cities). On the left panel, we have a star graph where the most populated node is the hub and on the right panel, we recover the minimum spanning tree. Figure taken from [6]

- $\beta \gg \beta^*$ the benefits term is negligible compared to the cost term. All new cities will connect sequentially to their nearest neighbour. If we select the node i such that the length of the link $i - j$ is the smallest, the algorithm is then equivalent to Prim's algorithm [11], and the resulting graph is a minimum spanning tree (see Chap. 18).

Figure 20.3 shows three graphs obtained for the same set of cities for three different values of β/β^* ($a = 1.1$, $\mu = 1.1$) confirming the discussion above about the two extreme regimes.

For $\beta \sim \beta^*$ we observe a different type of graph, which suggests the existence of a crossover between the star-graph and the MST. This graph is reminiscent of the hub-and-spoke structure that has been used to describe the interactions between city pairs [12, 13]. However, in contrast with the rest of the literature about hub-and-spoke models, we show that this structure is not necessarily the result of a global optimization and can emerge from the self-organization of the system.

Since the MST is characterised by a peaked degree distribution while the star graph's degree distribution is bimodal, we will monitor the crossover from the star graph to the MST with the Gini coefficient for the degrees that can be expressed as follows [14]

$$G_k = \frac{1}{2 N^2 \bar{k}} \sum_{i,j=1}^{N} |k_i - k_j| \qquad (20.14)$$

where $\bar{k}$ is the average degree of the network. The Gini coefficient is in $[0, 1]$ and measures the heterogeneity of the distribution: if all the degrees are equal, $G = 0$, and if all nodes but one have a degree 1 (as in the star-graph), we obtain $G \simeq 1$. Figure 20.4 displays the evolution of the Gini coefficient versus β/β^* (for different values of β^* obtained by changing the value of a, μ and N). This plot shows a smooth variation of the Gini coefficient pointing to a crossover between a star graph and the

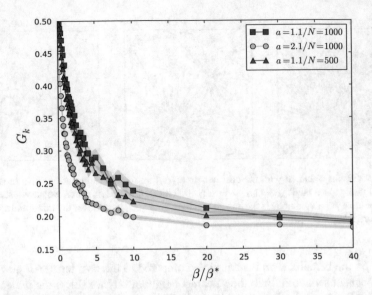

Fig. 20.4 Evolution of the Gini coefficient with β/β^* for different values of β^* and for different values of a and N. The shaded area represents the standard deviation of the Gini coefficient. Figure taken from [6]

MST, as one could expect from the plots on Fig. 20.3. Another important difference between the star-graph and the MST lies in the scaling of the total length L_{tot} with N. Indeed, in the star-graph case all the nodes are connected to the same node and the typical edge length is L, the typical size of the system the nodes are enclosed in. We thus obtain

$$L_{tot} \sim L\,N \tag{20.15}$$

On the other hand, for the MST each node is connected roughly to its nearest neighbour at distance typically given by $\ell_1 \sim L/\sqrt{N}$, leading to

$$L_{tot} \sim L\,\sqrt{N} \tag{20.16}$$

More generally, we expect a scaling of the form

$$L_{tot} \sim N^{\tau} \tag{20.17}$$

and on Fig. 20.5 we show the variation of this exponent τ versus β. For $\beta = 0$ we have $\tau = 1.0$ and we recover the behavior $L_{tot} \propto N$ typical of a star graph. In the limit $\beta \gg \beta^*$ we also recover the scaling $L_{tot} \propto \sqrt{N}$, typical of a MST. For intermediate values, we observe an exponent which varies continuously in the range $[0.5, 1.0]$ (see Fig. 20.5). This is rather surprising behavior and we will show in the following that it is a consequence of the hierarchical structure of the graphs.

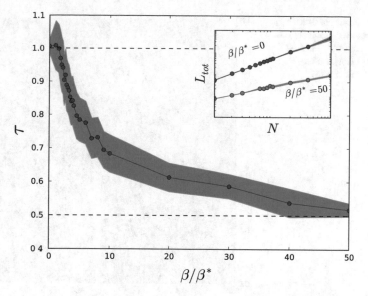

Fig. 20.5 Exponent τ versus β. For $\beta \ll \beta^*$ we recover the star-graph exponent $\tau = 1$ and for the other extreme $\beta \gg \beta^*$ we recover the MST exponent $\tau = 1/2$. In the intermediate range, we observe a continuously varying exponent suggesting a non-trivial structure. The shaded area represents the standard deviation of τ. Inset: In order to illustrate how we determined the value of τ, we represent L_{tot} versus N for two different values of β. The power law fit of these curves gives τ. Figure taken from [6]

20.2.3 Spatial Hierarchy and Scaling

The graph corresponding to the intermediate regime $\beta \approx \beta^*$ depicted on Fig. 20.3 exhibits a particular structure corresponding to a hierarchical organization, observed in many complex networks [15]. Inspired from the observation of networks in the regime $\beta/\beta^* \sim 1$, we define a particular type of hierarchy—that we call *spatial hierarchy*—as follows. A network will be said to be spatially hierarchical if (for a similar discussion, see Chap. 8 about spatial dominance):

1. We have a hierarchical network of hubs that connect to nodes less and less far away as one goes down the hierarchy;
2. Hubs belonging to the same hierarchy level have their own influence zone clearly separated from the others'. In addition, the influence zones of a given level are included in the influence zones of the previous level.

The relevance of this concept of hierarchy can be qualitatively assessed on Fig. 20.6 where we represent the influence zones by colored circles, the colors corresponding to different hierarchical levels. In order to go beyond this simple, qualitative description of the structure, we provide in the following a quantitative proof that networks in the regime β/β^* exhibit spatial hierarchy.

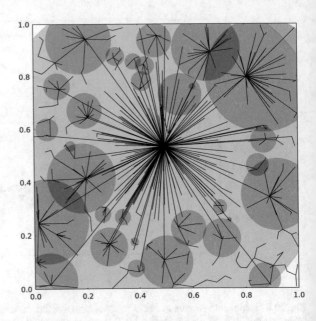

Fig. 20.6 Example of a graph where we represent the influence zones for the first two hierarchical levels. Figure taken from [6]

We propose here a quantitative characterisation of the part (1) in the definition of spatial hierarchy given above. The first step is to identify the root of the network which allows us to naturally characterize a hierarchical level by its topological distance to the root. We choose the most populated node as the root (which will be the largest hub for $\beta \ll \beta^*$) and we can measure various quantities as a function of the level in the hierarchy. In Fig. 20.7, we plot the average euclidean distance $\bar{d}$ between the different hierarchical levels as a function of the topological distance from the root node (for the sake of clarity, we also draw next to these plots the corresponding graphs). For reasonably small values of β/β^* (i.e. when the graph is not far from being a star-graph), the average distance between levels decreases as we go further away from the root node. This confirms the idea that the graphs for $\beta/\beta^* \simeq 1$ exhibit a spatial hierarchy where nodes from different levels are getting closer and closer to each other as we go down the hierachy. Eventually, as β/β^* becomes larger than 1, the distance between consecutive levels just fluctuates around $\ell_1 \sim 1/\sqrt{\rho}$ the average distance between nearest neighbours for a Poisson process, which indicates the absence of hierarchy in the network.

We now discuss the part (2) of the definition of spatial hierarchy, that is to say how the hubs are located in space. Another property that we can expect from spatially hierarchical graph is that of geographical separation between the respective influence zones of hubs belonging to the same level. We quantify this idea with the separation index that is equal to 1 if the nodes' influence zones do not overlap at all and 0 if they perfectly overlap. We plot this quantity averaged over the all the graph's levels for different values of β/β^* on Fig. 20.8. One observes on this graph that the separation index reaches values above 0.90 when $\beta/\beta^* \geq 1$, which means that the

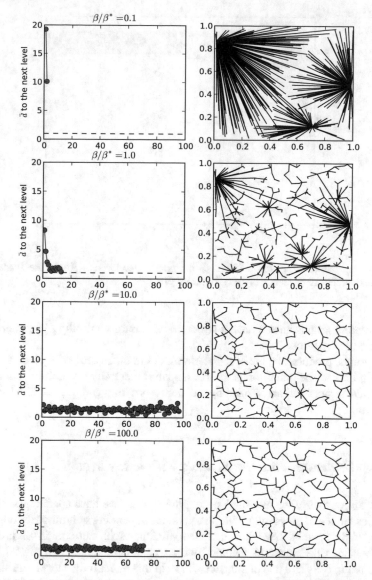

Fig. 20.7 Left column: Average distance between the successive hierarchy levels versus their topological distance from the root for different values of β/β^*, and on the right column, we show the corresponding graphs. The most populated node is taken as the root node. Figure taken from [6]

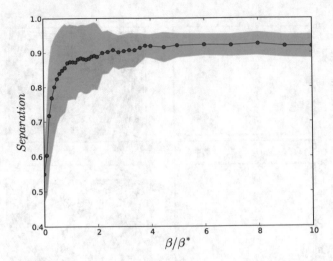

Fig. 20.8 Separation index averaged over all the graph's level versus β/β^*. The shaded area represents the standard deviation. Figure taken from [6]

corresponding graphs indeed have a structure with hubs controlling geographically well-separated regions.

The graphs produced by the cost-benefit model in the regime $\beta \sim \beta^*$ thus satisfy the two points of the definition given for a spatial hierarchy and exhibit a spatially hierarchical structure characterized by a distance ordering and geographical separation of hubs.

20.2.4 Understanding the Scaling with a Toy Model

We saw above that in the intermediate regime $\beta \sim \beta^*$ we have specific, non trivial properties such as L_{tot} scaling with an exponent depending continuously on β/β^*. Using a simple toy model, we show that the spatial hierarchy can explain this property.

We consider the toy model defined by the fractal tree depicted on Fig. 20.9 constructed recursively as a tree of connectivity z (in this figure only 3 levels are shown). For this model, the distance between the levels n and $n + 1$ is given by

$$\ell_n = \ell_0 b^n \tag{20.18}$$

where $b \in [0, 1]$ is the scaling factor. For a regular tree, each node at the level n is connected to z nodes at the level $n + 1$ which implies that the number of nodes at level n is

$$N_n = z^n \tag{20.19}$$

Fig. 20.9 A schematic representation of the hierarchical fractal network used as a toy model. Figure taken from [6]

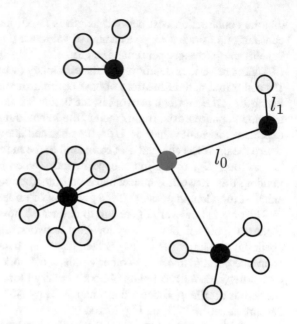

where $z > 0$ is an integer. A simple calculation shows that in the limit $z^g \gg 1$, the total length of the graph with g levels scales as

$$L_{tot} \sim N^{\frac{\ln(b)}{\ln(z)}+1} \qquad (20.20)$$

where $\frac{\ln(b)}{\ln(z)} + 1 \leq 1$ because $b \leq 1$ and $z > 1$. This simple model thus provides a simple mechanism where the exponent for L_{tot} varies continuously and depends on the scaling factor b. It provides a simplified picture of the graphs in the intermediate regime $\beta \simeq \beta^*$ and exhibits the key features of the graphs in this regime: the hub structure reminiscent of the star graph and where the nodes connected to each hub form geographically distinct regions, organized in a hierarchical fashion.

It is also interesting to note that the parameter z can be easily determined from the average degree of the network, and that the parameter b of the toy model can be related to the cost-benefit model by measuring the decrease of the mean distance between different levels of the hierarchy, as in Fig. 20.7. By plotting these curves for different values of β/β^*, we find that the coefficient of the exponential decays decreases linearly with β/β^* and therefore that $b \sim e^{\beta/\beta^*}$.

20.2.5 Efficiency

The question of the efficiency in transportation systems is of course of utmost importance. The cost-benefit analysis framework allows us to test the effect of various parameters and how efficient a self-organized system can be. For this, we can assume that the construction cost per unit length is fixed (i.e. the factor κ in Eq. (20.8) is constant), and since $\beta = \frac{\kappa}{K\varepsilon}$ a change of value for β is equivalent to a change in the benefits per passenger per unit of length.

There are various definitions of the efficiency (see for example Chaps. 4, 18 and 19) and a first natural measure of how optimal the network is, is given by its total length L_{tot}: the shorter a network is, the better for the company in terms of building and maintenance costs. In this model, the behaviour of the total cost is simple and expected: for small values of β/β^*, the obtained networks correspond to a situation where the users are charged a lot compared to the maintenance cost, and the network is very long ($L_{tot} \propto N$). In the opposite case, when $\beta/\beta^* \gg 1$ the main concern in building this network is concentrated on construction cost and the network has the smallest total length possible (for a given set of nodes).

The cost is however not enough to determine how efficient the network is from the users' point of view: a very low-cost network might indeed be very inefficient. A simple measure of efficiency is then given by the amount of detour needed to go from one point to another. In other words, a network is efficient if the shortest path on the network for most pairs of nodes is very close to a straight line. We will use the detour profile $\eta(d)$ defined in Chap. 4 (Eq. 4.40). We plot this detour profile for several values of β/β^* on Fig. 20.10a.

For $\beta/\beta^* \ll 1$, the function $\eta(d)$ takes high values for d small and low values for large d, meaning that the corresponding networks are very inefficient for relatively close nodes while being very efficient for distant nodes. On the other hand, for $\beta/\beta^* \gg 1$ we see that the MST is very efficient for neighboring nodes but less efficient than the star-graph for long distances. Surprisingly, the graphs for $\beta/\beta^* \sim 1$ exhibit a non trivial behaviour: for small distances, the detour is not as good as for the MST, but not as bad as for the star graph and for long distances it is the opposite. In order to make this statement more precise we compute the average of $\eta(d)$ over d, and plot it as a function of β/β^*. The results are shown in Fig. 20.10b and confirm this surprising behavior in the intermediate regime: we observe a minimum for $\beta/\beta^* \sim 1$. In other words, there exists a non-trivial value of β, i.e. a value of the benefits per passenger per unit of length, for which the network is optimal from the point of view of the users.

The existence of such an optimum is far from obvious and in order to gain more understanding about this phenomenon, we plot the Gini coefficient G_l relative to the length of the edges between nodes in Fig. 20.11. We observe that the Gini coefficient peaks around $\beta/\beta^* = 1$, which means that in this regime, the diversity in terms of edge length is the highest. The large diversity of lengths explains why the network is the most efficient in this regime: indeed long links are needed to cover large distances, while smaller links are needed to reach efficiently all the nodes. It is interesting to note that this argument is similar to the one proposed by Kleinberg [16] in order to explain the existence of an optimal delivery time in small-world networks.

We end this part on an empirical note. This cost-benefit model can be applied to railways and we estimate the value of β/β^* for these systems. For β, we use its definition (total maintenance costs per year divided by the total length and by the average ticket price per km), and in order to estimate β^* Eq. (20.13) is rewritten as

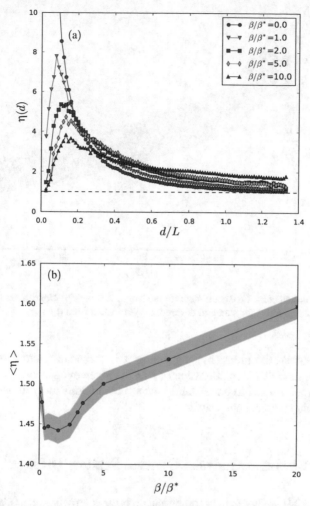

Fig. 20.10 (a) Detour function $\eta(d)$ versus the relative distance between nodes for different values of β/β^*. (b) Average detour index $< \eta >$ for several realizations of the graphs as a function of β/β^*. The shaded area represents the standard deviation of $< \eta >$. This plot shows that there is a minimum for this quantity in the intermediate regime $\beta \sim \beta^*$. Figure taken from [6]

$$\beta^* \simeq \frac{T_{tot}}{L_{tot}} \tag{20.21}$$

where T_{tot} is the total travelled length (in passengers·kms/year) and L_{tot} is the total length of the network under consideration.

Remarquably enough, the computed values for the ratio β/β^* shown in Table 20.1 are all of the order of 1 (ranging from 0.20 to 1.56). In the framework of this model, this result shows that all these systems are in the regime where the networks possess

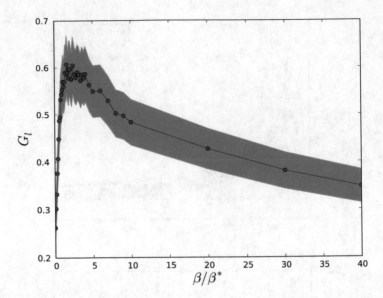

Fig. 20.11 Evolution of the Gini coefficient for the length versus β/β^* (for different values of β^*). The shaded area represents the standard deviation. Figure taken from [6]

the property of spatial hierarchy, suggesting it is a crucial feature for real-world networks. We note that here, the value of β/β^* is given exogeneously, and it would be extremely interesting to understand how we could construct a model leading to this value in an endogeneous way.

20.3   Cost-Benefit Analysis: General Scaling Theory

Subways [17–24] are not mere geometrical structures growing in empty space: they are usually embedded in large, highly congested urban areas and it seems plausible that some properties of these systems find their origin in the interaction with the city they are in. Previous studies [25, 26] have indeed shown that the growth and properties of transportation networks are tightly linked to the characteristics of urban environment. Levinson [25] for instance, showed that rail development in London followed a logic of both 'induced supply' and 'induced demand'. In other words, while the development of rail systems within cities answers a need for transportation between different areas, this development also has an impact on the organisation of the city. Therefore, while the growth of transportation systems cannot be understood without considering the underlying city, the development of the city cannot be understood without considering the transportation networks that run through it. As a result, the subway system and the city can be thought as two systems exhibiting a symbiotic behaviour. Understanding this behaviour is crucial if we want to

Table 20.1 Empirical estimates for β and β^*. Table giving the total ride distance (in km), the total network length (in km), the total annual maintenance expenditure (in euros per year) and the average ticket price (in euros per km). All the given values correspond to the year 2011. Table taken from [6]

Country	Total distance travelled (passenger·kms/year)	Total length (kms)	β^* (passengers/year)	Maintenance cost (euros/year)	Average ticket price (euros/km)	β (passengers/year)	β/β^*
France	$88.1 \cdot 10^9$	29,901	$2.94 \cdot 10^6$	$2.10 \cdot 10^9$	0.12	$5.85 \cdot 10^5$	**0.20**
Germany	$79.2 \cdot 10^9$	37,679	$2.10 \cdot 10^6$	$7.50 \cdot 10^9$	0.30	$6.60 \cdot 10^5$	**0.32**
India	$978.5 \cdot 10^9$	65,000	$1.51 \cdot 10^7$	$3.00 \cdot 10^9$	0.01	$4.61 \cdot 10^6$	**0.31**
Italy	$40.6 \cdot 10^9$	24,179	$1.68 \cdot 10^6$	$4.30 \cdot 10^9$	0.20	$8.89 \cdot 10^5$	**0.53**
Spain	$22.7 \cdot 10^9$	15,064	$1.51 \cdot 10^6$	$3.16 \cdot 10^9$	0.11	$1.91 \cdot 10^6$	**1.26**
Switzerland	$18.0 \cdot 10^9$	5,063	$3.55 \cdot 10^6$	$2.03 \cdot 10^9$	0.17	$2.36 \cdot 10^6$	**0.66**
United Kingdom	$62.7 \cdot 10^9$	16,321	$3.84 \cdot 10^6$	$12 \cdot 10^9$	0.16	$4.59 \cdot 10^6$	**1.19**
United States	$17.2 \cdot 10^9$	226,427	$7.59 \cdot 10^4$	$2.96 \cdot 10^9$	0.11	$1.18 \cdot 10^5$	**1.56**

gain deeper insights into the growth of cities and how the mobility patterns organise themselves in urban environments. More generally, this is the interesting problem of understanding evolving spatial networks in a substrate characterized by various scalar fields.

We discuss here a general coarse-grained approach, based on a cost-benefit analysis that accounts for the scaling properties of the main quantities characterizing transportation networks:

- the number of stations N_s,
- the total length L,
- the ridership R (number of passengers per unit time).

with the main features of the substrate that are:

- the population P
- the surface area A
- wealth characterized by the Gross domestic product (GDP) G for a country or the Gross metropolitan product (GMP for a city).

For subways the relevant unit is the city while for railways the natural scale is the country. In the following we will discuss a simple framework that allows to relate these different factors between them, and show that railways and subways can be discussed within the same framework but with some fundamental differences that we will highlight.

20.3.1   Theoretical Framework

The cost-benefit analysis seems to be the appropriate theoretical framework for understanding how network features and socio-economical indicators relate to one another. As discussed above, an iterative growth is considered where at each step an edge e is built such that the cost function

$$Z_e = B_e - C_e \tag{20.22}$$

is maximum. The quantity B_e is the expected benefit and C_e the expected cost of edge e. We consider networks after they have been built, and we assume that they are in a 'steady-state' for which we can write a cost function of the form

$$Z = \sum_e Z_e = B - C \tag{20.23}$$

where the summation is over all existing links. The quantity B is the total expected benefits and C the total expected costs, mainly due to maintenance (in the steady state regime). We further assume that, during this steady-state, operating costs are balanced by benefits. In other words, we assume that

$$Z \approx 0 \qquad (20.24)$$

Indeed, because lines and stations cost money to be maintained, we expect the network to adapt to the way it is being used. Therefore we can reasonably expect that at first order the cost of operating the system is compensated by the benefits gained from its use. We will apply this general framework to subway and railway networks in order to determine the behavior of various quantities with respect to the population and the GDP.

20.3.2   Subways

In the case of subways, the total benefits in the steady-state are simply connected to the total ridership R and the ticket price f over a given period of time. The costs, on the other hand, are due to the maintenance costs of the lines and stations, so that we can write (for a given period of time)

$$Z_{sub} = R f - \varepsilon_L L - \varepsilon_S N_s \qquad (20.25)$$

where L is the total length of the network, ε_L the maintenance cost of a line per unit of length, N_S the total number of stations and ε_S the maintenance cost of a station (for a given period time).

It is usually difficult to estimate the ridership of a system given its characteristics and those of the underlying city. Due to its importance for planning purposes, the problem of estimating the number of boardings per station given the properties of the area surrounding the stations has been the subject of many studies [27, 28]. Here we are interested in the dependence of global, average behavior of the ridership on the network and the underlying city. Very generally, we assume that the number R_i of people using the station i is a function of the surface area C_i serviced by this station—the 'coverage' [19]—and of the population density $\rho = \frac{P}{A}$ in the city and is of the form

$$R_i = \xi_i \, C_i \, \rho \qquad (20.26)$$

where ξ_i is a random number of order one representing the fraction of people that are in the area serviced by the station and who use the subway. The coverage depends, a priori, on local particularities such as the accessibility of the station, and should thus vary from one station to another. We take here a simple approach and assume that on average

$$C_i \sim \pi \, d_0^2 \qquad (20.27)$$

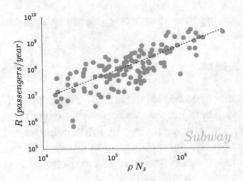
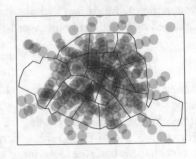

Fig. 20.12 Relationship between ridership and coverage. (Left) We plot the total yearly ridership R as a function of ρN_s. A linear fit on the 138 data points gives $R \approx 800 \, \rho N_s$ ($R^2 = 0.76$) which leads to a typical effective length of attraction $d_0 \approx 500$ m per station. (Right) Map of Paris (France) with each subway station represented by a red circle of radius 500 m. Figure taken from [29]

where d_0 is the typical size of the 'attraction basin' of a given station. If we assume that it is constant, the total ridership can be written as

$$R = \sum_i R_i \sim \overline{\xi} \pi d_0^2 \rho \, N_s \tag{20.28}$$

where $\overline{\xi} = \frac{1}{N_s} \sum_i \xi_i$ is of order $\mathcal{O}(1)$.

Using data gathered for 138 metro systems across the world (see [29] for details), we plot the ridership R as a function of $N_s \rho$ on Fig. 20.12(left) and observe that the data is consistent with a linear behavior. We measure a slope of $800 \, \text{km}^2/\text{year}$ which gives an estimate for d_0

$$d_0 \approx 500 \, \text{m} \tag{20.29}$$

This result is illustrated on Fig. 20.12 (right) by representing each subway stations of Paris with a circle of radius 500 m. The distance d_0 appears here an intrinsic feature of user's behaviors: it is the maximal distance that an individual would walk to go to a subway station. The average interstation distance ℓ_1 is another distance characteristic of the subway system. Rigorously, this distance depends on the average degree $<k>$ of the network so that $\ell_1 = \frac{2L}{N_s <k>}$. It has however been found [23] that for the 13 largest subway systems in the world, $<k> \in [2.1, 2.4]$, so that we can reasonably take $<k>/2 \approx 1$ and thus

$$\ell_1 \simeq \frac{L}{N_s} \tag{20.30}$$

The interstation distance depends in general on many technological and economical parameters, but we expect that for a properly designed system it will match human

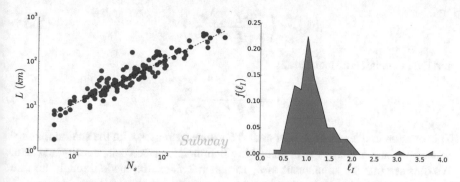

Fig. 20.13 Relation between the length and the number of stations. (Left) Length of 138 subway networks in the world as a function of the number of stations. A linear fit gives $L \sim 1.13 \, N_S$ ($R^2 = 0.93$) (Right) Empirical distribution of the inter-station length. The average interstation distance is found to be $\overline{\ell_1} \approx 1.2$ kms and the relative standard deviation is approximately 440 meters. Figure taken from [29]

constraints. Indeed, if $d_0 \ll \ell_1$, the network is not dense enough and in the opposite case $d_0 \gg \ell_1$, the system is not economically interesting. We can thus reasonably expect that the interstation distance fluctuates slightly around an average value given by twice the typical station attraction distance d_0

$$d_0 = \frac{\ell_1}{2} = \frac{L}{2 \, N_s} \tag{20.31}$$

It follows from this assumption that the interstation distance is constant and independent from the population size. In order to test our assumption, we plot on Fig. 20.13 (left) the total length of subway networks as a function of the number of stations. The data agrees well with a linear fit $L \sim 1.13 \, N_S$ ($r^2 = 0.93$). We also plot on Fig. 20.13 (right) the normalized histogram of the inter-station length, showing that the inter-station distance is indeed narrowly distributed around an average value $\overline{\ell_1} \approx 1.2$ km with a variance $\sigma \approx 400$ m, consistently with the value found above for $d_0 \approx 500$ m. The outliers are San Francisco, whose subway system is more of a suburban rail service and Dalian, a very large chinese city whose metro system is very young and still under development.

We can now express ℓ_1 in terms of the systems characteristics. Indeed, the total ridership now reads

$$R \sim \overline{\xi} \pi \rho \frac{L^2}{N_s} \tag{20.32}$$

If we assume to be in the steady-state $Z_{sub} \approx 0$, using the results from Eqs. (20.25, 20.32), the total length of the network and the number of stations are connected at first order in $\varepsilon_s / \varepsilon_L$ by

$$L \sim \left(\frac{4\varepsilon_L}{\pi \, \xi \, f \, \rho} + \frac{\varepsilon_s}{\varepsilon_L} \right) N_s \tag{20.33}$$

and the interstation distance reads

$$\ell_1 = \frac{4\varepsilon_L}{\pi \, \xi \, f \, \rho} + \frac{\varepsilon_s}{\varepsilon_L} \tag{20.34}$$

This relation implies that the interstation distance increases with the station maintenance cost, and decreases with increasing line maintenance costs, density and fare. We thus see that the adjustment of ℓ_1 to match $2\,d_0$ can be made through the fare price (or subsidies by the local authorities or national government).

So far, we have a relation between the total length L and the number of stations N_s, but we need another equation in order to compute their value. Intuitively, it is clear that the number of stations—or equivalently the total length—of a subway system is an increasing function of the wealth of the city. We thus assume a simple, linear relation of the form

$$N_s = \beta \frac{G}{\varepsilon_s} \tag{20.35}$$

where G is the city's Gross Metropolitan Product (GMP), and β the fraction of the city's wealth invested in public transportation. This relation can equivalently be interpreted as the proportional relation between the number of station per person and the city's development, as measured by its GMP per capita. On Fig. 20.14 (left) we plot the number of stations of different metro systems around the world as a function of the Gross Metropolitan Product of the corresponding city. A linear fit agrees relatively well with the data ($R^2 = 0.73$, dashed line), and gives $\frac{\varepsilon_s}{\beta} \approx 10^{10}$ dollars/station. However, the dispersion around the linear average behaviour is important: more specific data is needed in order to investigate whether differences in the construction costs and investments (or the age of the system) can explain the dispersion, or if other important parameters need to be taken into account.

Finally, we consider the number of different lines with distinct tracks. A natural question is how the number of lines N_{lines} scales with the number stations N_s, that is to say whether lines get proportionally smaller, larger or the same with the size of the whole system. We plot the number of lines as a number of stations on Fig. 20.14 (right) and find that the data agree with a linear relationship between both quantities ($R^2 = 0.93$). In other words, the number of stations per line is distributed around a typical value of 19, whatever the size of the system.

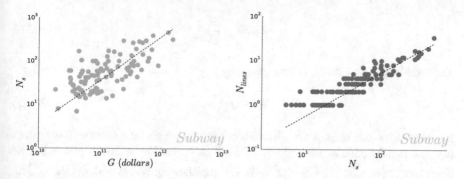

Fig. 20.14 (Left) Size of the subway system and city's wealth. We plot the number of stations for the different subway systems in the dataset as a function of the Gross Metropolitan Product of the corresponding cities (obtained for 106 subway systems). A linear fit (dashed line) gives $N_s = 2.51\,10^{-10}\,G$ ($R^2 = 0.73$). (Right) Number of lines and number of stations. We plot the number of metro lines N_{lines} as a function of the number of stations N_s. A linear fit on the 138 data points gives $N_{lines} \approx 0.053\,N_s$ ($R^2 = 0.93$), or, in other words, metro lines comprise on average 19 stations. Figure taken from [29]

20.3.3 Railways

At a different scale, railway networks answer a need for fast transportation between different urban centers, and we therefore expect their properties to be linked to the characteristics of the underlying country. An interesting question is to know whether subways and railway networks behave in the same way, but at different scales. In other words, we are interested to know whether subways are merely scaled down railway networks, or whether they are fundamentally different objects, following different growth mechanisms. Also, the existence of scaling between the system's output and its size is important as it suggests that very general processes are governing the growth of these networks [30, 31].

Even if it appears trivial, one could ask what is the main reason for the differences between subway and railway networks. As explained above, in the subway case, the interstation distance ℓ_1 is such that it matches human constraints: $\ell_1 \sim 2\,d_0$ where d_0 is the typical distance that one would walk to reach a subway station. For the railway network, the logic is different: while subways are built to allow people to move within a dense urban environment, the purpose of building a railway is to connect different cities in a country. In addition, due to the long distance and hence high costs, we assume that each city is connected to its closest neighbouring city. In this respect, the railway network appears as a planar graph connecting in an economical way, randomly distributed nodes (cities) in the plane. If we assume that a country has an area A and N_s train stations, the typical distance between nearest stations is

$$\ell_N = \sqrt{\frac{A}{N_s}} \qquad\qquad (20.36)$$

The total length $L \sim N_s \, \ell_N$ is then given by

$$L \sim \sqrt{A \, N_s} \qquad\qquad (20.37)$$

In order to test this relation for different countries, we plot the adimensional quantity $\frac{L}{\sqrt{A}}$ as a function of the number of stations N_s on Fig. 20.15. A power law fit gives an exponent 0.50 ± 0.08 ($R^2 = 0.87$), which is consistent with the theoretical discussion presented above.

At this point, we have a relation between L and N_s, but we need to find expressions for the other quantities. In contrast with subway systems, due to distances involved, the ticket price usually depends on the distance travelled and we denote by f_L the ticket price per unit distance. The relevant quantity for benefits is therefore not the raw number of passengers—as in subways—but rather the total distance travelled on the network T. Also, again due to the long distances spanned by the network, the costs of stations can be neglected as a first approximation, and we get for the budget the following expression

$$Z_{train} \simeq T \, f_L - \varepsilon_L \, L \qquad\qquad (20.38)$$

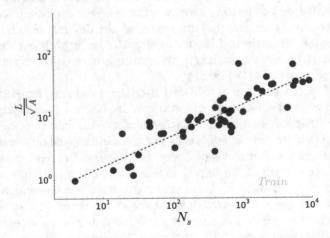

Fig. 20.15 Total length of the national railway network L rescaled by the typical size of the country $\sqrt{A}$ as a function of the number of stations N_s. The dashed line shows the best power-law fit on the 50 data points with an exponent 0.50 ± 0.08 ($R^2 = 0.87$). Figure taken from [29]

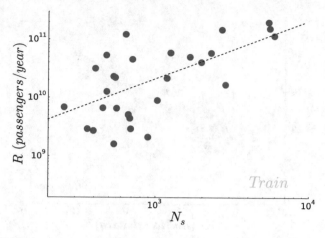

Fig. 20.16 Ridership and number of stations. The total yearly ridership R of the railway networks as a function of the number of stations. A linear fit on the 47 data points gives $R \sim 7.0\,10^8\,N_s$ ($R^2 = 0.86$). Figure taken from [29]

In the steady-state regime $Z_{train} \approx 0$, or in other words the revenue generated by the network use must be of the order of the total maintenance costs, which leads to

$$T \sim \frac{\varepsilon_L}{f_L} L \qquad (20.39)$$

In addition, if we assume that the order of magnitude of a trip is given by ℓ_N, the total travelled length is simply proportional to the ridership $T \sim \ell_N R$ leading to

$$R \sim \frac{\varepsilon_L N_s}{f_L} \qquad (20.40)$$

We thus plot the total daily ridership R as a function of the total number of stations N_s (Fig. 20.16), and despite the small number of available data points, a linear relationship between these both quantities seems to agree with empirical data on average ($R^2 = 0.86$). This result should be taken with caution, however, due to the important dispersion that is observed around the average behaviour, and the small number of observations.

According to the previous result, the total length and the number of stations are related to each other. We now would like to understand what property of the underlying country determines the total length of the network. That is to say, why networks are longer in some countries than in others. As in subway systems, economical reasons seem appealing. Indeed, the railway networks of some large african countries such as Nigeria are way smaller than that of countries such as France or the UK of similar surface areas. A priori, when estimating the cost of a railway network, one should take into account both the costs of building lines and the stations. However,

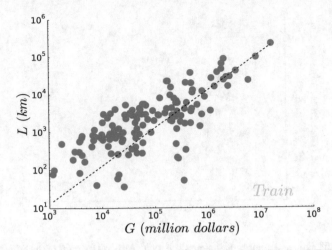

Fig. 20.17 Total length of the railway network L as a function of the country GDP G. The dashed line shows the linear fit on the 138 data points which gives $\varepsilon_L/\alpha \approx 10^4$ dollars.km^{-1} ($R^2 = 0.91$). Figure taken from [29]

Table 20.2 Summary of the differences between subways and railways

	Subway	Train
L/N_s	cste.	$\sqrt{\frac{A}{N_s}}$
R	$\frac{P}{A} N_s$	N_s
G	N_s	L

as stated above, considering the distances involved, the cost of building a station is negligible compared to that of building the actual lines. We thus can reasonably expect to have

$$L \sim \frac{\alpha\, G}{\varepsilon_L} \tag{20.41}$$

where G is here the country's Gross Domestic Product (GDP) used as an indicator of the country's wealth, and $\alpha < 1$ the ratio of the GDP invested in railway transportation. We plot L as a function of G on Fig. 20.17 and the data agree well ($R^2 = 0.91$) with a linear dependence between L and G. Again, the dispersion indicates that the linear trend should only be understood as an average behaviour and that local particularities can have a strong impact on the important deviations observed.

These different results for both subways and railways are summarized in the Table 20.2 where we show how the three main quantities (total length, ridership and gdp) vary for subway and railway networks respectively. The scaling of the length L of the network with the number of stations N_s reveals the different logics behind the growth of these systems. Another difference lies in the total ridership R: while it depends also on the population density P/A for subways, it only depends on the number of stations N_s for train networks. Finally, the size of both types of networks can be expressed as a function of the wealth of the region, represented here by the GDP G. However, because the interstation length is constant for subways, the size is expressed in terms of the number of stations N_s; in the case of railway networks, the cost of stations is negligible compared to the building cost of lines, and the size is expressed in terms of the total length L.

References

1. A. Fabrikant, E. Koutsoupias, C.H. Papadimitriou, in *Proceeding of the 29th International Colloquium on Automata, Languages, and Programming (ICALP)*, Lecture Notes in Computer Science, vol. 2380 (Springer, 2002), pp. 110–122
2. R. Pastor-Satorras, A. Vespignani, *Evolution and Structure of the Internet: A Statistical Physics Approach* (Cambridge University Press, Cambridge, 2003)
3. M.T. Gastner, M.E. Newman, J. Stat. Mech.: Theory Exp. **2006**(01), P01015 (2006)
4. M.T. Gastner, M. Newman, Phys. Rev. E **74**(1), 016117 (2006)
5. W.R. Black, Geographical Analysis **3**.3, 283–288 (1971)
6. R. Louf, P. Jensen, M. Barthelemy, in *Proceedings of the National Academy of Sciences*, vol. 110.22 (2013), pp. 8824–8829
7. M. Beesley, T. Coburn, D. Reynolds, *The London-Birmingham Motorway: Traffic and Economics, Road Research Laboratory* (Great Britain, 1960)
8. S. Erlander, N.F. Stewart, *The Gravity Model in Transportation Analysis: Theory and Extensions*, vol. 3 (Vsp, 1990)
9. F. Simini, M.C. González, A. Maritan, A.L. Barabási, Nature **484**(7392), 96 (2012)
10. M. Lenormand, S. Huet, F. Gargiulo, G. Deffuant, PLoS One (2012)
11. R.C. Prim, Bell Labs Tech. J. **36**(6), 1389 (1957)
12. M.E. O'Kelly, D. Bryan, D. Skorin-Kapov, J. Skorin-Kapov, Locat. Sci. **4**(3), 125 (1996)
13. M.E. O'Kelly, J. Transp. Geogr. **6**(3), 171 (1998)
14. P.M. Dixon, J. Weiner, T. Mitchell-Olds, R. Woodley, Ecology 1548–1551 (1987)
15. M. Sales-Pardo, R. Guimera, A.A. Moreira, L.A.N. Amaral, Proc. Natl. Acad. Sci. **104**(39), 15224 (2007)
16. J.M. Kleinberg, Nature **406**, 845 (2000)
17. L. Benguigui, J. Phys. I **2**(4), 385 (1992)
18. L. Benguigui, Environ. Plan. A **27**(7), 1147 (1995)
19. S. Derrible, C. Kennedy, Transp. Res. Rec.: J. Transp. Res. Board **2112**(-1), 17 (2009)
20. J. Sienkiewicz, J.A. Hołyst, Phys. Rev. E **72**(4), 046127 (2005)
21. D. Levinson, PLoS ONE **7**(1), e29721 (2012)
22. C. Von Ferber, T. Holovatch, Y. Holovatch, V. Palchykov, in *Traffic and Granular Flow'07* (Springer, 2009), pp. 709–719
23. C. Roth, S.M. Kang, M. Batty, M. Barthelemy, J. R. Soc. Interface **9**(75), 2540–2550 (2012)
24. B. Leng, X. Zhao, Z. Xiong, EPL (Eur. Lett.) **105**(5), 58004 (2014)
25. D. Levinson, J. Econ. Geogr. **8**(1), 55–77 (2008)
26. F. Xie, D. Levinson, Netw. Spat. Econ. **9**(3), 291 (2009)
27. R. Matsunaka, T. Oba, D. Nakagawa, M. Nagao, J. Nawrocki, Transp. Policy **30**, 26 (2013)
28. M. Kuby, A. Barranda, C. Upchurch, Transp. Res. Part A: Policy Pract. **38**(3), 223 (2004)
29. R. Louf, C. Roth, M. Barthelemy, PloS One **9**(7), e102007 (2014)
30. J.R. Banavar, A. Maritan, A. Rinaldo, Nature **399**(6732), 130 (1999)
31. R. Louf, M. Barthelemy, Sci. Rep. **4** (2014)

Index

A

Ad-hoc wireless networks, 295
Adjacency list, 6
Adjacency matrix, 5
Allometric scaling, 349
Alpha index, 54
Apollonian networks, 316
Area of faces, 167
Assortativity, 46
Average length of the Delaunay graph, 281
Average length of the Voronoi graph, 281
Average shortest path, 47

B

β-skeletons, 323
Betti numbers, 157
Betweenness centrality impact, 193
Biological networks, 174
Bluetooth graph, 309
Bouttier-Di Francesco-Guitter bijection, 178
Brandes algorithm, 67

C

Circuity, 182
Class first-passage times, 144
Cluster expansion, 302
Clustering coefficient, 45
Communication range, 295
Community detection, 146
Complete bipartite graph $K_{3,3}$, 10
Complete graph K_5, 10
Component, 24
    giant, 24
Computation geometry, 10

Congestion cost, 339
Contact network, 313
Continuum percolation, 295
Core and branches structure, 219
Crack, 288
Crossing number, 13
Crossing number, bounds, 15
Cycles, 23, 25

D

Degree of a face, 178
Delaunay graph, 278, 319
Dendricity, 182
Dependency, 67
Detour index, 58
Diameter, 47
Digraph, 23
Dual network, 6
Dual of the Voronoi, 278, 319
Dual, transportation case, 6
Dynamical euclidean minimum spanning
    tree, 358

E

Eccentricity, 47
Edge length ratio, 19
Edge orientation distribution, 56
Efficient transport network, 348
Entropy, 131
Erdos-Renyi graph, 235
Euclidean minimum spanning tree, 353
Euler relation, 12

© The Editor(s) (if applicable) and The Author(s), under exclusive license to Springer
Nature Switzerland AG 2022
M. Barthelemy, *Spatial Networks*,
https://doi.org/10.1007/978-3-030-94106-2

F

Feeder line, 386
First passage percolation, 109
Fitness model, 238
Forest, 180
Form factor, 58
Fragmentation models, 289
Fragmentation, multiscaling, 292
Full connectivity probability, 302

G

Gabriel graph, 320, 321
Gamma index, 54
General position assumption, 278
Geodesic, 112
Geographical Information System (GIS),
    188, 197
Gini coefficient, 413
Gravity law, 411
Gravity model, 156
Greedy algorithm, 350
Gross metropolitan index, 428

H

Headway, 386, 388
Hexagons, 280
Hidden variable model, 238
Hidden variable model with traffic, 240
Historical maps, 197
Homology, 157
Hub-and-spoke model, 339
Hub and spokes, 376
Hub-and-spoke structure, 339
Hub location problem, 377

I

Inclusion sequence, 321
In-degree, 24
Inverse degree, 48

J

Jensen's inequality, 307
Jordan curve theorem, 11

K

K-nearest neighbour model, 310
Kuratowsky theorem, 10, 15

L

Leaves, 174
Local optimization, 407
Longest link in the MST, 356
Loops, 25
Lune, 324

M

Map, 6, 14
Maximal planar graph, 13
Maximum BC, 72
Meshedness, 54, 213
Metabolic rate, 349
Minimum of random variables, 351
Minimum spanning tree, 350
Minimum spanning tree constant, 355
Mixed graph, 24
Mobile agents, 313

N

Navigability, 247
Neighbors in the Voronoi, 141
Nested inclusion relation, 115
Number of paths, 306

O

Order statistics, 352
Organic growth, 188
Out-degree, 24

P

Parallel transit lines, 386
Performance, 214
Peripheral node, 48
Persistent homology, 157
Planar fragmentation, 289
Planar graphs, 9
Planarity of street networks, 18
Planarization, 13
Planar map, 10
P-median, 373
Poisson-Voronoi tessellation, 279
Preferential attachment, 253
Preferential attachment and distance selec-
    tion, 255
Primal network, 6
Prim's algorithm, 350

R

Radial spanning tree, 358

Radial transit lines, 388
Radiation model, 157
Radius of a graph, 47
Railways, 429
Reachability, 24
Relative cost, 216
Relative neighborhood graph, 323
Root node, 408
Route factor, 58

S
Searchable networks, 247
Shape factor, 58, 169
Shape of faces, 167
Simplest paths, 131
Simplicity index, 124
Simplicity profile, 124
Skewness of graphs, 17
Soft random geometric graphs, 301
Spanning ratio, 322
Sparsification, 309
Spatial dominance, 140
Spatial hidden variable model, 239
Spatial hierarchy, 140, 415
Spatial networks, definition, 5
Spatial networks, representation, 6
Spatial planarity ratio, 18
Star graph, 68
STIT, 289
STIT tessellations, 288
Straightness centrality, 59
Street orientation distribution, 56
Stretch factor, 58
Subways, 208, 425

T
Template for subways, 327
Template for the subway structure, 219
Theta graph, 360
Thickness, 17
Thickness of graphs, 17
Topological data analysis, 157
Total disconnectivity threshold, 307
Total length of a graph, 285
Traffic effect, 240
Triangulation, 13, 319
Typology of planar graphs, 167
t-spanner, 112, 320, 359

U
Uncorrelated random graph, 236
Unit disk graph, 295

V
Veination pattern, 174
Vertex splitting, 17
Voronoi cell, 140, 277, 286
Voronoi tesselation, 140, 277

W
Watts-Strogatz model, 243
Watts-Strogatz model in dimension d, 244
Waxman model, 304
Witness Gabriel graph, 322

Y
Yao graphs, 360

Printed in the United States
by Baker & Taylor Publisher Services